T0255771

Socio-Economic Intervention in Organizations

The Intervener-Researcher and the SEAM Approach to Organizational Analysis

a volume in
Research in Management Consulting

Series Editor:
Anthony F. Buono
Bentley College

Research in Management Consulting

Anthony F. Buono, Series Editor

Challenges and Issues in Knowledge Management (2005)
edited by Anthony F. Buono and Flemming Poulfelt

Creative Consulting: Innovative Perspectives on Management Consulting (2004)
edited by Anthony F. Buono

Enhancing Inter-Firm Networks and Interorganizational Strategies (2003)
edited by Anthony F. Buono

Developing Knowledge and Value in Management Consulting (2002)
edited by Anthony F. Buono

Current Trends in Management Consulting (2001)
edited by Anthony F. Buono

Socio-Economic Intervention in Organizations

The Intervener-Researcher and the SEAM Approach to Organizational Analysis

edited by

Anthony F. Buono
Bentley College

and

Henri Savall
University Jean Moulin Lyon 3 and ISEOR

INFORMATION AGE
PUBLISHING

Charlotte, North Carolina • www.infoagepub.com

Library of Congress Cataloging-in-Publication Data

Socio-economic intervention in organizations : the intervener-researcher
and the SEAM approach to organizational analysis / edited by Anthony F.
Buono and Henri Savall.
     p. cm.
  Includes bibliographical references.
  ISBN 978-1-59311-621-7 (pbk.) — ISBN 978-1-59311-622-4
1. Organizational behavior. 2. Management. I. Buono, Anthony F. II.
Savall, Henri.
  HD58.7.S672 2007
  658—dc22

                                        2007003989

ISBN 13: 978-1-59311-621-7 (pbk.)
ISBN 13: 978-1-59311-622-4 (hardcover)
ISBN 10:   1-59311-621-7 (pbk.)
ISBN 10:   1-59311-622-5 (hardcover)

Printed in the United States of America

CONTENTS

INTRODUCTION

Anthony F. Buono

My first exposure to the Socio-Economic Approach to Management (SEAM) goes back to the mid-1990s when I met and worked with the ISEOR (Socio-Economic Institute of Firms and Organizations) team during annual meetings of the Academy of Management. Over a number of years, we collaborated on a professional development workshop series on consulting skills for the Management Consulting Division, sharing our thoughts and insights into organizational diagnosis and intervention. Key members of the ISEOR team—Henri Savall, Marc Bonnet, Veronique Zardet, Michel Peron, and Rickie Moore—became actively involved as workshop facilitators and members of the division. As I began to learn more about their particular approach to organizational analysis, I became increasingly intrigued by its integrative, dual focus on the qualitative and quantitative dimensions of organizational life, and the extended role of consultants as "intervener-researchers."

After working together over the next several years, Savall, ISEOR's founder and a professor at University Jean Moulin Lyon 3, invited me to study at ISEOR headquarters in Ecully, France during an upcoming sabbatical. I quickly accepted the offer and was able to immerse myself in the SEAM approach to organizational analysis, working closely with Savall,

Socio-Economic Intervention in Organizations: The Intervener-Researcher and the
SEAM Approach to Organizational Analysis, pp. ix–xii
Copyright © 2007 by Information Age Publishing

Bonnet, and a host of other ISEOR intervener-researchers. This volume is a result of that extended stay.

While my work at ISEOR headquarters was informative, thought-provoking, and highly enjoyable, a highlight of my time in France was the opportunity to visit what Savall described as the "living laboratory" of the SEAM approach—the headquarters and main factory of the Brioche Pasquier Group (BPG), one of France's leading bakeries. Serge Pasquier, chairman of the company, spent the day with us, touring the plant and reflecting on his company's strategy and his long-term relationship with Savall and ISEOR. In 1984, Pasquier's BPG consisted of a single plant with roughly 240 employees. At that point, he began a SEAM intervention that he would stay committed to for more than 2 decades. By the spring of 2003, BPG had expanded to 12 plants with over 3,000 employees and was viewed as one of the most successful bakery operations in France.

Reflecting on the success of his company, Pasquier credited much of its accomplishments to his devotion to the SEAM methodology, suggesting that throughout the organization SEAM was "like a religion." Pointing to the bakery system that he had created, he suggested "it is a complex process making something so simple," and the "strategic patience" associated with the SEAM approach had been a key component of the bakery's sustained level of high performance. Veronique Zardet's chapter, in the first section of this volume, provides an in-depth assessment of BPG and how socio-economic management had guided the company's growth and prosperity, and shaped its culture.

Pasquier is the perfect example of what Davenport and Prusak (2003, p. 18) referred to as an *idea practitioner*, those key individuals who make new management ideas "a reality within companies." These individuals, who embrace business improvement concepts and use them to bring about change in their organizations, are described as thoughtful, reflective managers—seasoned practitioners who have a basic sense of which ideas would truly benefit their companies.

As you read through the volume and learn about socio-economic analysis and implementation, skeptics might question the time commitment involved in a SEAM intervention, especially in the context of today's hypercompetitive environments and our increasing fascination with rapid change and fast-cycle organization development (see, for example, Anderson & Associates, 2000; Myer & Davis, 1998). The organizational world is increasingly characterized by reshaped and constrained notions of time—from reduced cycle time, simultaneous engineering, and faster meetings, to e-commerce on demand and anytime/anyplace exchanges. Management consultants, of course, must be sensitive to the resultant time-related needs and pressures faced by their clients. And part of the value-added dimension of consulting today, in no small measure, is reflected in a consultant's ability to facilitate the process of quickly getting things done. At the same time, another part of the real value that management consultants can add to the mix is reflected in their ability to "slow their clients down," enabling them to more fully reflect on, think through, and learn from their actions. As Savall has noted, there are times when a "mini"-analysis may be called for, but lasting, transformative change—which is the essence of the SEAM approach—cannot be implemented overnight. In fact, perhaps one of the reasons why so many of our organizational change programs fall well short of expectations (e.g., Beer, Eisenstat & Spector, 1990; Beer & Nohria, 2000; Kerber & Buono, 2005) is, in SEAM parlance, due to our lack of "strategic patience."

SOCIO-ECONOMIC INTERVENTION IN ORGANIZATIONS

The volume begins with a chapter by Henri Savall, founder and director of the ISEOR Institute and creator of the SEAM methodology, that presents an overview of the development of the socio-economic approach to management, and its guiding frameworks and methodology. The chapter's detailed explanation of the underlying thinking, tools, and techniques of socio-economic management serves as the primer for the remainder of the volume.

The book is then divided into three sections. The first part presents illustrations of SEAM interventions in different types of organizations, including industrial and service companies, and not-for-profit organizations, including cultural institutions and sports clubs. The next section looks at cross-cultural applications and assessments of SEAM experiments in Africa, Asia, Mexico, and the United States, with a concluding chapter on intervening in multinational corporations in general. The volume concludes with a section that examines different issues and challenges in SEAM intervention, ranging from the impact on and role of middle managers in the SEAM process, intervening in small organizations, SEAM's facilitative role in operationalizing and institutionalizing information technology, conceptualizing, and implementing organizational change, facilitating merger and acquisition integration, and the application of socio-economic management in sales and marketing. The book also contains a combined glossary and chapter index that provides a definition of key terms and concepts in the SEAM methodology and where they appear in the volume. These key terms are highlighted in **bold italics** throughout the volume, illustrating their application in different contexts.

It is our hope that the framework, applications and analyses in this volume will provide new insights into the theory and practice of socio-economic management. We would like to thank the book's contributors—true intervener-researchers—for their good natured colleagueship and willingness to accept our constant queries and proddings in moving this volume to completion. We also hope that this work has demonstrated the significance of thinking about organizations as socio-economic entities and the resultant demand for a truly integrative approach to management and organizational improvement efforts.

REFERENCES

Anderson, M. & Associates. (2000). *Fast cycle organization development*. Cincinnati, OH: South-Western College.

Beer, M., Eisenstat, R., & Spector, B. (1990). Why change programs don't produce change. *Harvard Business Review, 68*(6), 158-167.

Beer, M., & Nohria, N. (2000). Cracking the code of change. *Harvard Business Review, 78*(3), 133-141.

Davenport, T., & Prusak, L. (2003). *What's the big idea? Creating and capitalizing on the best management thinking*. Boston: Harvard Business School Press.

Kerber, K. W., & Buono, A. F. (2005). Rethinking organizational change: Reframing the challenge of change management. *Organization Development Journal, 23*(3), 23-38.

Myer, C., & Davis, S. (1998). *Blur: The speed of change in the connected economy*. New York: Perseus.

CHAPTER 1

ISEOR'S SOCIO-ECONOMIC METHOD

A Case of Scientific Consultancy

Henri Savall

I founded the Socio-Economic Institute of Firms and Organizations (ISEOR) in 1975, following my preliminary doctoral research work that laid the basis for a *socio-economic theory of organizations* (Savall, 1974, 1987, 2003a, 2003b). ISEOR is a management science research center that implements an *intervention-research* methodology (Savall & Zardet, 1987, 1992, 1995; Savall, Zardet, & Bonnet, 2000) of its own conception in enterprises and organizations. It constitutes a form of *scientific consultancy*. The ISEOR's intervention-research paradigm is based on the production of *generic knowledge* in the domain of management. Following a brief history of the origins of socio-economic theory and ISEOR's operating mode, the socio-economic model and intervention methodology in enterprises are presented (Péron & Péron, 2003). The characteristic concepts of the socio-economic theory of organizations are identified in bold italics and linked to an end-of-book glossary and applications throughout the volume that illustrate these principles and tenets with cases of *intervention-research*.

Socio-Economic Intervention in Organizations: The Intervener-Researcher and the SEAM Approach to Organizational Analysis, pp. 1–31
Copyright © 2007 by Information Age Publishing

THE ISEOR: A RESEARCH CENTER PRODUCING INNOVATIVE KNOWLEDGE FOR INTERVENTION IN ORGANIZATIONS

In the United States as well as in Europe, teacher-researchers tend to have a dichotomist conception of research and consultancy (Savall, Zardet, Bonnet, & Moore, 2001). Although research is aimed at elaborating and producing new knowledge, consultancy largely consists of applying preestablished knowledge obtained through former research in enterprises and organizations (International Labor Organization, 1998). The ISEOR vision is very different, envisioning consultancy as management research technology—which is why we employ the term scientific consultancy, explicitly including consultancy in the research process. Thus, intervention-research and *scientific consultancy* are used interchangeably. Our stance is based on the firm conviction of the absolute necessity in management science to scientifically observe the research object. Thus, the work of researchers present in the field—referred to as *intervener-researchers* (or interveners for short)—is justified by the will to observe the object being studied in a scientific manner.

Enterprises and organizations constitute the *field for scientific observation* (Savall & Zardet, 2000). Methodological opportunism enables intervention-research to increase the quality of scientific information, since researchers penetrate the organization, allowing up-close observation of their research object, without settling for more superficial observation from the outside (Savall & Péron, 2003). They negotiate their "geopolitical" position within the enterprise, organizing a dialectic confrontation and a *cooperation*-based system with company actors to coproduce knowledge of scientific intent. Each intervention-research permits pursuing two types of knowledge: (1) knowledge specific to the organization following case-study rationale; and (2) generic knowledge that contributes to the renewal of knowledge in the discipline of management sciences. Destructuring context-bound material makes it possible to identify specific knowledge, and then *generic knowledge*. The objective of this book is to present both generic knowledge and specific knowledge, through illustrations of case studies developed in each of the following chapters.

Generic contingency emerges based on three epistemological procedures created and developed at ISEOR: *cognitive interactivity, contradictory intersubjectivity*, and *generic contingency* (Savall, 2003a, 2003b; Savall & Zardet, 2004). Intervention-research thus permits practicing *integrated epistemology* as part of in situ intervention-research actions inside enterprises. The socio-economic model presented in the following section is a generic representation progressively constructed through successive experiments carried out over the past 30 years in more than 1,150 enterprises and organizations in 34 different countries.

The ISEOR Research Laboratory

In its 34 years of existence, ISEOR has trained more than 450 doctoral and postdoctoral researchers who have spent between 9 months and 25 years working at the institute's laboratory. It can be likened to a scientific "meta-consultancy" firm. It is a consultancy firm to the degree that it sells services to enterprises and organizations; and at the same time, it is a fundamental research laboratory since every organization also constitutes a field of scientific observation, useful for the advancement of knowledge.

ISEOR representatives negotiate with each potential client, with the goal of crafting an intervention that aims to create innovative management methods. Each *socio-economic intervention* can thus be considered a "machine for negotiating" innovative solutions, with the underlying goal of reducing the *dysfunctions* experienced by the enterprise.

Historically, intervention-research contracts have provided the economic *resources* necessary for ISEOR to self-finance 97% of its budget. The remainder is covered by doctoral research grants attributed by the university to several members of the research team. Intervener-researchers who take part in contractualized interventions are remunerated by ISEOR for their participation in the intervention-research work. This operating model guarantees the laboratory's self-financing status, which enables it to maintain its scientific independence and avoid submitting it to the fluctuations of "fashionable" research themes and financing schemes. However, this also forces the laboratory to negotiate intervention-research every year to cover its budget, but without recourse to canvassing. The team's reputation, its publications, the dissemination of intervention-research results and the innovative character of its proposed management methods are factors of attraction that activate spontaneous contracts from both public and private enterprises and organizations.

The process of creating new knowledge and expertise grounded in intervention-research, which applies previously-experimented knowledge in different contexts, can be summarized as follows (see Figure 1.1). The socio-economic intervention approach is practical, operational, and can be applied both by research teams and professional consultants. This interaction among researchers, management practitioners and consultants bears witness to a conception of the university's societal role, founded on *strategic interactivity*, as opposed to the scenarios prevailing in most academic communities in management sciences (Savall & Zardet, 2001), that of *academicism* and of *alignment*. In the *academicism* scenario, researchers consider themselves above practitioners in the analysis and comprehension of the world of professional activity. Conversely, in the *alignment* scenario, the university imitates the highly-competitive market of management training, thus losing any *value-added* differential.

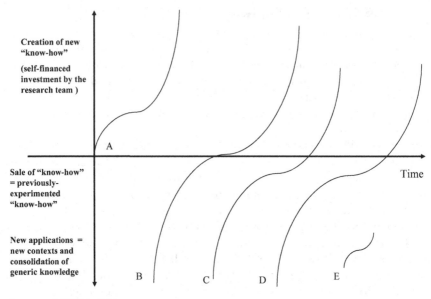

Creation of new
"know-how"

(self-financed
investment by the
research team)

A

Sale of "know-how"
= previously-
experimented
"know-how"

Time

New applications =
new contexts and
consolidation of
generic knowledge

B   C   D   E

Source: © ISEOR 2004.

Figure 1.1.   Process of creating new *generic knowledge* based on *intervention-research*.

The *strategic interactivity* scenario, in contrast, asserts the university's specific mission in contract with enterprises and consultancies, i.e., constructing and diffusing new generic knowledge while organizing interaction and joint-action among these partners (Tabatoni, 2003).

ISEOR has conceived innovative training programs at the university based on its research results, which are published by its researchers in reports, doctoral theses, articles, communications and books for teacher-researchers, practitioners and management consultants. As for ISEOR *intervener-researchers*, some of them pursue university careers, some of them choose management consultant careers, and some join enterprises and organizations. Most of them stay on the research team for 3 or 4 years before electing a university or professional career, maintaining working, even cooperative, relations with ISEOR after they leave.

SOCIO-ECONOMIC THEORY AND THE INTERVENTION MODEL

The theory underlying ISEOR's intervention strategy was constructed in stages beginning in 1973 to 1974, initially elaborating on a methodology to evaluate *hidden costs* and a subsequent approach to conducting *socio-*

economic projects focused on the creation of value-added. ***Socio-economic management tools*** were then created and experimented. Finally, intervention theory and change management procedures were elaborated, capitalizing on knowledge acquired from interventions in over a thousand organizations.

The Socio-Economic Theory of Enterprises and Organizations

The theory stems from two theoretical refutations and one experimental verification in the domain of economic sciences and accounting. Neoclassic and Marxist theories, enemy "sisters" born of the same mother—the classic school of political economy—as well as their ersatz successors (neoliberal, Keynesian and regulation schools) all made the same two critical mistakes. First, they consider that technical (or financial) capital is a factor in the production of economic value of the same nature, level and importance as human input. Capitalism and Marxism also consider that humans exercise their activity in situations of subordination and submission, and that they accept either hierarchical authority in enterprises, organizations and democratic institutions, or party dictatorship in other regimes.

Elementary observation shows (one would be tempted to say "It is obvious ..." if a rigorous epistemological viewpoint did not summon our prudence) that only human activity is intelligent and active, adapting itself to its environment and creating products-goods and services, marketable and nonmarketable, by detecting emerging needs and by innovating. So-called intelligent machines were conceived and maintained by humans who incorporated into them some of the knowledge that only humans are capable of creating, producing, reproducing, diffusing, transforming, perfecting, and questioning. Fixed or variable, material or financial capital is therefore a precious tool developed by ***human potential*** and constructed or destroyed, exploited or wasted by them.

The same elementary observation shows that humans are anything but spontaneously obedient, submissive, or even subordinated. They are agents moved by their contradictory drives: conflict/cooperation, individualism/teamwork, and autonomy/cooperation. Human and ***social performance*** is thus the result of an eminently dialectical and unstable system of behavior. Although norms attempt to adjust these behaviors and their effects on individual, family, professional and institutional performance, they typically fall short of totally succeeding to do so.

The second theoretical refutation amounts to highlighting the relative ineffectiveness and inefficiency of decision-making and economic-information classification systems, at the level of enterprises and organizations, whether this concerns accounting standards, budgetary procedures or financial analysis and decision-aid models. The decision to invest or divest should take into account the *integral performance* of an organization and evaluate it in order to ensure *sustainable development*. Yet, accounting and financial systems, in their dominant conception, maintain the ephemerality of performance. They do not encourage maintaining or developing performance in the medium- and long-term (more than a year). Private capitalism, a creator of enterprises, products and jobs, shrinks and becomes discouraged for lack of funds, because financial flows are diverted toward successions of relatively-sterile speculative bubbles generated by a surfeit of floating capital that engenders hypertrophied spheres and speculative mechanisms. Conversely, lack of capital to be truly and durably invested in productive circuits, provoked by the exaggerated attraction of speculative markets, condemns enterprises to underdevelopment or downsizing, for lack of funding for genuinely productive activities.

Within this context, any practitioner or reality-oriented theorist can make two observations on business operation. First, the primary factor of success is human, individual, team, and collective behavior. Second, the principal *immediate-result indicators* for a given period and in a given environment (e.g., earnings, *value-added*, level of financial results, thicker order books, delivery time, customer and user service) depend on numerous, decentralized, fluctuating behaviors, and not on theoretical instructions—little or poorly followed. The level of an organization's performance depends on its degree of *cohesion*, on its capacity for durable *cooperation* among its members, that is, their know-how in the *negotiation* of productive, effective and efficient relations, which are ineluctably deconstructed and which must unceasingly be reconstructed.

Socio-Economic Theory: Concepts and Variables

The *socio-economic theory of organizations* can be summed up as follows. An organization is a complex whole made up of five types of *structures* (physical, technological, organizational, demographic and mental) that interact with five types of human behavior (individual, trade group, occupational group, affinity group, and collective). This on-going and complex interaction creates activity thrusts that constitute the enterprise's *functioning*. Such functioning, however, is permeated with anomalies, disturbances, and discrepancies between the desired *functioning* (referred to as *orthofunctioning*), and the observed functioning (i.e., dysfunctions). These *dysfunctions* can be classified into six families: *working conditions, work organization, communication-coordination-cooperation (3C), time management, integrated training,* and *strategic implementation*. These six families constitute both explicative variables of the enterprise's actual functioning and domains of improvement actions meant to correct the dysfunctions identified during the diagnostic of the organization.

Regulations engendered by these *dysfunctions* generate costs for the organization that are usually hidden costs, and which taint the relevance, *effectiveness* and *efficiency* of decisions. The socio-economic method of *hidden cost* analysis includes five headings of socio-economic indicators, three predominantly social (*absenteeism, work accidents*, and occupational sickness and *personnel turnover*) and two predominantly economic (product/ service *quality* and direct *productivity* (product quantities).

Hidden costs constitute the *destruction of value-added*, which affects the overall economic performance of the enterprise. Such hidden costs have six components. The first three constitute *charges* the organization could avoid, at least partially, if its dysfunction level was not so high—*excess salaries, overtime*, and *overconsumption*. The last three components are particular in nature, for they do not constitute actual *charges*; they represent *nonproducts*, that is, a loss of production or activity engendered by dysfunctions, or a lack of earnings, which is an opportunity cost. These are *nonproduction, noncreation of potential* and *risks* endured by the enterprise. Hidden cost-performance thus affects both the enterprise's *immediate results* and its *creation of potential* (i.e., actions currently carried out to ensure the *immediate results* of future years). Thus, they constitute potential reserves, budgetary maneuver margins for improving the enterprise's economic performance.

Given these dynamics, an *overall approach to the enterprise* is required to (1) explain the level and mechanism of a firm's economic performance and (2) inspire effective and efficient improvement actions for *sustainable performance*. Socio-economic analysis and the management of organizations constitute a conceptualization of the enterprise's functioning and its capacity for *survival-development*, aimed at increasing its *socio-economic performance*, notably by means of the fundamental factor of *confidence*.[1] Indeed, the notion of confidence, which is at the heart of the socio-economic approach, is at once a *value*, a *doctrinal component* and *operational lever for increasing performance*. The quest for performance without *confidence* is outside the field of socio-economic analysis, for it is characterized by *the simultaneous, integral search for both economic and social performance*. The absence of performance and confidence leads inexorably to the disappearance (through decline, liquidation or merger) of the organization or to an artificially maintained *survival*. As for the problematics of confidence without goals of economic performance, they pertain to other disciplinary fields of human sciences such as psychology, sociology or anthropology. Thus, *socio-economic management* institutes socio-economic behavior as playing the part of pilot and referee in the dialectical movement between *ethics* and *deontology* on one hand, and between *effectiveness* and *efficiency* on the other.

According to the socio-economic theory of organizations, the spontaneous evolution of the enterprise does not permit a spontaneously adaptation to its environment susceptible to preserve, or even develop, its capacity for *short-, medium-, and long-term* survival-development as well as its *competitiveness* (Perez, 2004). The multiple changes that accompany a more deliberate evolution, that is, chosen and anticipated change, imply a structured, methodic process of *socio-economic intervention*.

Socio-Economic Intervention

Socio-economic intervention is a method for *engineering* change actions and implementing strategies. The process is aimed at improving the *integral quality* (internal and external) of organizations and the *piloting of social and economic performances* through two axes: a *cyclical problem-solving process* and *management tools*. The first axis is a *change conduct framework* that makes it possible to improve quality by simultaneous developing structure and behavior quality. *Integral quality* is composed of three levels: product quality (goods or services), functioning quality and management quality (see Figure 1.2).

The second axis consists of setting up a *piloting structure* and a management method, which are based on *stimulating tools* that are characterized by their capacity to increase the involvement of human potential and improve management analysis and decision-making quality. The socio-economic problem-solving method comprises four stages (the detailed characteristics of which are discussed later in this chapter):

- *Diagnostic* of dysfunctions and evaluation of hidden costs;
- Cooperative conception and development of innovative organizational solutions;
- Structured *implementation* of *improvement* actions; and
- *Evaluation* of *qualitative*, *quantitative*, and *financial performances*.

In order to involve all company actors in the change process or *metamorphosis*, socio-economic intervention procedure includes two simultaneous and synchronized actions:

- *Horizontal action* involving the executive management team and middle management, organized into *collaborative-training clusters*. Each cluster is composed of a manager and his or her immedi-

Module source : N 99 (7)

Source: © ISEOR 1999.

Figure 1.2.   *Internal quality* of an organization.

ate collaborators. The number of *clusters* depends on the size of the enterprise and the size of the management personnel; collaborative-training in the use of *management tools* is carried out in each one of the clusters.

- *Vertical action* in at least two units (e.g., departments, services, agencies) involving each unit's management team and line personnel.

This double horizontal and vertical action, referred to as the *Horivert process*, helps to ensure better articulation of and link between the socio-economic intervention and the company's strategy, and expose and resolve dysfunctions of a strategic nature as well as those linked to daily operations (see Figure 1.3). This *intervention architecture* derives from the principle of embedding change within the enterprise, from the strategic apex all the way down to the shop floor. This architecture also promotes collective learning and team *integration*, sources of effective and efficient cohesion.

The *Horivert process* architecture thus involves training all clusters of an enterprise and selecting at least two sectors where in-depth socio-

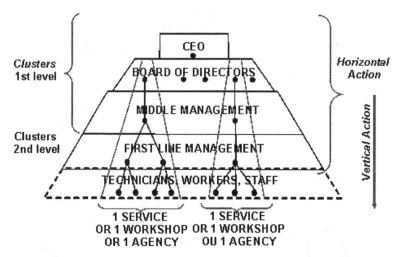

Clusters 1st level

Clusters 2nd level

CEO

BOARD OF DIRECTORS

MIDDLE MANAGEMENT

FIRST LINE MANAGEMENT

TECHNICIANS, WORKERS, STAFF

Horizontal Action

Vertical Action

1 SERVICE OR 1 WORKSHOP OR 1 AGENCY

1 SERVICE OR 1 WORKSHOP OU 1 AGENCY

Source: © ISEOR 1982.

Figure 1.3.   The *Horivert process*.

economic innovation actions will be carried out. As the applications throughout the book illustrate, however, there is a great diversity of concrete situations, underscoring the highly-adaptable nature of the Horivert process. Thus, the Horivert model can be applied at both the enterprise level (see, for example, chapter 2 by Veronique Zardet, chapter 3 by Vincent Cristallini, chapter 5 by Olivier Voyant, and chapter 7 by Nathalie Krief; see also Foucart, 2003; Millet, 2005; Vangénéberg & Zanta, 2005) and interfirm level with a group of small enterprises. In the latter case, the *Multi-SB (small business) Horivert* model entails organizing an inter-enterprise horizontal action—for example, bringing together five or six company heads and one or two of their immediate collaborators—and as many *vertical actions* as there are small companies in the group. This scaled-down Horivert process was designed and applied (1985) during an operation organized by the French city of Lille's Chamber of Commerce and Industry, made available to SBs in Belgium (1995), applied in Mexico (2002), and then again in Belgium (2004). As illustrated in Laurent Cappelletti's chapter, the Multi-SB (small business) Horivert model was adapted to the sector of French notary public offices (1998 to the present). This application to notary public offices permits disposing of a large statistical database, since 350 notary public offices have been involved in process of this type.

This variety of applications shows that the general model is easy to contextualize. The case of the opera (chapter 4 by Philippe Benollet) illustrates the implementation of a *Horivert process* with only one vertical action, even though the opera's initial request exclusively concerned a vertical sector, without *horizontal action*, which was finally accepted by the director of the opera during negotiation with the ISEOR.

Setting Up a Piloting Structure: Management Tools

There are a number of management tools that facilitate the intervention and create a pilot *structure* for the initiative. The *periodically negotiable activity contract* formalizes priority objectives and the means made available for each person in the organization through personalized, bi-annual *double dialogue* with the immediate hierarchical superior. A *remuneration* bonus is attached to attaining collective, team, and individual objectives. The bonus is self-financed through *recycling* **hidden cost** *into* **value-added**.

The *competency grid* is a synoptic tool that visually displays the competencies actually available in a team and its organization. It facilitates the development of an *integrated training* plan, particularly well-adapted to each person and to the evolving needs of the team.

The *priority action plan* is a *cooperative* inventory of actions to be carried out biannually to attain priority objectives, after arbitration of priority objectives and feasibility tests, notably in terms of time.

The *strategic piloting logbook* is made up of **qualitative, quantitative, and financial** indicators utilized by every member of the management team to concretely pilot people and activities in their zones of responsibility. The logbook permits *measuring, evaluating, following* the execution of actions, and *surveying* the sensitive parameters of operational and strategic activities, in view of *piloting* decisions to be made with regards to these activities. Line personnel are also instructed in recording piloting indicators.

The *self-analysis of time grid* or time management permits researching a more efficient structure of time *scheduling* by developing individual and collective scheduling, as well as *cooperative delegation* as appropriate across organizational members.

The *internal-external strategic plan* clarifies company strategy for the coming 3 to 5 years *as regards all actors involved*, that is, external targets (clients, suppliers, partners, institutions) and internal clients (from the chief executive officer to staff and workers as well as governance). It is updated every year to take into account evolutions in its relevant *external* and *internal environments*.

Overall, *socio-economic intervention* makes it possible to become better acquainted with a company's dysfunctions while proposing a reorganization project of its functioning, which is adapted to its particular constraints thanks to the *involvement* of relevant actors (Boje & Rosile, 2003;

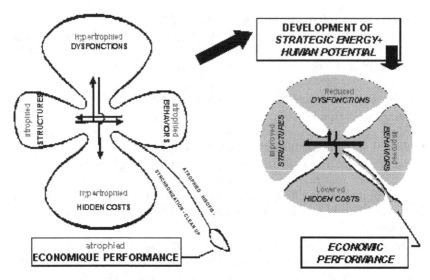

Figure 1.4.   **Strategic development** through **socio-economic intervention**.

Buono, 2001, 2003; Harris, 2001; Hayes, 2001). Figure 1.4 presents a schema of **strategic** company **development** obtained through socio-economic intervention.

The tools of **socio-economic management** set up over the course of the structured change process with company actors, permit better mastering of operating costs, in particular through the partial conversion of **hidden costs** into the creation of value-added. They have also been applied in conducting **projects** that involved a wide diversity of organization within a given territory, both enterprises and institutions with problems that go well beyond the perimeters of an enterprise, such as youth training and insertion, illiteracy, organization of public health partners, and promotion of an economic territory (cf. Savall & Bonnet, 1988, 1989, 1997; Savall & Zardet, 2005a, 2005b).

Setting up this change process permits developing operational and **piloting acts** within the enterprise by activating three levers: **synchronization, clean-up** and **HISOFIS**. These levers facilitate the **synchronization** of actions, services and people, regular clean-up of the organization, structures, behaviors, and **game rules**, and developing stimulating information (**HISOFIS effects**[2]), i.e., an information system made up of warning signals, information captors and indicators that better integrated and more stimulating, cultivating automatic action-taking reflexes among company actors.

Implementation and Evaluation

In order for the *implementation* of the *management tools* to be effective, certain rules must be respected. First of all, the process should *sustainably involve the full range of appropriate internal actors* in setting up the socio-economic tools. At the summit of the enterprise, it is essential that the board of directors fix strategic orientations and validate action plans defined at subordinate levels, especially those that involve transversal action across several company sectors. The method calls for the *mirror-effect* technique in every department or service, thus permitting everyone to become aware of the importance of dysfunction costs. Once this occurs, they will be better able to design and implement improvement solutions. The mirror effect is also used in projects of territorial development and interenterprise cooperation. Second, the board should announce certain *policy-making options* that affect strategic and organization choices, as well as the game rules of company operation. Finally, the enterprise needs the *contribution of external energy* to mobilize its hidden costs and performances, not only because an external eye helps it to recognize its major resource pools, but also because it is necessary to introduce specific socio-economic methods for conducting the process of change and organizational innovation in the enterprise.

According to the *socio-economic intervention* model, there are *three key forces of change* that drive the process (see Figure 1.5). The *Improvement-Process Axis* symbolizes the dynamics of the four change sequences; the *Management-Tool Axis* reflects the contribution of permanent management tools to all management personnel during the *horizontal action*. Finally, the *Policy-Decision Axis* represents the will to change at the summit of the enterprise, which, beyond discourse alone, is expressed through transformation decisions in different domains and functions of the enterprise. Indeed, if it is true that the socio-economic theory of organizations aims to reinforce the enterprise at its heart, through the principles of *piloting* and shared management by the entire management team, it also includes active principles in the different domains and functions represented by the branches of the *SEAM star* (see Figure 1.6). *Socio-economic management* reinforces internal team *cohesion* and the external cohesion of company strategy as well as the interaction between the two (see the chapter 7 by Nathalie Krief). It also drives innovative strategic choices, as illustrated in the chapter by Veronique Zardet (chapter 2).

ISEOR's numerous *research-intervention* findings have led to a fundamental criticism of the economy of scale theory, because the hidden costs of dysfunctions created by large organizations, with their excessive fragmentation of tasks and specialization of functions, cancel out most of the expected gains (Zardet & Voyant, 2003; see also Benollet's chapter 4). The *effectiveness* and *efficiency* of automation, if unaccompanied by a

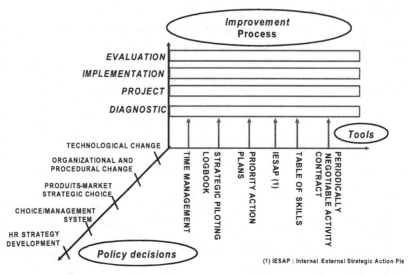

Figure 1.5.   The *three forces of change*.

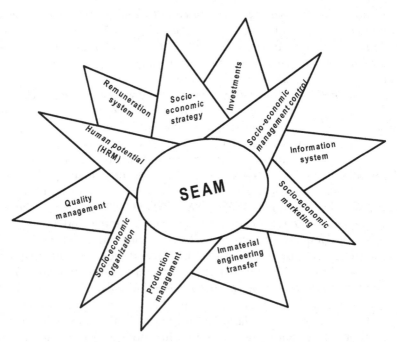

Figure 1.6.   *The SEAM star.*

simultaneous intervention with employees, are thus questioned (see Benollet in this volume). The crucial importance of *strategic implementation*, beyond mere strategic choices, is demonstrated by numerous cases where strategy remained no more than an intention or a dream on the part of top management—unless those choices are embedded in concrete participation of all company actors. In other words, beyond a company's *superstructure* (see Figure 1.2), its *infrastructure* (i.e., the organization's management and operation), plays a determining, but often underestimated, role.

In the domain of marketing and sales, numerous intervention-research findings show the vital character of the sales function throughout the entire organization. The enterprise's capacity for *survival-development* should therefore be consolidated and developed, considering *every* organizational actor as a salesperson, regardless of their position or job, or whether their clients are internal or external. *Socio-economic marketing* is based on an interactive conception of commercial relations between the salesperson and the client, in order to (1) better define the needs of the latter, which are often poorly articulated, and (2) better respond to them.

The Critical Role of the Human Factor

The management of people is at the heart of the socio-economic theory of organizations. The innumerable dysfunctions and hidden costs often found in organizations reflect the depletion of human energy without *creation of value-added*, as well as fatalism on the part of organizational actors, who are often resigned to the repetition of dysfunctions. The theory of the fragmented person (Savall, 1974, 1979) states that every individual possesses a triple status: citizen, consumer, and producer. The individual is typically coddled as a *consumer*, respected as a *citizen*, but less appreciated as a *producer*. This dynamic explains certain passive or apathetic chronic behaviors. Action against change resistance thus requires finding levers that activate individual and collective energy, in order to modify behaviors, which, interacting with structures, generate proper functioning (*orthofunctioning*) but also numerous dysfunctions. The *SEAMES® software* program created by the ISEOR and utilized by *intervener-researchers* for more than 20 years identifies approximately 3,700 types of dysfunctions. The principles of the conception of this software program and its utilization in the intervention-research framework are presented in Appendix 1 of the present chapter.

One of these levers is negotiating and periodically renegotiating commitment to more active, more effective, and more efficient behavior, transforming "negative" energy (devoted to regulating dysfunctions) into active energy, with the intent of creating more value-added. However, this transformation also implies modifying activity *structures*, especially *working conditions* and *work organization*, to improve transversal and vertical relations, and to incite every actor to develop his or her creation of value-added. Actions include those aimed at obtaining appropriate multiskill capacity within teams, *synchronized decentralization*, and *cooperative delegation*, all accompanied with precise specification manuals and *game rules*. The focus is on improving everyone's "contribution-retribution" couple. Participative management, especially if embedded with socio-economic values, facilitates the best performance, generating the *creation of value-added*.

Within this context, socio-economic intervention constitutes a *collaborative learning* process involving all actors, the results of which are personal *development*, intensified teamwork, reinforced cohesion, and sustainable economic performance of the firm.

Socio-Economic Intervention Procedures

Establishing contact with an organization triggers the *negotiation* process between that enterprise and ISEOR, a process that is completed with the signature of an *intervention agreement*. This agreement stipulates, in a highly detailed manner, the intervention's objectives, the methodological framework that will be deployed, the services the ISEOR team will provide, the calendar of the intervention's development, the cost of the intervention, and the payment schedule. These details comply with the principle of transparency vis-à-vis the enterprise and ensure the effectiveness of the intervention process. The intervention negotiation is a critical phase that can last several months and will have a decisive influence on the quality of the intervention.

The Negotiation Phase of a Socio-Economic Intervention

Starting with the negotiation phase, an internal (client-based) intervention team is set up. The agreement gives a full account of the team, its size, the training it will receive and the contribution it will make to the accomplishment of the services to be provided, under ISEOR's quality-control and methodological steering.

The initial contact between the ISEOR and an organization is always at the initiative of the latter. Indeed, the ISEOR employs neither canvassing nor hard-sell techniques and only intervenes when solicited by a business or an organization, or at the request of an intermediary third-party such as a professional association or a government department that expresses the desire to prescribe or recommend an intervention in one of its affiliated enterprises or organizations.

Establishing Contact

Contact thus must be initiated by a member of the organization, referred to as the *introducer*. This person has heard about socio-economic theory—although real knowledge may be limited—either by word-of-mouth during a training session, at a conference, in a book, or by an intervention that the individual had either heard about or witnessed in an enterprise where he or she was formerly employed. The introducer plays a critical role, for the first stage of the negotiation phase is conducted with that person. This process entails introducing the socio-economic method and testing its relevance to the objectives and problems of the enterprise. The introducer's role is a delicate one, because the introducer must quickly understand that negotiation, if it is to continue, should directly involve top management and enable ISEOR representatives to reach a key character—the *decider-payer*. This person is the authority who steers the organization (e.g., CEO, general manager, director of the enterprise) and has the financial authority to allot budgetary means to the intervention and to sign the agreement.

Accessing the Decider-Payer

Access to the decider-payer is thus organized with the introducer, who usually remains involved in the second stage of negotiation. A semi-structured, in-depth interview with the *decider-payer* focuses on identifying and formalizing his or her objectives for the intervention, the stakes involved, and expectations and major concerns regarding managing the enterprise. This interview makes it possible to draw up the first schematic document of the intervention project, which includes:

- A diagram or "architectural drawing" of the enterprise, according to the *Horivert model*, defining the number and content of *clusters*, as well as two or more sectors of the enterprise where vertical actions could take place (see Figure 1.3).

- A rough schedule identifying the major phases of the intervention and their positioning in the annual plan. Except in the case of small enterprises, complete *socio-economic management implementation* requires 1 year for enterprises with 1,000 employees, and 2 to 3 years for organizations employing up to 30,000 people.

- A methodological specifications manual, referred to as the *OMSP* (*Objective-Products/Method-Products/Service-Products*) *Tool*. *Objective-products* are the objectives assigned to the intervention reflecting the enterprise's viewpoints, as they were understood by ISEOR representatives during the first interview appointment. *Method-products* are the intervention methods that the ISEOR proposes to apply in this particular enterprise to attain these *objective-products*.

Finally, *service-products* designate the concrete services to be carried out by the interveners in application of each *method-product*. These services enable the enterprise to clearly identify ISEOR intervention modes within the organization, such as conducting collaborative-training sessions or conducting diagnostic qualitative interviews. The *OMSP Tool* also clearly specifies which services are to be performed by *internal-intervener* members of the enterprise, such as providing personal-assistance to middle management and conducting *diagnostic* interviews.

These three documents are presented and discussed with the decider-payer during a second meeting. This continued discussion attempts to create a *mirror-effect* to ensure the client-enterprise's objectives are well understood (project/objectives). This is the point in the negotiation process when the decider-payer acknowledges the objectives, stakes, and problems to be solved that he or she expressed during the first meeting. During the second meeting the *socio-economic theory of organizations* is explained in more depth and questions about the methodological specificities of socio-economic management are explained. This second meeting also makes it possible to focus on the content of the proposed OMSP TOOL and to engage the decider-payer's involvement in modifying the initial proposals as needed. At the end of this meeting, a first agreement-in-principle is obtained from the director, regardless of any financial consideration, since no budget has yet been presented.

Preparing and Negotiating the Cost Estimate

Following the second meeting, the three key documents mentioned above are adjusted and a financial estimate of the intervention is developed, based on the volume of services to be provided. A third meeting is then held to validate the intervention content and present the *cost estimate*. In cases where the proposed costs exceed the company's willingness or ability to pay, a second round of negotiation attempts to adjust the volume of services to fit the available budget. As an example, the client organization's internal intervention team can constitute a major resource, since it can take over certain services, replacing ISEOR intervener-researchers and thus reducing the costs of the intervention.

Signing the Contract

The negotiation phase is completed when the intervention agreement has been signed and the first installment has been paid. The *intervention agreement*, complete with the outline of the *Horivert* action, the overall plan, the OMSP (*Objective-Method-Service-Products*) tool specifications, the budget and the payment schedule, constitutes a fundamental, legally binding document that is regularly referred to throughout the intervention to provide evaluation and progress reports and the results of each stage.

The Chronology of a Socio-Economic Intervention

The *socio-economic management implementation* process unfolds progressively. It can be broken down into four major phases:

1. The *initial setting-up of socio-economic management* (Phase A) generally lasts a year (8 months as in the case of very small enterprises with between 3 and 20 employees). It consists of applying the *Horivert process* and securing the first utilization of the *socio-economic management tools* (Phase A is explained in detail in the following paragraphs; phases B, C and D are introduced more briefly and illustrated in the chapters in the volume).

2. *Territorial extension* (Phase B) is applicable to large firms. It consists of developing new *vertical actions* (diagnostic, project, implementation, evaluation) and extending *horizontal action* to members of the intermediate supervisory staff who have not yet been involved in the intervention (e.g., *collaborative training*, *personal assistance*).

3. The *in-depth, integrative development phase* (Phase C) consists of choosing specific actions to consolidate the initial setting-up (Phase A) and extension (Phase B), working toward a new threshold of socio-economic performances. Phases B and C can take place simultaneously.

4. Finally, the *maintenance phase* (Phase D) entails very light intervention and mainly consists of conducting *collaborative training* sessions for the team of directors to ensure continuity and to optimize the use of *socio-economic management tools* and concepts. It also involves conducting audits. In order to avoid losing long-term socio-economic management *effectiveness*, it is recommended that the maintenance phase be renewed every year, similar to the annual maintenance of the organization's computer system. In medium-size enterprises, maintenance takes approximately 1 day per semester and 1 day per year in small firms.

Phase A: Initial Set Up

The setting-up process begins by applying the Horivert process, simultaneously conducting horizontal actions throughout the enterprise involving directors and top managers and, at the same time, conducting at least two vertical actions that involve lower-level workers. The sequence of the intervention is a key factor of success and Figures 1.7 and 1.8 illustrates its key principles. *Horizontal action* starts with the first session of *collaborative training* in all *clusters*. This initiative triggers the *horizontal diagnostic*. Collaborative training action continues at strictly regular intervals (once a month), in parallel with the horizontal diagnostic, and then the *horizontal project*. Once the latter has begun, *vertical actions* are then initiated, focusing on the diagnostic followed by the improvement project. Thus, by the end of the first year, *horizontal* and *vertical diagnostics* and *projects* have been undertaken and completed, the main tools are set up and the first *periodically negotiable activity contracts (PNAC)* have been drawn up.

A typical initial setting-up phase begins with Horizontal Action, mobilizing top and middle teams for *development* of *social and economic performance*. As Figures 1.7 and 1.8 illustrate, these horizontal actions include collaborative training sessions combined with *personal assistance* to coach every member of top and middle management teams in the use of *socio-economic management* tools and in their application to their own professional situation (see Figure 1.5, the *management-tool axis*). As Figure 1.5 also illustrates, the problem-solving actions begin with a *horizontal diagnostic* and continue with a *project* and its *implementation*.

The collaborative training action is carried out in the form of group sessions lasting half a day, once a month for every *cluster*. In the interval between two collaborative training sessions, every participant receives one hour of *personal assistance*, adjusting the content of group sessions to his or her own case. There are six themes for the *collaborative training* session: (1) the general introduction of socio-economic analysis and intervention in the enterprise, (2) *time management* and *competency grid*, (3) presentation of the horizontal *mirror-effect* diagnostic, (4) *internal-external strategic action plan (IESAP)*, *priority action plan (PAP)* and the expert-opinion of the diagnostic, (5) *strategic piloting indicator* logbook, and (6) periodically negotiable activity contracts (PNAC). The following sessions are followed by additional personal assistance, so that by the end of the initial setting-up phase all members of all clusters are equipped with the requisite tools to improve the piloting of their department activities and employees, including priority action plans and biannual periodically negotiable activity contracts.

The *collaborative training* program is similar for all clusters to ensure that the entire enterprise will progressively be instilled with common language and management tools. The training sessions in each cluster are called collaborative training sessions. Indeed, they are based on an active pedagogical approach: time devoted to exchange among group members

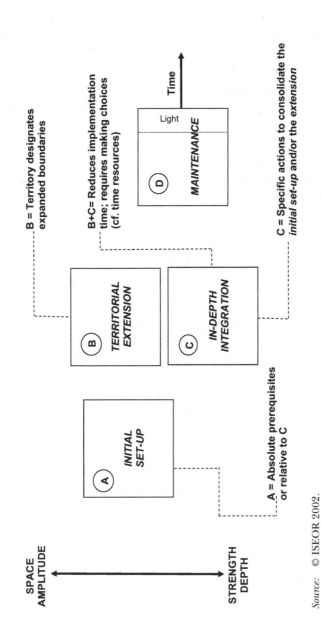

SPACE
AMPLITUDE

STRENGTH
DEPTH

A = Absolute prerequisites
or relative to C

INITIAL
SET-UP

Ⓐ

B = Territory designates
expanded boundaries

TERRITORIAL
EXTENSION

Ⓑ

IN-DEPTH
INTEGRATION

Ⓒ

B+C = Reduces implementation
time; requires making choices
(cf. time resources)

C = Specific actions to consolidate the
initial set-up and/or the extension

MAINTENANCE

Ⓓ

Light

Time

Source: © ISEOR 2002.

Figure 1.7. The SEAM *set-up process*.

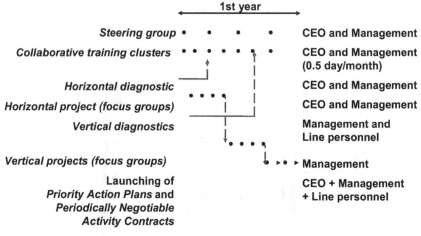

Source: © ISEOR 1982.

Figure 1.8.   The *chronobiological* process: Progressive rhythm of *SEAM intervention*.

(i.e., members of top and middle management teams for a given sector) is at least equal to time devoted to training by the instructor. The objective of these exchanges is, of course, adjusting socio-economic management concepts and tools to the specific reality of the enterprise, but they also serve to make resolutions so that training sessions all contribute to engendering changes in the team's professional practices. These resolutions are written down during each session on *resolution charts* that are distributed to every member of the team at the end of the session. Thus, following collaborative training and personal assistance on *self-analysis of time* and *time management*, decisions are frequently made to modify and simplify meeting procedures and to reduce mutual, spontaneous interruptions between company members. The *IESAP* and PAP collaborative training sessions provide the occasion for exchanges that clarify which managers will be responsible for the *PAP* and the PAP breakdown chart in the enterprise.

Personal assistance is further prepared for each manager by completing a first application of the tools, such as carrying out a self-analysis of the individual's time management over a 3-day period, drawing up the list of department operations in preparation for the team's *competency grid*. Thus, personal assistance permits verifying the assimilation of training content and counseling the employee on the continuation of his or her appropriation of the proposed tools. Personal assistance evaluations are drawn up at the start of the following collective collaborative-training session, through a roundtable discussion on work accomplished and lessons learned by the team.

When an enterprise includes numerous levels of management, *collaborative training* action is reinforced for first level clusters (e.g., board of directors and department heads), it is lighter for supervisor clusters (three or four sessions).

The Horizontal Diagnostic

In an effort to list the *dysfunctions* expressed by the entire top and middle management, in order to determine a general and transversal diagnostic of dysfunctions, an initial horizontal diagnostic is carried out, which will eventually be continued through focused vertical diagnostics. During the horizontal diagnostic, every member of the top and middle management cluster, who are also members of a collaborative training cluster, is individually interviewed during a one-hour session and documented by *exhaustive fieldnote quotes*. Once all interviews have been completed, each interview is processed by analyzing the *fieldnote quotes* (as expressed by the interviewed person in his or her natural language) that point out the dysfunctions. The fieldnote quotes from all interviewees are classified according to six themes (categories of dysfunctions), then subthemes and finally synthesized into *key-ideas*, with the assistance of the *SEAMES® software* program developed by ISEOR for processing qualitative data. The diagnostic, which is thus structured in the form of key-ideas, illustrated by *fieldnote quotes*, is presented to the entire group of interviewees in each collaborative training cluster. These presentations attempt to make the participants recognize the dysfunctions. This technique is referred to as the *mirror effect*. Once this mirror effect has been presented and assessed in each cluster, ISEOR intervener-researchers develop their *expert opinion*, which (1) identifies major dysfunctions expressed in the mirror-effect and (2) expresses the *non-dit (unvoiced comments)* in the form of key-ideas (i.e., the dysfunctions that were not expressed, but nonetheless perceived by the intervention team). The expert opinion thus constructed is presented in each *cluster*. This permits distinguishing between profound dysfunctions requiring in-depth transformations and those considered as *misunderstandings* that essentially call for communication and clarification action. The *expert opinion* constitutes the chief platform for the *horizontal project* to develop the horizontal *socio-economic performance improvement project*.

Developing the Horizontal Project

The *horizontal project* designates all improvement actions aimed at reducing dysfunctions identified during the horizontal diagnostic and fosters attaining strategic company objectives. The project is developed under the responsibility of the client organization's manager, referred to as the *project leader*, with the support of a *focus group* whose mission is to propose cooperative solutions. The focus group is composed of three sub-groups: the *core group*, the *plenary group*, and *task groups*, which meet alternately in four sessions spread over a period of 4 to 5 months. The core group is composed of the head of the project together with his/her two or three closest collaborators. As a general rule, the *plenary group* comprises the entire top management team.

The *core group* defines the major objectives/constraints that direct the quest for solutions. It then examines the solutions put forward in the *plenary group*, one after the other. In the intervals between plenary group meetings, specific *task groups* pursue in-depth research for solutions on a portion of the diagnostic, steered by a member of the plenary group to ensure *coordination* between the plenary group's work and that of the task groups. The latter can include executives and experts beyond the perimeter of the plenary group. The maximum size of each group is 15 persons. To structure research solutions, the *pivotal ideas* (*idées-force*) in the *expert opinion* are grouped into three or five "*dysfunction baskets*," each of which is entrusted to a task group. The themes of these "baskets" represent the *root causes of dysfunctions*.

When the *focus group* has accomplished its work, the selected solution proposals constitute the actions that fuel the *internal-external strategic action plan* (*IESAP*) and are scheduled over several consecutive semesters.

Implementation of the Horizontal Project

The actions proposed by the focus group and validated by the board of directors are carried out progressively. Written into the *IESAP,* they then appear in the *priority action plan* (*PAP*) of the first semester at the overall company level. These actions are then broken down and distributed into the priority action plans of the various divisions and departments. A more precise breakdown of actions to be accomplished is carried out at this level. These actions, in turn, are written into the PAPs of the different levels of responsibility in the enterprise.

The Vertical Diagnostic

The main objective of *vertical action* is to involve lower-level personnel in the metamorphose process and to enhance performance. Vertical action is based in at least two departments; it entails carrying out a diagnostic, followed by a *project*. The underlying process involves drawing up the inventory of *dysfunctions* within the diagnosed sector and evaluating the resulting *hidden costs*. *Competency grids* for the entire sector are also part of the vertical diagnostic, enabling an in-depth analysis of the sector's training-employment appropriateness, a perpetual cause of a number of dysfunctions.

The first phase of the *vertical diagnostic* is similar to that of the horizontal diagnostic—in-depth semistructured interviews focusing on dysfunctions are carried out with every member of the department's managerial team. The individual interviews are also undertaken with small group interviews (3 to 4 persons) of lower-level personnel. Overall, 30 to 60% of all shopfloor personnel are interviewed, with special care devoted to ensuring variety in the sample (e.g., trades, status, seniority, age, gender). These qualitative interviews are processed in the same manner as the horizontal diagnostic.

The evaluation of *hidden costs* is then prepared, drawing on complementary interviews with the department's managerial staff, to determine the precise frequency and modes of the *regulation of dysfunctions* identified during the qualitative interviews. The intervention team then proceeds to a monetary valuation of the hidden costs, by evaluating the costs linked to *absenteeism, industrial injuries, personnel turnover, nonquality* and *direct productivity gaps*. Time spent on dysfunction regulation (*overtime* [excess time]) and *nonproduction* [missed production] is evaluated using the *hourly contribution to margin on variable costs (HCMVC)*, also referred to as the *hourly contribution to value-added on variable costs (HCVAVC)*.

The vertical diagnostic *mirror-effect*—comprising the *qualitative diagnostic*, the *hidden costs* evaluation and the *competency grids*—is first presented to the department's management team, then to the shopfloor personnel who were interviewed, in the presence of the management team. The *expert opinion*, which is drawn up by the intervention team following the same principles as the horizontal expert opinion, is then presented to the management team and the board of directors.

The Vertical Project

The *focus group* is charged with developing improvement solutions under the supervision of the department head, who as noted above is also the project leader. The *core group*, including the *project leader* and his or her immediate superior, is assigned to set priority objectives-constraints for researching *improvement* solutions. The quest for these solutions is the responsibility of the *plenary group*, under the guidance of the project leader, the department's entire managerial team and managers representing the other company departments, which makes it possible to deal with dysfunctions at the very interfaces between neighboring departments. *Task groups* continue in-depth research of solutions during the intervals between plenary group meetings. Piloted by a member of the plenary group, this effort draws in other participants at all company levels not represented in the plenary group. This focus group, like the horizontal focus group, concentrates on four of five *pivotal ideas (idées-forces)* in *dysfunctions baskets*.

At the conclusion of the *project phase*, solution proposals are gathered together and the project's *economic balance* is calculated, evaluating (1) the cost of the proposed actions and (2) the expected performance in terms of increased *value-added* based on a reduction of dysfunctions. The overall balance (performances minus costs) allows the identification of whether the project will be economically profitable. The complete inventory of proposed solutions, as well as the economic balance, is then submitted to the board of directors for validation.

Implementation of the Vertical Project

The selected actions are broken down and distributed into the *priority action plans* of all the departments concerned. A detailed plan for accomplishing these actions is then developed, with certain actions becoming targets for specific objectives and indicators, stipulated in the *periodically negotiable activity contracts* of the personnel (managers and participants) responsible for these target actions.

Socio-Economic Evaluation

A comparative evaluation, carried out one year after launching the intervention, is important because it allows the organization to (1) identify significant accomplishments, (2) estimate the subsequent gains in value-added, and (3) recognize the efforts that are still required. In order to construct this *evaluation*, interviews are again carried out in the team of directors and in a small sample of managers and shopfloor personnel, noting points of view from different categories of actors (principle of *contradictory intersubjectivity*). A comparative evaluation of hidden costs is also carried out, based on the list of dysfunctions identified during the initial diagnostic. The gains in value-added actually identified are compared to the cost of the action, which permits establishing the actual, not provisional, *economic balance* as well as calculating the profitability of change actions achieved by the enterprise.

This evaluation has an important impact. It allows the enterprise to become aware of the improvements accomplished, which encourages actors to continue their improvement efforts, thus generating change energy, which ineluctably deteriorates.

The Extending and Rooting Phases of Socio-Economic Management

Upon the optional demand by the enterprise, which has registered growth in its socio-economic performance at the end of the first year, the *extension phase* continues the *initial setting-up* in a new geographical area during a second year of intervention. In particular, new vertical actions are launched in different departments of small and medium-sized enterprise (see chapter 4). At this point, the enterprise can request additional, *in-depth intervention*, with or without territorial extension, in order to reinforce the integration of socio-economic management concepts and tools in its principal management functions and processes.

Thus, the *socio-economic management control* method is aimed at transforming the role of management controllers toward functions of expertise, support, and assistance for operational managers in preparing projects to be inserted in their priority action plans, their *budgeted action plans*, and their *economic balances*. Other method-products, such as *socio-economic marketing*, technological project piloting, the *vital sales function,* and creating a network of *internal-intervener* are frequently developed as necessary (as examples, see chapters 5 and 7).

The Maintenance Phase

During the *maintenance phase* the client organization is provided with a framework of services that provides light but regular assistance to the enterprise in an effort to maintain and build on the initial successes of the intervention. The best frequency is one session per semester (1 to 2 days depending on the size of the enterprise) with the board of directors and management teams. The focus is placed on launching *priority action plan* campaigns, *periodically negotiable activity contracts*, and updating *internal-external strategic action plans (IESAP)*. For ISEOR interveners, these maintenance services provide excellent opportunities to audit the evolution of the enterprise's qualitative, quantitative, and financial performances, allowing them to periodically assess the enterprise's social and economic outcomes, which results in sustaining the dynamics of *progress*.

The Critical Role of Internal-Interveners

Internal-interveners are members of the enterprise who, during the setting-up phase of socio-economic management, devote part of their work time (approximately 25%) to collaborative support activities, under the supervision of ISEOR intervener-researchers. They benefit from 10 days of in-depth training dispensed by ISEOR, covering the internal intervener's skills, diagnostic and innovative *socio-economic project* methodology, and *socio-economic management tools*. This training takes place within the premises of ISEOR headquarters and brings together directors, managers, and consultants from different companies, thus providing an opportunity for exchange of experience and discussion with other enterprises that are also engaged in setting up socio-economic management.

There are two facets to the internal intervener's role: to carry out specified services and to ensure that *socio-economic management* is sustained and anchored within the company. ISEOR thus attempts to achieve a veritable transfer of expertise to the enterprise through these internal interveners, which makes it possible to sustain the management system.

The services performed by the internal interveners involve cointervention participation with ISEOR intervener-researchers in the vertical *socio-economic diagnostics*, including conducting individual and small group interviews, processing the interviews with the *SEAMES® software* program, conducting hidden cost interviews, and calculating the hidden costs. In addition, these individuals are responsible for part of the personal assistance provided to managers and supervisors on socio-economic team-management tools (e.g., *self-analysis of time*, competency grids, piloting indicators logbook, priority action plans, periodically negotiable activity contracts). After the setting-up phase, the enterprise may give them a lighter role, assisting management teams in the maintenance of tools during biannual campaigns of competency grid updating and PAP and PNAC renewals. During the extension (or anchoring) phase, once their competency has increased, they are sometimes entrusted with carrying out entire *socio-economic diagnostics*, with regular, distant supervision by the ISEOR, including leading focus groups and participating in *socio-economic evaluations*.

It is important that *internal-interveners* come from different departments and trades in the enterprise (e.g., quality assurance, human *resources*, management control, audit) and that some of them are managerial staff that perform operational functions such as production or sales. Indeed, this variety in the network of internal interveners increases their credibility among the managers they advise, and permits a better appropriation of socio-economic management throughout the enterprise's different departments.

THE IMPORTANCE OF SOCIO-ECONOMIC MANAGEMENT

The work done by the ISEOR is, in our opinion, characterized by a number of specific traits that stand out in comparison to other change management methods. Economic and social aspects of organizational life are embedded and constitute an *integrated* management system: *integration* of the enterprise's functions, interconnection of socio-economic management tools, and their integration into the change process. The following chapters bear witness to the adaptability of this integrated management system to a wide range of situations and contexts, including different countries, business sectors and sizes, organizational types and corporate objectives and missions.

The findings revealed by studies of ISEOR interventions, of which this book provides only a sample, demonstrate the erroneous character of certain theories that, nonetheless, continue to top the management science "box office" in universities and business schools around the world. Commonly accepted concepts—for example, economies of scale, job specialization, strategies of domination—are called in to question by the theory of *hidden costs*. It is our hope that twenty-first century business education programs will be rethought in light of these findings—a goal that we have been striving to accomplish at the University of Lyon for more than 20 years.

NOTES

1. For a fuller discussion of the idea of confidence, see the *Management Sciences* special issue (1998), especially the articles by Bidault, Charreaux, Devillebichot, Jameux, Marchesnay, Urban, Allouche, Capet, Igalens, Trépo, Albouy, Hirigoyen, Vailhen, Evraert, Teller, Filser, and Peaucelle.

2. *HISOFIS* is an acronym for Humanly Integrated and Stimulating Operational and Functional Information System.

REFERENCES

Boje, D., & Rosile, G. A. (2003). Comparison of socio-economic and other transorganizational development methods. *Journal of Organizational Change Management, 16*(1), 10-20.

Buono, A. F. (Ed.). (2001). *Current trends in management consulting*. Greenwich, CT: Information Age.

Buono, A. F. (2003). SEAM-less post-merger integration strategies: A cause for concern. *Journal of Organizational Change Management, 16*(1), 90-98.

Foucart, M. (2003). Responsabilités partagées et partenariats territoriaux [Shared responsibilities and regional partnerships]. In ISEOR (Ed.), *L'université citoyenne* (pp. 65-87). Paris: Economica.

Harris, M. (2001). Evaluation is value-added in management consulting. In *knowledge and value development in management consulting*. Lyon, France: ISEOR—HEC—University of Lyon.

Hayes, R. (2001, March). Using real option concepts to guide the nature and measured benefit of consulting interventions involving investment analysis. *Proceedings of the First International Conference of the Management Consulting Division of the Academy of Management* (pp. 531-540). Lyon, France.

International Labor Organization. (1998). *Le conseil en management: Guide pour la profession* [Management consulting: A guide for professionals] (3rd ed.). Geneva, Switzerland: ILO Editions.

Millet, N. (2005). Méthode de construction de projets territoriaux [Method for constructing regional projects]. In ISEOR (Ed.), *Enjeux et performances des établissements sociaux: Des défies surmontables* (pp. 209-216). Paris: Economica.

Perez, R. (2001). *Stratégie de la compétitivité et emploi* [Strategy of competitiveness and employment]. Paris: Harmattan.

Péron, M,. & Péron, M. (2003). Postmodernism and the socio-economic approach to organizations. *Journal of Organizational Change Management, 16*(1), 49-55.

Savall, H. (1974, 1975). *Enrichir le travail humain dans les entreprises et les organisations* [Work and people: An economic evaluation of job enrichment]. Paris: Dunod.

Savall, H. (1979). *Reconstruire l'entreprise: Analyse socio-économique des conditions de travail* [Reconstructing the enterprise: Socio-economic analysis of *working conditions*]. Paris: Dunod.

Savall, H. (1987). Les coûts cachés et l'analyse socio-économique des organisations [Hidden costs and the socio-economic analysis of organizations]. *Encyclopédie du management* (pp. 599-628). Paris: Economica.

Savall, H. (2003a). An updated presentation of the socio-economic management model. *Journal of Organizational Change Management, 16*(1), 33-48.

Savall, H. (2003b). International dissemination of the socio-economic model. *Journal of Organizational Change Management, 16*(1) : 107-115.

Savall, H., & Bonnet, M. (1988). Coûts sociaux, compétitivité et stratégie socio-économique [Social costs, competitiveness and socio-economic strategy], *Encyclopédie de la Gestion*. Paris: Editions Vuibert.

Savall, H., & Bonnet, M. (1988). *Outils de pilotage socio-économique des projets industriels* [Socio-economic piloting tools for industrial projects]. Paper delivered at the 2nd International Conference "Génie Industriel, facteur de compétivitié des entreprpises," Nancy, France.

Savall, H., & Péron, M. (2003, August). *How to negotiate possible obstacles between academics and practitioners: action research and consultancy as a scientific research tool in management. Case studies.* Paper presented at the Academy of Management Meeting, Seattle.

Savall, H., & Zardet, V. (1987). *Maîtriser les coûts et les performances cachés: Le contrat d'activité périodiquement négociable* [Mastering hidden costs and performances: The periodically negotiable activity contract]. Paris: Economica

Savall, H., & Zardet, V. (1992). *Le nouveau contrôle de gestion: Méthode des coûts-performances cachés* [New management control: The hidden cost-performance method]. Paris: Éditions Comptables Malesherbes-Eyrolles.

Savall, H., & Zardet, V. (1995). *Ingénierie stratégique du roseau, souple et enracinée* [Strategic engineering of the reed, flexible and rooted]. Paris: Economica.

Savall, H., & Zardet, V. (2000, October). *La valorisation de la recherche en sciences de gestion dans l'environnement économique et social. Enquête auprès des enseignants-chercheurs* [Valorization of management science research in the social and economic environments. Inquiry with teacher-researchers]. Paper delivered at the 9th International Conference on Strategic Management (AIMS), Doctoral Research Commission—FNEGE Research Group, Montpellier, France.

Savall, H., & Zardet, V. (2001). L'évolution de la dépendance des acteurs à l'égard des dysfonctionnements chroniques au sein des organisations: Résultats de processus de métamorphose [The evolution of actor dependency regarding chronic dysfunctions in organizations: Results of a process of metamorphosis]. In T. de Swarte (Ed.), *Psychanalyse, management et dépendance au sein des organisations* (pp. 179-212). Paris: Harmattan.

Savall, H., & Zardet, V. (2004). *Recherche en sciences de gestion: Approche qualimétrique. Observer l'objet complexe.* [Unpublished English translation: Research in management sciences: The qualimetric approach. Observing the complex object.] Paris: Economica.

Savall, H., & Zardet, V. (2005a). *Tétranormalisation: Défis et dynamiques* [Competitive challenges and dynamics of tetra-normalization]. Paris: Economica.

Savall, H., & Zardet, V. (2005b). Le potentiel humain: Source primordiale de création de valeur ajoutée. La théorie socio-économique des organisations [Human potential: The all-important source of creation of value-added. The socio-economic theory of organizations]. *La revue des Ressources Humaines et du Management, 19,* 6-7.

Savall, H., Zardet, V. & Bonnet, M. (2000). *Releasing the untapped potential of enterprises through socio-economic management.* Geneva: International Labor Office - ISEOR.

Savall, H., Zardet, V., Bonnet, M. & Moore, R. (2001). A system-wide, integrated methodology for intervening in organizations: The ISEOR approach. In A. F. Buono (Ed.), *Current trends in management consulting* (pp. 105-125). Greenwich, CT: Information Age.

Tabatoni, P. (2003). *Prospective: l'Université creuset de l'innovation* [Prospects: The university as crucible for innovations?]. In ISEOR (Ed.), *L'université citoyenne* (pp. 237-252). Paris: Economica.

Vangénéberg, R., & Zanta, R. (2005). Horivert multi-PME. In ISEOR (Ed.), *Enjeux et performances des établissements sociaux: Des défies surmontables.* Paris: Economica.

Zardet, V., & Voyant, O. (2003). Organizational transformation through the socio-economic approach in an industrial context. *Journal of Organizational Change Management, 16*(1), 56-71.

APPENDIX 1

SEAMES®

A Professional Knowledge Management Software Program

Véronique Zardet and Nouria Harbi

ISEOR has accumulated a large quantity of data representing experience and know-how acquired through its extensive *intervention-research* work (Savall, 1974, 1975). The collected and stored data has incrementally increased over the years, as has the need for scientific understanding and application of that data. Over the past 18 years, our networked use of *SEAMES® software* (Socio-Economic Approach to Management Expert System) (Harbi, 1990), a professional application for business diagnostic, has enabled us to accumulate this large volume of information about enterprises, their operations and their *improvement projects*. This raw data contains a certain amount of knowledge, codependencies and correlations, implicit and useful, only waiting to be exploited.

Data warehousing and data mining are currently very active fields of research and development: *Knowledge Discovery in Databases* (KDD). This term covers the entire cycle of knowledge discovery, from the selection of data to the conception of large database networks that process information extraction from this data, one of whose stages is data mining.

Socio-Economic Intervention in Organizations: The Intervener-Researcher and the SEAM Approach to Organizational Analysis, pp. 33–42
Copyright © 2007 by Information Age Publishing
All rights of reproduction in any form reserved.

The estimated time spent performing actual data mining (which is the knowledge discovery stage itself) represents less than 20% of project time. Thus, more than 80% of project time is dedicated to operations of data selection, cleaning, enrichment and coding. Data mining signifies the intelligent interpretation of warehoused data, using data processing tools, to uncover operational knowledge. This is also referred to as automatic extraction of knowledge, a term that stresses the final product of this data mining process—*knowledge*. Indeed, information is considered data when it is void of defined usefulness, but it becomes knowledge when it acquires the power to predict.

THE ORIGIN OF DATA

ISEOR conceived and implemented the original method of **socio-economic diagnostic** the principles of which are presented in the first chapter of this book. Since 1986, the ISEOR team, in collaboration with the computer team of INSA Lyon, under the direction of Professor Jacques Kouloumdjian, has worked jointly on a program of modeling knowledge on business **dysfunctions**, their costs as well as their solutions. The first expert system in socio-economic diagnostic was created in 1988 (Figure A.1). **SEAMES® software** is made up of a "factbase" imbedded in a database. Part of this data is permanent (possibility of adding specific data from businesses), and the other part contains data that is linked to the context of each socio- economic diagnostic (**fieldnote quotes**). The rule-base construction is based on the following principle: "*If* this condition, *then* that action."

SEAMES thus facilitates performing qualitative **socio-economic diagnostics** by respecting the interviewed actors' "natural" language (fieldnote quotes), and by crossing contextual data with **generic key-ideas** regarding dysfunctions, which are part of the permanent database. This software program makes it possible to gain in **productivity** while guaranteeing the **exhaustive** quality of identified problems. It also illustrates the complementary nature of the technology and **human potential** in data processing—abundant **fieldnote quotes** (12 to 15 per person interviewed) are expressed in the very words of each person, and then linked to the **generic key-idea** displayed in the program. This operation cannot be done automatically, given the strong polysemic and contingent nature of "natural" words employed by actors. Indeed, the actual key words of fieldnote quotes are always different than those employed formulating the defined key idea. Furthermore, taking key ideas up to an even higher generic level—referred to as **idée-forces** (**pivotal ideas**)—it done without any technological tools.

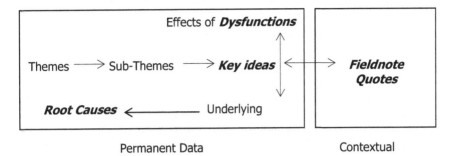

Figure A.1. *SEGESE® Expert System software:* Structure of *generic knowledge* on *dysfunctions*.

Utilization of the SEAMES® software program for more than 18 years in more than 1,000 enterprises, in addition to the expertise of interveners, has permitted the creation of a large volume of datasets on enterprises. Data includes all information concerning the companies where *socio-economic diagnostics* were performed. This information represents:

- data about the intervention (e.g., type of intervention, starting date, cost, number of interveners, number of interviews);
- profile of the company (e.g., name, address, sector of activity, sales figure, number of employees, distribution of personnel);
- results of the *qualitative diagnostic* (e.g., personnel interviewed, fieldnote quotes classified by theme, sub-theme and generic ideas on *dysfunctions*, elements concerning *hidden costs*); and
- elements on innovation projects to reduce dysfunctions.

THE FUSION OF DATA

Data comes from disparate sources. Starting with divergent and separate forms of information, the data fusion process produces a composite object while maintaining the integrity of the initial data. Data fusion is a recent field of research, multidisciplinary and growing fast. Theoretical research in this domain calls on data mining, mapping and visual form reconnaissance techniques, as well as learning from digital and symbolic data. However, this domain, with its multiple applications, still suffers from insufficient overall methodological order.

Figure A.2 captures the complete knowledge extraction process, limited to the processing of structured data. It is not a question here to search through texts or images, but only to process data, knowing that informa-

Sources of data (Data Warehouse)

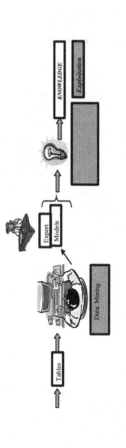

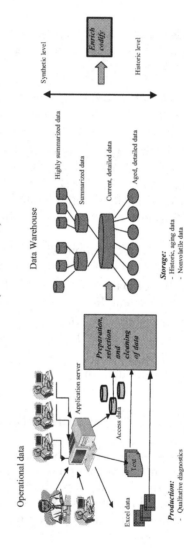

Source: © ISEOR 1988.

Figure A.2. Stages of the Knowledge Extraction from Data Process

tion can be extracted from all forms of data and that the existence of warehouses reduces the time needed to complete a project.

The Information Extraction Process

This process is made up of six stages: preparation, selection, cleaning, enrichment, coding, mining and validation. The flow of these different stages is illustrated in Figure A.2. The project does not unfold in linear fashion. It can become evident during the validation stage, for example, that the obtained performances are insufficient or that users in the domain find the information unusable. It then becomes a question of choosing another data mining method or of reconsidering the coding, or again seeking to enrich data.

ANALYZING DATA IN THE SEAMES DATA WAREHOUSE

One of the analyses consists of submitting qualitative data produced by the SEAMES application to quantitative processing. For example, counting the occurrences of citations by theme, subtheme and key ideas within each subpopulation of the same company (e.g., directors, managers, line personnel), or counting by groups of companies (e.g., by industry, service, public service) (See Table A.1 and Figure A.3; Savall, Zardet, & Harbi, 2004).

Three types of analysis can be performed:

- **Intraorganizational comparative analysis** entails studying the gap in perception of dysfunctions between the different populations of a given company, in order to identify sensitivity gaps and expression contrasts. Figure A.4 illustrates this by comparing managers' comments with line personnel comments. A high degree of proximity can be observed regarding *communication-coordination-cooperation, time management, integrated training* and *strategic implementation*. However, a sharp contrast becomes evident concerning *working conditions* and *work organization*.
- **Transorganizational analysis** illustrates the positions of several enterprises belonging to the same group. In a sample of 2,622 employees in 10 enterprises (Table A.2), for example, sensitivity contrasts can be observed among the three subpopulations. These

Table A.1.   Examples of Results Produced Through Quantitative Processing by Company

| | | Direction | Manage- ment | Shopfloor Personnel | |
		fod2	foe8	fop5	Total
Number of interviews (A):		47	319	44	410
Total number of persons interviewed (B):		47	319	198	564
Total number of fieldnote quotes (C) :		507	2407	581	3495
Working	Nb quotes =	42	8%	89%	89%
conditions	(D) Nb quotes / C =	365	15%	114%	144%
	(E) Nb quotes / B =	161	28%	81%	366%
	(F) Nb quotes / A =	568	16%	101%	139%
Work	Nb quotes =	122	25%	260%	260%
organization	(D) Nb quotes / C =	535	22%	168%	168%
	(E) Nb quotes / B =	110	19%	56%	250%
	(F) Nb quotes / A =	767	22%	136%	187%
Communication-	Nb quotes =	102	202%	217%	217%
Coordination-	(D) Nb quotes / C =	452	19%	142%	142%
Cooperation	(E) Nb quotes / B =	96	17%	48%	218%
	(F) Nb quotes / A =	650	19%	115%	159%
Time	Nb quotes =	38	8%	81%	81%
management	(D) Nb quotes / C =	212	9%	66%	66%
	(E) Nb quotes / B =	36	6%	18%	82%
	(F) Nb quotes / A =	286	8%	51%	70%
Integrated	Nb quotes =	38	8%	81%	81%
training	(D) Nb quotes / C =	228	9%	71%	71%
	(E) Nb quotes / B =	70	12%	35%	159%
	(F) Nb quotes / A =	336	10%	60%	82%
Strategic	Nb quotes =	165	33%	357%	351%
implementation	(D) Nb quotes / C =	615	26%	193%	193%
	(E) Nb quotes / B =	108	19%	55%	245%
	(F) Nb quotes / A =	888	25%	157%	217%

Source: © ISEOR 2004.

differences indicate potential zones of conflict, since they under-score the insensitivity of one group with regards to the strong sensitivity of another. Feeble contrast, on the other hand, is a sign of greater consensus.

- **Benchmarking analysis** permits situating an enterprise within the group to which it belongs. For example, Figure A.5 compares Organization F and Organization G belonging to the same branch of activity. G presents an absolute insensitivity toward time management and integrated training dysfunctions, in contrast with F which demonstrates "ISO sensibility" about *time management*.

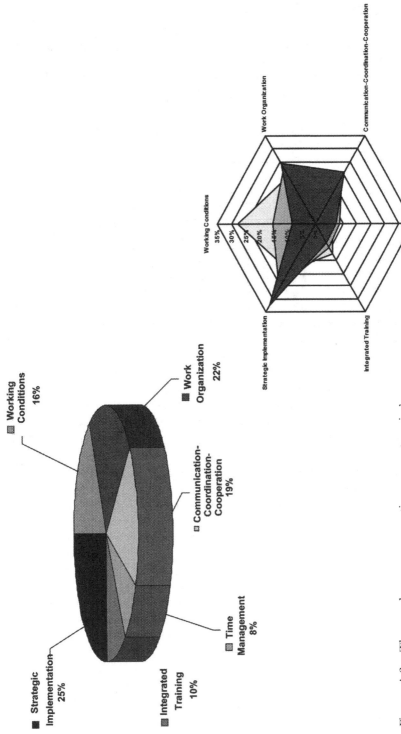

Figure A.3. Three schemas representing an enterprise's results (*dysfunction* themes) (continues on next page).

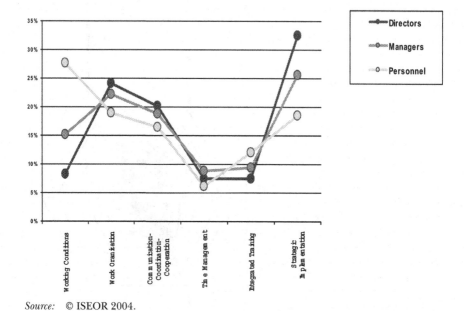

Source: © ISEOR 2004.

Figure A.3.   Continued.

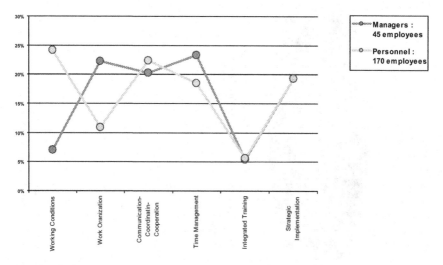

Source: © ISEOR 2004.

Figure A.4.   Comparing expression profiles between managers and line personnel in an enterprise.

Table A.2. Degrees of Sensitivity per Population for the Entire Sample

	Managers/ Directors	Managers/ Line Personnel	Directors/ Line Personnel
Working conditions	+50%	−30%	−37%
Work organization	+17%	+16%	− 6%
Communication- coordination- cooperation	0	+ 7%	+ 4%
Time management	−14%	+50%	− 4%
Integrated training	+51%	+ 6%	−24%
Strategic implementation	−31%	+46%	+40%

Source: © ISEOR 2004.
Note: N = 2,622 persons: 177 directors, 710 managers, 1,735 line personnel.

SEAMES: A GROUPWARE APPLICATION

SEAMES is a type of groupware that through networking, makes it possible to perform collaborative types of work involving several users: ISEOR interveners, private consultants and company ***internal-interveners***. All of these users contribute to:

- *The enrichment of SEAMES® software with new data*: With every utilization, the database is made more complete with new information concerning businesses, their dysfunctions and the specific solutions implemented. It is thanks to this mechanism that the application has evolved since 1986. For example, the initial database of typical dysfunctions has gone from 100 its first version to more than 3,500 in 2005.

- *The fusion of results:* Users can participate in the same interventions without geographical constraints. They can merge and consolidate their work. Thus, internal company interveners contribute to developing diagnostics with experienced interveners from the ISEOR. Fusion of results originating from different enterprises can be performed, with the goal of extracting knowledge on those enterprises.

REFERENCES

Harbi, N. (1990). *Système expert en diagnostic d'entreprise* [Expert system in business diagnostic]. Doctoral thesis in computer sciences. Lyon, France: INSA .

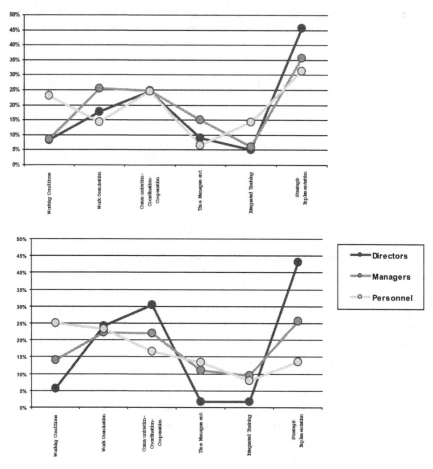

Figure A.5.   Organization F and Organization G graphics in six domains.

Savall, H. (1974, 1975). *Enrichir le travail humain dans les entreprises et les organisa-tions* [Work and people: An economic evaluation of job enrichment]. Paris: Dunod.
Savall, H., Zardet, V., & Harbi, N. (2004, March). *Spectral analysis of socio-economic diagnostics: Qualimetric treatment of qualitative data*. Paper presented at the First International Conference Cosponsored AOM-RMD-ISEOR "Crossing Fron-tiers in Qualitative and Quantitative Research Methods," Lyon, France.

PART I

SEAM INTERVENTIONS IN ORGANIZATIONS

CHAPTER 2

DEVELOPING SUSTAINABLE GLOBAL PERFORMANCE IN SMALL- TO MEDIUM-SIZED INDUSTRIAL FIRMS

The Case of Brioche Pasquier

Véronique Zardet

The Brioche Pasquier Group (BPG) is an independent family-run industrial concern that was founded in 1974. In 1984, the company adopted the socio-economic approach to management (Savall, 1974, 1975, 1979, 2003a, 2003b; Savall & Zardet, 1987, 1992, 1995; Savall, Zardet, & Bonnet, 2000; Zardet, 1982; Zardet & Voyant, 2003), and has since experienced steady, sustained growth in its business, human and *social performance*. The chapter examines BPG's growth and expansion, focusing on the way the company has been managed and controlled, and the contribution that *socio-economic management* has made to the enterprise.

After briefly presenting the company, its *development* and the characteristics of the *socio-economic intervention* it has undergone, the chapter

*Socio-Economic Intervention in Organizations: The Intervener-Researcher and the
SEAM Approach to Organizational Analysis*, pp. 45–70
Copyright © 2007 by Information Age Publishing

focuses on BPG's strategic decisions and the guiding principles underlying them, both in the firm's markets (products and services) and approach to human *resources* management. The chapter concludes with a discussion of the company's governance, strategic and operational management systems, and how they have evolved over time.

THE BRIOCHE PASQUIER GROUP: 1984 TO 2004

The Brioche Pasquier Group (BPG) was founded in 1974, in the west of France, to make and market prepackaged industrial pastries (e.g., brioches, croissants, doughnuts) for large- and medium-sized supermarkets. Since 1980, BPG has been the leading manufacturer in the French pastry market. In 1984, when the company introduced the principles of socioeconomic management, it had 240 employees located at one site. ISEOR's research-intervention program involved several years (1984-1988) of intensive intervention with BPG and its management team. The program, which was then significantly scaled down, has been ongoing ever since. It has constantly accompanied the company's *strategic development*, facilitating BPG's ability to confront a range of issues and challenges that have emerged along the way (see Pasquier, 1992, 1993, 1995, 1998).

The company was founded with a clear product concept. Brioche Pasquier would offer high quality breakfast and teatime products, with product freshness and low prices as its key success factors. All the company's industrial and organizational strategy has been based on constantly improving product quality in terms of taste and freshness, and on maintaining low prices.

Governance and Synchronized Decentralization

The Pasquier Group's governance system played a major role in the company's extensive expansion and its overall performance. As the company grew, it created industrial sites positioned as close as possible to its customers to cut delivery times while preserving the advantages and flexibility of a small organization. BPG is now structured into 14 subsidiaries divided into two fields of activity (bun-making and pastry-making). Three of the subsidiaries, which are located in Spain, Belgium and the United Kingdom, are exclusively commercial. The 11 others are both industrial and commercial sites. The salient features of the system draw on the principle of *synchronized decentralization*, which is evident in the company's general organizational principles, in its strategic thinking and decision-making system, and in its *management tools*.

An Organization Founded on "Fully-Fledged Site" Principles

The company's founding organizational principle rests on proximity and decentralization. Each of the company's 14 sites is considered a fully-fledged operation: each site has its own workforce, with 200 to 300 employees, its own comprehensive budget, and a two-person management team, the site's chief executive officer, who is also the company's official representative, and his deputy managing director, who represents the company's two main business activities. Depending on the site, if the managing director has industrial management duties, his deputy is the sales manager (and vice versa).

The typical organization of a site is characterized by:

- Approximately 20 departmental managers heading the production, sales and staff departments (3 in production: production management, methods and maintenance, quality, and 2 in sales: sales administration and transport-delivery). Thus even on sites with relatively few of people, there is at least one employee in each of the five staff departments to ensure that the site can operate as an independent entity. This strategy helps to reduce the dysfunctions and hidden costs caused by a lack of operational and functional autonomy.

- Commercial functions organized into regional areas. Each site is allotted an area in which it canvasses and manages its existing customer base; for example, each bun-making site produces all the company's products for its own clientele, as does each pastry-making site. As a result, BGP has discarded the specialized production principle in favor of site autonomy in terms of commercial relations, meeting delivery deadlines, and direct and close relations between sales and production.

- A sales department on each site run by a sales manager, who is responsible for his or her own customer portfolio, orders, the "freshness" goal in commercial relations, managing the team of salesmen, and budget.

- Few hierarchical levels, with only four levels from the group level down to the worker or employee level.

- Production is organized into product lines. A manufacturing line, which is run by a line manager, is responsible for the entire manufacturing process from the unloading of raw materials up to the shipping of finished products.

- Preventive maintenance integrated into the lines, both by entrusting all production operatives with the responsibility of level-one maintenance of their machines (e.g. oven, packing machine) and by including a maintenance engineer on the line team. That individual is under the direct authority of the line manager and is in charge of level-two maintenance for all the machines assigned to that line. Each line also has *productivity* and quality indicators, which are printed out every morning for the previous day's activity, and posted up every day at the head of the production line.

A Small Head Office and Strategic Decentralization

The structure of the Pasquier Group includes a relatively small head office, with approximately 60 people—only 2% of the company's workforce. The remaining 98% work in the 14 sites that operate as fully-fledged subsidiaries. In addition to the obvious task of accounting and financial consolidation, the head office's exclusive mission is to formulate methods and work tools, develop the decentralized and consolidated *piloting* indicators, and provide training on the sites to enable the site staff departments to be as autonomous as possible. Group synchronization is thus assured by rules and policies that are common to all the sites.

It is important to note that BPG's structure was based largely on historical reasons rather than to the socio-economic theory of organizations. The company opted not to have either a human resources (HR) director or a marketing director, instead planning to integrate these functions into the duties of appropriate line managers (e.g., HR functions integrated into the duties of each company manager, marketing integrated into the sales function) to avoid the drawbacks of critical functions that were out of touch with the realties of the organization and its customers. In addition, information technology plays a key role in the company, with an information system that is accessible in real time from any location in the company. In this way, each site's financial indicators can be accessed remotely, as can the consolidated figures per activity. "On-board" computing has been used for more than 20 years, and all delivery salesmen have access to a computer in their trucks allowing them to transmit updated orders from the customers whose orders they are delivering.

Internal and external communication is also decentralized to each site management department. Each year, after the group's financial statements have been released to shareholders and to the stock market, each site management department organizes a verbal presentation of site and group results for all personnel, at which each employee is given a copy of the annual financial statements. This form of decentralized organization into profit centers, featuring the widespread implementation of the principle of internal customer-supplier relations and internal invoicing and transfer prices, nonetheless allows consistency and *synchronization* to be maintained, both through the application of common organizational principles and through the introduction of *cooperation* mechanisms and *management tools*. The company's control function is polymorphous; it includes "traditional" budgetary control, cost accounting featuring monthly closings that are available in a very short space of time (the 8th of the following month), and an assessment of strategic achievements every 6 months, which is qualitative as well as quantitative and financial.

Company Growth Patterns

A study of the company's growth reveals three distinct periods:

- **1974 to 1990:** The major challenge facing the company was to control a high rate of internal growth (20% annual growth rate for the period) as the company became the market leader. The period from 1984 to 1987 was more specifically characterized by the company's high rate of internal growth, an increase in workforce and pay scales, productivity gains in the workforce, a sharply increasing rate of self-financing investments, and a consolidation of owners' equity.
- **1988 to 1989:** A second *development* phase began, and the rate of growth remained high based on both internal and external expansion.
- **1990 to 2000:** The group embarked on a diversification strategy in the industrial pastry business, successively buying out four small firms. The major challenge then became that of controlling the dual diversification in products and markets, since industrial pastry is primarily aimed at nonhousehold catering customers (e.g., schools, hospitals, armies, local authorities). Over this 10-year period, BPG also embarked on a strategy of international expansion, both by extending its marketing operations outside France to neighboring countries and by conducting external growth operations, two in Spain and one in Italy.

Since 2001 the growth rate in both BPG's domestic and international markets appeared to slow down, confronting the company with new challenges characterized by low growth and high liquid assets (400 million euros cash, no debt).

The Appendix, which retraces the variation in the group's main economic and financial indicators, indicates the success of the company's strategy vis-à-vis its environment. Over a 25-year period, the company has successfully pursued a strategy of economic and commercial growth by paying close attention to consumer needs. Operationally, BPG has also significantly reduced the lead time between product manufacture and sale (further supporting its commitment to product freshness). The group's high growth rate has been achieved at the expense of its competitors on a low-growth market. As an illustration of this impressive growth rate, Brioche Pasquier's pastry revenues grew by 18% in 2001 compared to only 8% for the entire French market. In 2002, Brioche Pasquier was the leader in France in the pastries segment, with a 44% share of a highly competitive market (the number two company, the Harry's group, had 18% market share).

ISEOR'S 20-YEAR RESEARCH AND INTERVENTION PROGRAM

In 1984, ISEOR conducted an initial 1-year *intervention-research* at Brioche Pasquier. Although the company was experiencing high growth and satisfactory business and social results, its management foresaw long-term risks related to the growing size of the company and its sustainability. In essence, the company was faced with the question of how to prepare its future managers over the next 15 to 20 years to take over from the founding managers, who at the time were barely in their early 40s.

1984-1986: The First Horivert Intervention

The initial *Horivert intervention* (see Figure 2.1) included only one *vertical action*: the production plant, which at the time employed 135 people (nearly two-thirds of the total workforce). The *horizontal action* consisted of (1) carrying out the *horizontal diagnostic* of the enterprise, through interviewing all members of the board of directors and senior management, and (2) providing *collaborative training* for all managers in *socio-economic management tools*. The *vertical action* in the production factory entailed carrying out a *socio-economic diagnostic* of *dysfunctions*

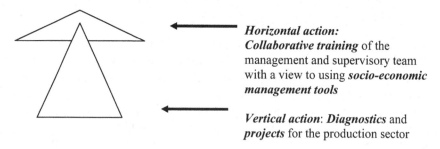

Horizontal action:
Collaborative training of the management and supervisory team with a view to using *socio-economic management tools*

Vertical action: *Diagnostics* and *projects* for the production sector

Source: © ISEOR 1984.

Figure 2.1.   The *Horivert intervention* (1984-1986).

Table 2.1. *Main Dysfunctions*:
Initial *Socio-Economic Diagnostic* (1985)

Working conditions
- Irregular working hours in the packaging department
- Cluttered premises, lack of space

Work organization
- Uninteresting work in packaging and lack of motivation
- Compartmentalization: barriers separating the dough/baking/packaging departments

Time management
- Delays and backlogs in maintenance work

Communication-coordination-cooperation
- Lack of communication-coordination-cooperation between manufacturing and support services (maintenance, shipping, etc.) and the sales department

Integrated training
- Lack of flexibility of production operatives

Source: © ISEOR 1985.

Table 2.2. *qQfi Evaluation of Hidden Costs* per Person per Year (1985)

Indicators	*Qualitative* Assessment	*Quantitative* Assessment	*Financial* Assessment
Absenteeism and work accidents	High rate in packaging	5.8%	1,600€
Personnel turnover	Not assessed	Not assessed	Not assessed
Nonquality	• Scrapped dough • Unsold recalled products • Breadcrumbs	21 tons per year 1% of products 75 tons per year	1,500€
Direct productivity *gaps*	Frequent excess weight Production losses	50 tons per year 4% per year	2,000€
Total *hidden costs* per person per year			5,100€

Source: © ISEOR 1985.
Note: All figures have been converted to euros at 2004 value.

and evaluating *hidden costs*, and then conducting a socio-economic innovation *focus group* to look for solutions to these dysfunctions.

The socio-economic diagnostic in the plant created an inventory of the main dysfunctions in terms of how work processes, *communication-coordination-cooperation* and integrated training were organized (summarized in Table 2.1). The analysis also revealed *hidden costs* amounting to €5,100 per person per year (see Table 2.2).

The *collaborative training* and consultation program, which involved the company's senior management and the managerial staff in charge of the administrative, industrial and commercial teams, provided the foundation for the adoption of the *socio-economic management method*. All managers from production, commercial and administrative departments were trained in *time management*, improvement of *communication-coordination-cooperation*, multiannual strategic action plans and biannual *priority action plans*, establishing personalized objectives, and setting up *socio-economic organization*.

As part of ISEOR's "toolkit," *priority action plans* (**PAP**) and *periodically negotiable activity contracts* (**PNAC**) were introduced throughout the company (1986), and have been used on a permanent basis every since. These tools are now viewed as "commonplace" and used by all organizational members. In 1985-1986, a focus group corun by the company chairman and CEO and ISEOR *intervener-researchers* proposed a socio-economic innovation action plan, which included among other things an organizational model for the plant based on product lines organized as microplants. Following the deployment of this plan in 1987, a socio-economic *evaluation* revealed that senior management, managerial staff, and line personnel perceived substantial *improvements*—and evaluated a gain of €30,100 per person per year (see Table 2.4).

Among the major actions, setting up production lines organized as microenterprises generated significant results. One single supervisor was now responsible for manufacturing and packaging a line of products, with a dozen employees under his or her supervision carrying out tasks of kneading, baking, packaging and maintenance. This type of multitasking capacity on the production lines made it possible to more easily compensate for absences and vacation leave. Under the former organization, kneading, baking, packaging and maintenance were separate services that never felt responsible for product quality, and even less for *productivity*. Under the new organization, every morning the supervisor posts productivity, quality and rejects results from the day before for the production line as a whole and for each individual operator. In connection with this increased responsibility of the operators, the role of supervisors evolved considerably toward BPG's mission of development and improvement in security, quality and operator training.

The fact that the reduction in *hidden costs* of 30,100€ was greater than the amount of hidden costs calculated in the initial 1985 diagnostic is not unusual. The difference indicates that calculated hidden costs are typically underestimated at the diagnostic stage, because the staff that helped work out the figures had not yet mastered the hidden cost detection technique. They acquire these skills in the subsequent stages of the change

Table 2.3. Main *Innovation Actions* Introduced in the First Plant (1985)

- Organization into product line units: A line manager was placed in charge of a team, from kneading through to packaging, including a maintenance technician

- Occupational flexibility was developed in each line

- Production schedules and working hours were stabilized

- Managerial staff's role was enhanced by entrusting them with additional development tasks: Improvements in safety, health, etc.

- A loss-making product line was shut down

Source: © ISEOR 1985.

Table 2.4. *qQfi Evaluation* of the *Conversion of Hidden Costs Into Value-Added* per Person/per Year: First Plant (1987)

Indicators	*Qualitative* results	*Quantitative* results	*Financial* results
Absenteeism	• Greater motivation • Better working hours	Absenteeism at packaging down 0.98%	1,000€
Work accidents	• Better training and awareness of the risk of accidents	Not assessed	Not assessed
Personnel turnover	Not assessed	Not assessed	Not assessed
Nonquality	• Better handling of faults and defects • Fewer scrapped articles	• Losses down 0.5% • Improvements in quality (recall rate down 0.39%)	8,500€
Direct productivity gaps	• Socio-economic organizational model introduced based on product lines	• 2% of gains per day • Removal of a loss-making line relating to improved management tools • Time-savings in the organization of production • Less maintenance work	20,600€
Total conversion of **hidden costs** per person per year			30,100€

© ISEOR 1987.

process, a veritable learning process involving new management methods and tools.

Hidden costs are essentially evaluated on the basis of interviews with the management, which serve a double objective: estimating dysfunction frequency (e.g., average number of breakdowns per day) and describing *regulations* that respond to *dysfunctions* (e.g., preassessing breakdowns, calling and waiting for the maintenance service, stopping production). Until company actors become more familiar with such dysfunction analysis, they tend to forget to mention certain dysfunctions, underestimate their frequency and/or the time spent regulating dysfunctions. All of these factors combined mean that the first hidden cost estimation in an enterprise tends to be underestimated by at least one third.

1987-1988: Horivert Intervention in the Second Plant

In 1986, the company built a second plant in the east of France, and soon noticed a number of dysfunctions on the new site. A second *Horivert intervention* was conducted, including a complete *socio-economic diagnostic* of the site and its 28 workers. The dysfunctions it revealed (see Table 2.5) indicated a striking similarity with the initial 1985 diagnostic, but with the *hidden costs* (Table 2.6) assessed as being much higher.

Launching a new enterprise or a new unit frequently generates major costs. Certain costs are related to the learning process of organizational members who lack experience, often leading to product *nonquality* or insufficient *productivity*. Other costs are linked to organizational actors

Table 2.5.   *Dysfunctions* in the New Plant (1987)

Working conditions	• Bad layout of the premises
	• Lack of safety on certain machines
	• Frequent changes in production personnel working hours
Work organization	• A high degree of compartmentalization between manufacturing and packaging
Communication-coordination-consultation	• Lack of general information on the company for site personnel
	• Strained relations between the new site and the parent plant
	• Excessively informal mode of management on the part of supervisory staff
Integrated training	• Unqualified workers

© ISEOR 1987

Table 2.6. *Hidden Costs* per Person/per Year:
New Plant (1987) Partial *Evaluation* Based on Two Indicators

Nonquality	
• Breakdowns	2,500€
• Adverse effects to sales	1,500€
Direct productivity gaps	
• Overproduction on one line	5,000€
• Underproduction on another line	3,400€
Total *hidden costs* per person and per year	12,400€

Source: © ISEOR 1987.

not being acquainted with one another and who are just beginning to learn to work together, which causes high *communication-coordination-cooperation* costs.

1987-1988: Intervention in Sales

The next two interventions involved a training program on *socio-economic management* for senior management and the sales staff, and an in-depth analysis of the *dysfunctions* and assessment of *hidden costs* in production. In view of the difficulties involved in developing the company's sales, however, a broader socio-economic diagnostic and innovation project were implemented, concentrating more particularly on the recently established site in Paris. On the basis of the dysfunctions (Table 2.7) and hidden costs (Table 2.8) observed, improvement actions were worked out, organized and put in place.

The principal *improvement* actions that were implemented involved:

- establishing a more regular system of *communication-coordination-cooperation* among the heads of sales, salespersons and sales promoters;
- devising a framework for welcoming and integrating new employees, designed to reduce turnover, featuring visits to the home offices and the production sites for all new salespersons;
- a mentor/sponsorship system, in which each new salesperson was assigned a mentor/sponsor (a company employee who accompanies him or her over the first few months on the job); and
- clearer delegation of tasks between the head of the division and the head of sales.

Table 2.7.   *Dysfunctions* in the Sales Area (1987-1988)

Work organization	• The function of sales representative vaguely defined • Organizational problems due to a high rate of staff turnover
Time management	• The difficulties managerial staff had in getting organized when faced with commercial imperatives
Communication-coordination-cooperation	• Poor relations between head office and the sales areas • Lack of consultation between salesmen and sales promotion staff
Integrated training	• Managerial staff insufficiently trained in recruitment techniques and sales training • Insufficient induction and *integration* of new employees
Strategic implementation	• Pay scales not in line with the cost of living in the region

Source: © ISEOR 1988.

Table 2.8.   Partial *Hidden Costs* Analysis per Person/per Year: The Sales Area (1987-1988)

Absenteeism and *work accidents*	4%	Not assessed
Personnel turnover	40%	2,500€
Nonquality and *direct productivity gaps*: • Loss of goods • Promotional rate exceeded • Late deliveries • Disputes		8,500€
Total *hidden costs* per person and per year		11,000€

© ISEOR 1988.

Finally, a new stock and order management system was set up to reduce the numerous errors in ordering stock, overstocking and depleted stock, which were identified during the diagnostic. It consisted particularly of reinforcing *communication-coordination-cooperation* between sales personnel and the sales management department, setting up indicators relative to errors and making each salesperson aware of the importance of following the ordering procedure.

The year 1988 marked the end of the period of intensive intervention aimed at introducing the socio-economic management system. The next steps involved the *"maintenance" phase* of the *socio-economic management system*, in which the company's senior managers and two ISEOR *intervener-researchers* participated in a 1-day session twice a year. The basic objective of these sessions was to identify and analyze ongoing-problems, with the intent of responding with appropriate changes in strategic investment, structural organizational decisions, and selection of new management tools as needed. These sessions have been held twice a year since 1999 (the 34th session took place in April 2004).

1995-1998: Follow-Up and Support

As a follow-up to the initial work done, ISEOR intervener-researchers worked with BPG's senior management to undertake an *evaluation* of the company's economic *vigilance* system, including a precise diagnostic of the *dysfunctions* apparent in this system with a focus on creating and marketing new products. Emphasis was placed on reactivating the *strategic ambitions* of the *Internal/External Strategic Action Plans* (*IESAP*), which the CEO considered too focused on company growth to the detriment of strategic innovation, particularly in terms of products and markets. Decentralized strategic alerts were implemented throughout the entire enterprise by lines of products, major raw material purchases, and technology. These strategic alerts were entrusted to binomial teams made up of the head of a plant and an operational group leader. Following this strategic alert stimulation action, the company created and commercialized new products, broke into new markets, in particular by intensifying its presence throughout Europe. The company bought a factory in Spain and created plants in the United Kingdom, Belgium and Italy.

In 1988 ISEOR carried out a *socio-economic evaluation* of the flexible working hour policy the company had voluntarily implemented as early as 1995, which was well before the relevant French laws that were promulgated in 2000 and 2001. The assessment revealed the positive achievements of the policy and measured the *improvements* the company had achieved in *economic* and *social performance*.

The results, in terms of *social performance*, showed that the enterprise increased the number of jobs by 15% (compared to its initial objective of 10%), maintained salary levels (despite a 15% reduction of weekly working hours), and made major training investments in its employees, both old and new. Interviews carried out with both management and shopfloor employees pointed to personnel satisfaction and brought to light the factors that made it possible for the flextime and worktime program to generate *economic performance*. Sales increased by 27% and a high-level customer service was maintained, even on employees' days off, thanks to augmented multitasking capacity. These outcomes were made possible through careful job redeployment, in-depth study of tasks that could be eliminated, and delegation of managerial responsibilities to personnel. The *hourly contribution to margin on variable costs* (**HCMVC**) increased by 38%.

BPG Performance Context

The Pasquier Group is a perfect example of *SEAM implementation* over a long period, more than 20 years, supplemented by regular light ISEOR interventions. In the early years, performance gains following *in-depth intervention*, when the objective was to set up socio-economic management throughout the firm, consisted of converting hidden costs into *value-added* in a context of steady growth. From 1990 onward, the objective shifted to consolidate the *integration* of socio-economic management tools and concepts though *maintenance*/follow-up efforts (two ISEOR intervener-researchers working one day every six months with top management). Currently, a new emphasis has emerged, with questions about how BPG could avoid energy waste and management system entropy, and structure and organize the enterprise in the context of its growth. During the company's growth, its labor force went from 250 employees in 1984 to 3,000 in 2006. In this type of high-growth context, there is high *risk* of an experiencing an exponential rise in *dysfunctions* and *hidden costs*. At the same time, pressures from a highly competitive environment and with extremely aggressive behaviors from clients (major distributors, superstores) also exerted significant pressure on the firm.

BPG's Stock Performance

Founded with only 20,000 French francs (3,000 euros) of owners' equity, the company is now listed on the stock exchange on France's second market. In 1985, Brioche Pasquier went public, offering 20% of its capital to staff members and 20% to the public. In 1994, the French economic journal *Usine nouvelle* rated BPG the top French industrial concern among those firms with revenues less than 3 billion French francs. In December 1997, a study conducted by the Boston Consulting Group identified the company as having created the most *value-added* on the French market between 1987 and 1997, with a stronger stock market performance than that of well-known multinational companies. All these various citations testify to the financial stability of a successful company that has withstood significant pressure and change in a highly competitive environment.

In 2002, the value of the company's stood at 400 million euros. In 2002, however, the market price dropped 18%, largely due to the short-term expectations of financial analysts, who criticized the "low" rate of return of BPG's European expansion strategy. In response, the following year the company decided to buy back 90 million euros of the publicly floated shares, increasing the Pasquier family's share of capital back up to 85% (compared to 57% in 2002). The proportion of shares owned by organizational members was approximately 10%.

Human and Social Performance

Brioche Pasquier's strategic management of human *resources* over the last 20 years was guided by a number of policies and commitments, including:

- A high level of investment in *integrated training* for all personnel (roughly 10% of total payroll), a commitment that has lasted for more than 20 years.
- A strategy focusing on developing personnel skills, responsibilities and internal promotion.
- An executive recruitment policy, searching for a combination graduates, postgraduates and self-taught personnel.
- A flexible working hour policy (working hours reduced to 33.75 hours a week), creation of new jobs (15% increase), and expansion of customer delivery days from five to six per week, all of which has significantly improved product and service quality.
- A policy of employee share-ownership and financial profit-sharing based on *periodically negotiable activity contracts*.

Staff turnover and *absenteeism* rates in the group have also been low (except for turnover in sales staff), in large part due to the opportunity internal promotion made possible by BPG's steady growth.

STRATEGIC CHOICE, DEVELOPMENT OF VALUE-ADDED AND HIGH QUALITY CUSTOMER SERVICE

The company's major strategic decisions, which are characterized by its internal structure (i.e., choices regarding organization and management methods), facilitated its attempt to enact its *external environment*. BPG believed that its own *strategic force*, facilitated by the application of ISEOR's methods, would allow it to exert influence on its market stakeholders—customers, suppliers, and competitors.

A Strategy Based on Innovation and Strategic Patience

Brioche Pasquier's strategy was to distinguish itself through a high degree of innovation in terms of services, products and markets. One of the founding principles of the company was ecology, with an emphasis on making healthy products using high quality raw materials. The company pursued an own-brand strategy (Brioche Pasquier) combined with the image of a traditional, quality product. To consolidate this brand strategy, BPG had to confront its supermarket resellers, who always tried to boost the sales of their own brands rather than producer brands. This strategy has proved to be effective, especially in comparing BPG's rate of growth and earnings to that of the supermarket brands. As an example, the company's share of supermarket brands fell from 20% in 1996 to 10% in 2001; in 2001, the company's total revenues rose 18% from the previous year, while the share of supermarket brand products fell by 34%.

Another feature of BPG's strategy is its *strategic patience*, giving priority to controlling what existed internally before pursuing expansion, in particular expansion through external growth. As an example, although the company had numerous takeover opportunities it only followed through on a selected few because BPG is very demanding in terms of its financial criteria, quality of technology, the human resource potential of the acquired work force, and the high degree of *creation of potential* generated by the target company. After each acquisition, a detailed examination of the target firm is carried out, focusing on its products, organization, and tools and management principles in order to ensure consistency for all the sites.

Strategic innovation results from strategically analyzing and acting on the *hidden infrastructure* of the organization. In analogy to building terminology, the hidden infrastructure refers to the foundation components required to carry out the company's characteristic business activities, which make up the *superstructure* of the company. For instance, in the context of a potential acquisition, the *infrastructure* of the target enterprise can be examined through the *competency grid* of management team members and by studying internal team *cohesion* through an examination of it new product (or forthcoming) product portfolio. Similarly, the acquiring firm can assess the degree of autonomy, responsibility and technical competency of production operators. Theses elements can be detected by observations during company visits and interviews with key members of the management team.

Strategic innovation has also manifested itself through a highly interactive conception of the commercial supply-demand relationship. In BPG's marketing vision, the company has favored the idea by which the solvent need of the customer-consumer progressively emerges and takes shape through a process of *interaction* between the company and its customers. This emerging concept in marketing research (see Filser, 1994) involves literally all organizational members as "active" strategists who exert influence on their environment, gather information of strategic scope, and use it "in real time" (i.e., short spaces of time ranging from 1 day to 1 week).

In essence, the company engages in *active and participative vigilance* of its environment. In particular, marketing and sales functions were drawn closer together and the role of the salespeole has been explicitly integrated (with qualitative and quantitative objectives) with the reporting of information on the competition. Consumer behavior was also observed in real time in sales outlets. In this latter instance, sales personnel are in the best position to check whether frontline shelf space (where products are displayed) conforms to the negotiated requirements, or whether the distributor has reduced the shelf space to the possible benefit of a competitor that is leading an aggressive commercial campaign.

Finally, the company's main strategic decisions were innovative in comparison with the strategic behavior of its competitors and the recommendations typified in prevalent strategic models. Focusing its commercial strategy on the freshness of its products, for example, BPG has, at times after numerous hesitations, made a number of strategic industrial and commercial decisions, the most characteristic of which are discussed below.

Competing Through Price

The supermarket sector is notorious for ruthlessness price negotiation with suppliers. The Brioche Pasquier Group is admittedly no exception to the rule. At the same time, however, BPG relentlessly pursues a strategy of cutting its own costs by reducing its *hidden costs*, converting them into *value-added*. As a result, product prices regularly fall (in constant currency), but the company's earning capacity continues to steadily grow. This reallocation process concerns not only production and higher sales volumes, but also the development of the company's *potential to create*, that is, intangible investments throughout the organization with the participation of many actors, from internal focus group brainstorming sessions to improve delivery times and product use-by dates (fresh products), and carefully preparing the start-up of new sites (e.g., the entire managerial team of a newly created site is recruited two years before the site starts operating), to investing in highly developed training schemes (e.g., nearly 10% of the payroll is regularly allocated to training). The reduction of *hidden costs* and their *conversion into value added* has therefore led to a strategy of controlled growth in terms of the workforce and pay scales.

Decentralizing the Sales Force

A long-term trend in supermarkets is the high concentration of supply, especially in terms the competition. Yet, despite such concentration, BPG decentralized its sales force to ensure greater service and support for its customers. In this way, general terms of sale are not centrally negotiated with leading customers (e.g., 8 customers make up 80% of group sales) at the head office. Instead, they are negotiated by different site managers in close consultation with the Group General Management.

Opting for Multipurpose Plants

Although many companies opt for specialized sites per product, BPG decided to produce a wide range of products in one location. Despite the significantly higher level of initial investment, the company has succeeded in cutting the hidden costs generated by specialization (see the discussion of fully-fledged sites later in this chapter).

Growing Market Share Through Quality Initiatives

In an extremely competitive market with relatively low-growth for several years, Brioche Pasquier took a gamble to boost its sales through improvements to its products and services, attempting to grow its market rather than merely attempt to take market share from its competitors. BPG's strategy was to continually offer higher quality through constant and careful monitoring of end-user needs, above and beyond those fulfilled by supermarkets, as a way of increasing demand. As a result, the firm had become a driving force in the growth of the French pastry market for a number of years, whereas its competitors were mired in low rates of growth.

Such strategic decisions contributed to high levels of performance—while the company's short-term earning capacity obviously improved, this has gone hand in hand with a high level of activity in creating potential. The proportion of potential creation in relation to *immediate results* can judiciously be evaluated by the magnitude of the following ratio:

$$\frac{\textit{Indirectly productive time} \text{ (charged to intangible investments)}}{\textit{Directly productive time}}$$

When applied to the entire workforce, this ratio is the expression of the company's steady innovation drive, including social innovation, product innovation, technological innovation, and innovation in management by taking *dysfunctions* and *hidden costs* into account.

With regard to industrial and commercial organization, the company's major challenge was to reduce the lead time between manufacturing and product consumption as much as possible. Over approximately a 20-year period, BPG was able to cut this lead time from 10 days to 24 hours in 80% of cases, an achievement due primarily to controlled flows of production, enhanced information between supermarket customer production sites, and the creation of production units throughout the country to reduce the distances with customers. Production ultimately shifted to a just-in-time basis, 6 days a week with two 7- to 8-hour shifts, throughout the year. As an example, 90% of orders that received electronically every day at 1 p.m. are delivered to the points of sale the following day at 9 A.M. and are available for purchase by 10 A.M.

The process of systematically listening to the end consumer has been accompanied by a quality-rating system. Quality marks are given every day by a cross-section of the company's staff on the basis of specific criteria (e.g.,, taste, color, shape, texture). These marks are an integral part of the production management control charts.

A Strategic Approach to People at Work

Brioche Pasquier has fully integrated the inherent logic of *hidden costs* into its budgetary, strategic and business decisions. Each member of the company's managerial staff and all the employees are fully alerted to any dysfunctions they detect and to their resulting hidden costs. Such assessment has been elevated to a basic principle of BPG management. As an example, the number of bad tele-transmissions to customers is recorded in *piloting* indicators, as are the immediate or subsequent impacts of these problems on product turnover and the risk of worsening commercial relations with consumers. This type of assessment led the company to discover that a different technological option, which was initially viewed as more costly, was actually more economical in that it helped to convert *hidden costs* into greater *value added*. The improved performance of the company (see the Appendix) clearly shows that the reduction in the company's hidden costs resulted both in lower production and distribution costs and, above all, in improving product turnover and increasing sales. In essence, hidden costs have been converted into increased *value added* and cash flow.

Within this context, BPG's personnel are considered a strategic lever for creating *value-added*. Recognition, a sense of responsibility and contractualization are the major principles underlying the firm's human resources management. The *socio-economic management method* is founded on a principle of *internal marketing*, where socio-economic diagnostic facilitates the ability of the organization to capture and focus on fulfilling employee needs and to create a periodic inventory of management failings and dysfunctions. The methods and know-how of the company's workforce are thus formalized and circulated on a regular basis—to the point where preventive actions can take up the equivalent of 30% of company working time. For instance, several years ago a large focus group met on a regular basis to develop a standardized packaging and palleting system for all the company's various products. The project had an important knock-on effect financially, cutting the cost of packaging materials, stock reference management and packaging labor costs.

In the firm's industrial units, production was organized around product lines: a team of operatives headed by a line manager was responsible for making the entire product, from mixing the dough right through to the palleting machine. Each line was evaluated every day according to two indicators—quality and productivity. This process was put in place in 1985, following a *socio-economic diagnostic* that revealed the numerous dysfunctions inherent in the highly compartmentalized organizational model (dough—manufacturing—packaging; see Table 2.3).

ORGANIZATIONAL VISION AND STRATEGIC VIGILANCE

Strategic vigilance, thinking and decision-making are generally considered the preserve of headquarters in the corporate world. The Pasquier Group has opted for a totally different form of organization, which was inspired by the desire to remain close to the marketplace and to maintain a minimal structure at headquarters. For ten years it has put in place a focus group mechanism, which is referred to as "file" (guideline) as a way of ensuring economic *vigilance* and appropriate strategic decisions.

The term "file" is an abbreviation of an ISEOR concept, that of "chef de file" (leader) for strategic vigilance. When the Pasquier Group wanted to revitalize its *strategic vigilance* system in 1995, ISEOR suggested that the company follow the *synchronized decentralization* principle it had already adopted for its operational activities by extending it to strategic vigilance. This led the company to abandon the idea of setting up a dedicated strategy department and to share the various domains (products, markets, technologies, human potential) across the 14 sites.

A "file" is led by one of the 14 sites and run by the site manager (or his or her deputy) assisted by a functional manager from the head office. There are roughly 10 "files" in the company, with individual sites leading a product "file" (e.g., doughnut file, milk roll file) and a customer file (corporate customer) for the entire group. One of the sites is responsible for purchasing raw materials (e.g., chocolate) for the entire group, and is in charge of strategic vigilance regarding potential suppliers, world prices, and strategic decisions for the group management committee. The latter meets four times a year, bringing together all site managers, the three directors of BPG's activity centers, and the group's senior management. This committee assesses the work prepared by the "files" and arbitrates and defines new strategic orientations. The final strategic decision falls to the group's chief executive officer and managing director.

This strategic consultation mechanism is supported with a strategic and operational *communication-coordination-cooperation* mechanism involving all group executives. Every 6 months, team operational plans (*priority action plans*) are created, bottom-up, for each department on each site. The site managers convene their middle management team (line managers, sales managers, heads of staff departments) for the purpose of identifying the site's major dysfunctions and setting priority goals for the site for the coming 6 months. Each member of middle management then puts forward five main points of the next *priority action plan* for the site during a planning session, which ensures the synchronization of the *PAPs* for a given site. A 2-day seminar organized by the Group then takes place, bringing together all the two-person site management teams. Each member presents the priority actions proposed in their site PAP, further facilitating the *synchronization* of the PAPs between sites.

The *priority action plans* hold a central place in strategic planning. First, they complement long-term strategic thinking (a timeframe of several years) by proposing half-yearly action plans including actions on two environmental targets involving *external* and *internal actors*. Second, they allow all of middle management, including line management, to be involved in preparing for *strategic implementation* since the plans are drawn up by the units. Finally, they serve as a basis for negotiating half-yearly aims-means pairs for people or small groups in the units.

The collective *cooperation* mechanism, based on the PAPs and half-yearly seminars, is completed with personal consultation mechanisms. The priority action plans are scaled down by the PAP leaders for each of their employees in the form of activity contracts. These contracts are periodically negotiable, which the company refers to as improvement contracts and which feature team (e.g., improving the quality of a production line) as well as personal (e.g., training, skill development, participating in a focus group) goals. The *improvement* contracts are appraised every six months, giving rise to a bonus equal to approximately one month's annual pay. In parallel, monthly steering meetings, known as "recommendations," bring together all the PAP leaders and their immediate superiors to examine the *progress* of all the PAPs and to confer with one another regarding future actions.

The preparatory stage of the priority action plans (once every 6 months) and the appraisal-*negotiation* of the *periodically negotiable activity contracts* facilitate the development of a highly articulated organization, combining strategic and operational decision making. As a result, the company's strategic goals find concrete expression in the actions carried out by the units. In this sense, *strategic implementation*—that is, choosing a combination of actions to achieve one's goals—has a retrospective effect on the nature and level of the company's *strategic ambi-*

Table 2.9.   BPG Major *Innovative Strategic Decisions* (1984 to 1994)

1. Construction of a relocated plant in eastern France instead of doubling the size of the parent plant.	1984
Abandonment of economies of scale and size.	
2. Nonspecialization of the 8 plants. Abandonment of cost-cutting minimization by adjusting production runs	1990
3. Plants became fully-fledged sites (subsidiaries): Global production and service costs minimized rather than the sole cost of industrial production	1987
4. Rational management of product portfolios: better product mix new/old "refreshed" products rather than a headlong rush toward new products	1989
5. Decentralization to corporate account portfolio sites in the face of the concentration of supermarkets.	1992
Marketing and sales force functions merged.	

Source: © ISEOR 1994.

tion. This retrospective effect occurs over a very short space of time (a few weeks), in step with the process of preparing the priority action plans, and over the 6-month period during which the strategic implementation actions take place.

CONCLUSION

The Pasquier Group has created a management system founded on decentralization, organization and ISEOR-based management tools, together with real synchronization and combined strategic and operational planning. Since 1985, the strategic leader mechanisms, PAPs and improvement contracts have enabled the company to steadily evolve without experiencing significant setbacks. This long-running successful performance has more specifically been achieved on the basis of strategic decisions (see Table 2.9) that have moved BPG away from traditional management principles that have been called into question by the socio-economic theory of organizations. The underlying process has also experimentally endorsed the innovative strategic choices that the Pasquier group has successfully implemented in its growth strategy.

The role and contribution of the ISEOR and *SEAM* approach in the success of the Pasquier Group can be summed up as follows. First, BPG and its management team have developed a strategic action dynamic (the interval between the decision and achievement of the decided action), including its major strategic decisions. Second, ISEOR has fulfilled a mission of *scientific consultancy*, and not merely that of an advisor recommending quick fixes. Solutions are suggested to management based on socio-economic management concepts connected with the *socio-economic theory of organizations*. In essence, there are never "ready-made" solutions applied by the consultant. In the Pasquier Group, this scientific consultancy enabled the CEO to make innovative decisions, both in terms of the organization (see Table 2.9) and its product-market strategy. The biannual intervention *maintenance* follow-up by the ISEOR with the group's board of directors includes a first half-day, during which the participants discuss the current major *dysfunctions* of the firm. The second half-day consists of a participative analysis of those dysfunctions and their origins. This analysis is based on concepts which enable the enterprise to outline solutions and decisions, and the follow-up *implementation* is regularly checked during the next biannual session. Among the concepts that have been drawn on in recent years, there has been strategic *cohesion* with the firm's *strategic ambitions*, the assets and constraints of *synchronized decentralization* have been assessed (in comparison to centralization), visible costs versus hidden costs have been analyzed, and the rate of *intangible investment* has been monitored, all of which are indispensable to the making of *la bonne brioche*.

Evolution of Brioche Pasquier's Economic and Financial Indicators (1986 to 2005)

		1986	1990	1996	2000	2001	2002	2003	2004	2005*
Earnings	French Francs	178MF	497MF							
	Euros	27M	76M	193M	332M	436M	479M	497M	514M	401M
Percent average annual growth			45.4%	25.6%	34.8%	31.3%	9.8%	3.8%	3.4%	6.6%**
Net results		6MF	32MF	15M	26M	33M	31M	16M	31M	8M
Rate of net results/earnings		3.3%	6.4%	7.7%	7.8%	7.6%	6.5%	3.2%	6%	2%
Self-financing capacity					37M	46M	50M	42M	43M	35M
Progression of capital		36MF	53MF		114M	158M	181M	79M	84M	115M
Industrial investments					38M	35M	62M	32M	24M	29M
Overall number of employees		350	604	1,333	1,685	2,031	2,444	2,501	2,438	2,830
Annual increase %						20.5%	20.3%	2.3%	-2.5%	+16%

Earnings between 1986 and 2005:1,385%   or an average of +72%/year

Employees between 1986 and 2005: 708%   or an average of +37%/year

Results between 1986 and 2004*** : 416%   or an average of +23%/year

Source: © ISEOR 2005

Notes: *Figures processed according to IFRS norms. **% calculated after processing of 2004 data according to IFRS norms. ***The year 2005 is not included, in application of IFRS norms for 2005.

REFERENCES

Filser, M. (1994). *Réflexions sur les perspectives de l'étude des marchés* [Reflections on the perspectives of market study]. Research paper. Institut de l'administration des entreprises, Dijon, France.

Pasquier, S. (1992). Preface. In H. Savall & V. Zardet, *Ingénierie stratégique du roseau* (pp. 21-23). Paris: Economica.

Pasquier, S. (1993). *Le management socio-économique dans une entreprise très performante cotée au second marché boursier.* Paper presented at the colloquium "Evolution de l'expert comptable: Le conseil en management" conference proceedings, ISEOR, Economica, Paris.

Pasquier, S. (1995). La communication réalisée dans un cadre conceptuel global, stimulant et multiple: Les réalisations performantes [Communication carried out through a stimulating, multiple and global conceptual framework: Achievements in performance]. *Revue Brises, INIST, CNRS, 17,* 27-36.

Pasquier, L.M. (1998). *Le rôle du dirigeant: moteur du changement et de la croissance interne et externe de la PME-PMI.* Paper presented at the colloquium "PME-PMI: Le métier du dirigeant et son rôle d'agent de changement," conference proceedings, ISEOR, Economica, Paris.

Savall, H. (1974, 1975). *Enrichir le travail humain dans les entreprises et les organisations* [Work and people: An economic evaluation of job enrichment]. Paris: Dunod.

Savall, H. (1979). *Reconstruire l'entreprise: Analyse socio-économique des conditions de travail* [Reconstructing the enterprise: Socio-economic analysis of working conditions]. Paris: Dunod.

Savall, H. (2003a). An updated presentation of the socio-economic management model. *Journal of Organizational Change Management, 16*(1), 33-48.

Savall, H. (2003b). International dissemination of the socio-economic model. *Journal of Organizational Change Management, 16*(1), 107-115.

Savall, H., & Zardet, V. (1987). *Maîtriser les coûts et les performances cachés: Le contrat d'activité périodiquement négociable* [Mastering hidden costs and performances: The periodically negotiable activity contract]. Paris: Economica

Savall, H., & Zardet, V. (1992). *Le nouveau contrôle de gestion: Méthode des coûts-performances cachés* [New management control: The hidden cost-performance method]. Paris: Éditions Comptables Malesherbes-Eyrolles.

Savall, H., & Zardet, V. (1995). *Ingénierie stratégique du roseau, souple et enracinée* [Strategic engineering of the reed, flexible and rooted]. Paris: Economica.

Savall, H. & Zardet, V. (2004). *Recherche en sciences de gestion: Approche qualimétrique. Observer l'objet complexe* [Research in management sciences: The qualimetric approach. Observing the complex object]. Paris: Economica.

Savall, H., Zardet, V., & Bonnet, M. (2000). *Releasing the untapped potential of enterprises through socio-economic management.* Geneva, Switzerland: International Labor Office-ISEOR.

Zardet, V. (1982). Vers une gestion socio-économique de l'hôpital. Cas d'expérimentation [Towards socio-economic management of hospitals. Cases of experimentation]. Unpublished doctoral dissertation, University of Lyon, France.

Zardet, V. (1985). Des systèmes d'information vivants: étude des conditions d'efficacité à partir d'expérimentation [Living information systems: A study of efficiency conditions through experimentation]. *Sciences de Gestion, Economies et Sociétés, 6,* 229-266.

Zardet, V. (1986). Contribution des systèmes d'information stimulants à l'efficacité de l'entreprise. Cas d'expérimentations [Contribution to information systems stimulating for enterprise efficiency. Cases of experimentation]. Unpublished doctoral dissertation, University of Lyon, France.

Zardet, V., & Voyant, O. (2003). Organizational transformation through the socio-economic approach in an industrial context. *Journal of Organizational Change Management, 16*(1), 56-71.

CHAPTER 3

SEAM
IN A SERVICE COMPANY

Developing Vigorous, Disciplined and Empowering Management[1]

Vincent Cristallini

The management literature places heavy emphasis on the role of leadership in a company's workings and its influence on organizational performance. And yet the capacity to stimulate, organize and guide individual and team actions through leader-based energy typically presupposes that this can be accomplished methodically through the application of management methods. To a certain degree, consultants can help to fulfil this role, which consists of a combination of providing energy for transformation (stimulating the desire to change) and providing methods for **implementation** (setting up the move to action). Yet, although the idea of energy and method may appear paradoxical, literally to the point of cancelling each other out, in practice they can be combined very effectively when the energy provided is methodical and when a method's application releases, organizes and maintains energy. In this sense, the *energy-method* combination can be a powerful performance generator.

Socio-Economic Intervention in Organizations: The Intervener-Researcher and the SEAM Approach to Organizational Analysis, pp. 71–98
Copyright © 2007 by Information Age Publishing

Each year ISEOR runs three to four long-term (1 year) consultancy missions, which focus on improving the target companies' *socio-economic performance*. These actions involve introducing the *socio-economic management method* (Savall & Zardet, 1987, 1992, 1995) throughout the company and advising top and middle managers with regard to three major challenges:

- improving the company's *organization* and working
- improving management of people and activities
- improving *strategic implementation* and the information system.

Mobilizing people in this manner always produces encouraging results. But when people scrupulously implement new ways of working (methods) with determination (energy), the results can be spectacular. This is the case for the company that is the focus of this chapter.

This chapter is based on the case of an electronic surveillance company with 700 employees, Générale de Protection. The organization is a European subsidiary of a large American company in the electricity sector. It was founded in the 1990s and has enjoyed strong growth since then. This growth was supported by a strong sales and business focus, aimed at winning as many customers as possible through aggressive sales techniques and a well-designed financing concept. This rather simple *strategy*, however, came into serious question when the company began to accumulate substantial losses.

In retrospect, two phenomena accounted for these losses. First, the enterprise had lost control over its *charges* because of some unchecked practices of its members (e.g., unjustified wage hikes, disproportionate traveling expenses, lost or stolen materials). Second, sales personnel, loosely supervised by their hierarchy but highly interested in increasing revenues, allowed themselves to propose exorbitant solutions that were not adapted to customer needs.

Top management decided to base the company's recovery on ISEOR's *socio-economic intervention* framework. This decision reflects a bold strategic move, as the conventional wisdom suggests that during crises companies should focus on the short term rather than a change program that is intended to produce effects in the medium to long term. The case also demonstrates impressive improvement—in terms of both speed and extent—in the firm's financial results, that were attributed to profound changes in organizational *structures* and *behaviors*. The guiding premise in the chapter is that although management may have the energy for change (strength in action), the fact remains that if the *engineering* concept itself (i.e., formalization and use of principles, methods and tools by the actors) is disregarded, the expenditure of human energy will not necessarily result in enhanced performance.

GÉNÉRALE DE PROTECTION'S STRATEGIC SITUATION

Générale de Protection operates in a very strong growth sector with markets that have enormous development potential thanks to the introduction of new technologies, a widespread use of the precautionary approach, and new forms of petty crime. Indeed, in France, faced with a marked tendency toward "judiciarization" of societal affairs, numerous actors (companies, organizations, individuals) are forced to demonstrate that they did behave responsibly when anticipating possible hazards. These actors are thus led to protect their premises against vandalism, breaking and entering, malevolence so as to avoid problems with insurance companies in the event of claims. The electronic surveillance business is essentially an unregulated profession where numerous players offer services that vary in nature and quality. The simplicity of the basic electronic surveillance product facilitates the diversity of operators on the market. The electronic surveillance market has also attracted the attention of powerful institutional business concerns, such as banks and insurance companies, either with the goal of adding to their range of services or attempting to claim a stake in a promising sector.

The American group that took control of Générale de Protection made what turned out to be a bad deal, buying the company for more than its actual value. Shortly after the buyout, the company began to show significant losses, which led stockholders to inject over $15 millions of liquid assets into the firm to avoid voluntary liquidation. The stockholders' strategy then shifted toward reestablishing **immediate results** in an attempt to remove this "dead weight" from their portfolio. Générale de Protection was suffering from being a mere "cash cow," in which getting customers was all that mattered. There was a frantic race for sales, with sales teams subjugated by an "emotionally threatening" style of management, literally alternating the "carrot" with the "stick," but constantly exerting mental pressure on employees. A conflict between the top directors of the firm further heightened the malaise in the company. One director left Générale de Protection, taking several of the best sales people with him to set up another company. There was very high **personnel turnover** among the sales group.

Setting a Foundation for Intervention: Gaining Client Buy-In

Based on ISEOR's observations of client behavior, the signing of an **intervention agreement** only reflects the "tip of the iceberg." In reality, a consulting agreement only takes on full meaning during the course of the action itself. In fact, it is striking to note that potential clients often do not really believe that their organizational performance can be totally transformed through acts of management and by mobilizing ***human potential***.

The Traditional Demand for a Magic Wand

Most clients approach ISEOR when their performance—be it social or economic—is already rather poor. Their state of disarray seems to cause them to lose their perspective when it comes to the conditions and methods required to turn their situation around. The underlying dynamic reflects a psychoanalytic tendency, as if the mere fact of bringing in expertise is itself the right solution because they have finally decided to act. In cases such as Générale de Protection, in which the firms' management have been unable to overcome organizational difficulties, consultants are often views as potential "saviors" that will enable the organizations to succeed. Yet, when an organization becomes a client, its management is not always consistent in their trust (or distrust), at times even dismissing the possible influence on the consultant who is considered as being anything but a savior.

Générale de Protection's top management called on ISEOR after having apparently accepted that socio-economic development was a useful method for improving overall performance. As illustrated in Table 3.1, the objectives of the intervention *suggest* acceptance on this approach. However, although Générale de Protection's top management bought into a traditional ***HORIVERT approach*** to introducing ***SEAM***, they implicitly thought that the approach in itself would not be sufficient to turn around their poor marketing performance. They did not believe "culturally" in the potential of mobilizing people through management.

Table 3.1.   Objectives of the *Socio-Economic Intervention*

Contributing to the Development of Corporate Culture: Creating Customer Loyalty to Ensure Development
• Assist the company's top management in strategic thinking on the evolution of its business activities and structure.
• Restore sustained marketing performance.
• Reinforce managerial staff skills by training staff to use *socio-economic management tools* for steering activities and managing people to develop *transversality* in the organization.
• Support the company in setting up qualitative, quantitative and financial indicators to develop cost control discipline.
• Restructure services, in particular the company's marketing, technical and administrative network, while basing the changes on the existing organization;
• Transfer sustainable skills through intervention and a *socio-economic project* with a team of internal interveners from the company.

Source: © ISEOR 2001.

ACTIONS / Month	May 01	June 01	July 01	Sept 01	Oct 01	Nov 01	Dec 01	Jan 02	Feb 02	Mar 02	Apr 02	May 02
APPRAISAL OF MARKETING NETWORK												
• Appraisal of marketing network	X	X										
• Marketing network focus group			X	X	X	X						
HORIZONTAL ACTION												
• **Steering group**	X			X			X			X		X
• *Collaborative-Training cluster* A	X		X		X	X			X			
• Collaborative-Training on *socio-economic management tools* for 7 other clusters	X		X		X	X			X	X		
• *Personal assistance* of cluster A members		X	X		X		X		X		X	
• Personal assistance of clusters B & H		X	X		X		X		X		X	
• Horizontal diagnostic				X	X	X						
• Senior management *focus group*								X	X	X	X	
• Collaborative-Training on *socio-economic management tools* for clusters from horizontal level 2							X	X		X	X	X
VERTICAL ACTION												
• *Vertical diagnostics*							X	X	X			
• *Vertical focus groups*										X	X	X X
ENGINEERING TRANSFER												
• In-depth training of 8 *internal interveners* and 3 members of top management	X		X		X		X					

Source: © ISEOR 2001.

Figure 3.1.   Projected intervention schedule during the first year.

As a result, they also demanded appraisal and specific marketing *projects* (see Figure 3.1), which ISEOR agreed to for several reasons. First, the stockholders and top management were in a state of emergency, and not receptive to the speed in which *dysfunctions* and **hidden costs** could be reduced. Second, the specific marketing actions requested involved adapting ISEOR's intervention expertise to a particular theme and time-scale. This step, in effect, involved a rapid diagnostic and project processes specific to the marketing function (see introductory chapter on the *SEAM star points*). Finally, ISEOR's expertise reflects marketing know-how through such improvement processes as *socio-economic marketing*, strategic watch, and *vital sales function*. We will see later that the actual deployment of the socio-economic method in this company confirmed the fact that there was fundamentally no need to attach any "bumps" to the intervention process, since the HORIVERT action fulfilled its goal of turning around marketing performance, despite the company's rapid abandonment of the specific work on marketing.

Indeed, socio-economic management is integrative to such a degree that it deals globally with the problems of a given organisation. Clients, before going through socio-economic intervention, are so focused on the fragmented approach to company operations that they cannot imagine that hard and fast marketing problems can be dealt with through the overall *socio-economic management* process. This is the reason why our client tried to ensure that part of the intervention should be specially earmarked for marketing and realized later on that is was not necessary with a *socio-economic intervention*.

The Reality of Intervening in the Field:
Dealing With Unexpected Change

The appraisal of the sales force began as planned. It was designed as a *mirror-effect* diagnostic, with both qualitative and quantitative components (Savall & Zardet, 2004), focused on calculating the marketing network's hidden costs. It was during this initial period that the company's stockholders decided to stop backing Générale de Protection's top management and appointed the Administrative and Financial Director, Bernard Richerme, as the company's new chief executive officer. Given the enormous tensions in the company, the new CEO and ISEOR jointly decided to quickly present the qualitative assessment in order to mobilize the sales force and enter into a new constructive logic (Richerme, 2003). This new constructive logic consisted of drastically restraining previous practices, turning over a new leaf and rebuilding the management of the company on new grounds.

At that very point in time, there was another dramatic turn of events. The new CEO decided to recall the firm's charismatic ex-marketing director in an effort to boost the sales force and launch the *HORIVERT* action while exempting the sales force. The appraisal of the sales force and the *project* that was to follow were suspended sine die. It appeared that the company's top management understood that its salvation and rapid and sustainable improvement in performance depended on effective management of the company rather than solely on any marketing action, no matter how vigorous.

The CEO exempted the sales force from the Horivert process so as to prevent the demise of the company, whose finances were at rock bottom for lack of sufficient sales. Yet, at the same time, while resigning himself to boost the sales force with a classical "One-Shot" method, the CEO seemed to have understood that reactivating a salesperson's network could not sustainably solve the serious and far-reaching dysfunctions experienced by the company.

As a result of these changes, it took several months for the sales staff to begin the **HORIVERT** approach. While their efforts resulted in a significant contribution (e.g., salespeople obviously longed to be taken care of, they always attended and actively participated in face-to-face or collective meetings), such timing setbacks are not conducive to effective action, either symbolically or practically. Given the dynamic realities of organizational life, however, they are still manageable from a methodological viewpoint, as long as the **HORIVERT** framework continues to work harmoniously.

The intervention project **negotiation** was headed by the CEO, his director general and the administrative and financial director (who became CEO at the start of the intervention). It became evident that the CEO and director general were keeping the administrative and financial director at "arm's length" and that there were points of contention between them. As noted earlier, the CEO and director general were dismissed by the stockholders. The administrative and financial director became CEO, and could have completely changed course, if only for the symbolic reason of making a break with the past.

In such critical situations of sudden change, the **intervener-researcher** must be highly proactive in ensuring the continuity of the **intervention agreements**, because many of the actors will have vested interests in the status quo and it is often too easy to simply stop an action. This proactive orientation should be clearly expressed in private meetings with managers and publicly through group sessions, conveying the message that everyone is working for the good of the company, its jobs, prosperity, and permanence. It is important that these values are fully understood by clients, who interpret such behavior as a strong sign of professionalism.

The Intervention Process

The first days of any intervention are decisive because they shape the initial image of the consultant and can influence the client's faith in the consultant over the long term. It is during this period that the client sees the level of **value-added** the consultant can provide.

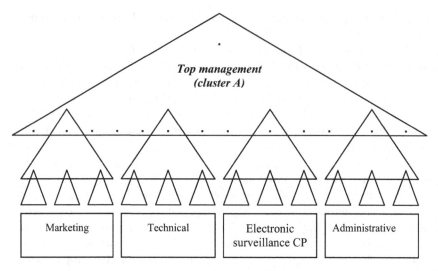

Source: © ISEOR 2001.

Figure 3.2.   Simplified *intervention architecture* by functions.

In the Générale de Protection case, a ***horizontal diagnostic*** involving approximately 90 managers was undertaken (see Figure 3.2). The diagnostic revealed nearly 1,000 ***dysfunctions*** quoted by people, some of which were very substantial. Besides detailing the symptoms these numerous dysfunctions represent, the horizontal diagnostic pointed to serious conceptual defects with regard to current management knowledge and thinking. Générale de Protection, for example, relied on excessive centralization, which can readily disempower organizational members. In such instances, the consultant's value added consists of demonstrating that beyond the detail there are shapes and pictures charged with meaning that emerge. For example, when many people express themselves about fragmentation of tasks, "deresponsibilization" of employees and excessive control leading to guilty feelings, the explicit criticism of Taylorism is conspicuous. This phenomenon may be compared to impressionism, where it is not the details that count but the overall picture we perceive as we step back.

An initial step toward in-depth analysis consists of summarizing special and varied statements in a generic way. The consultants thereby identified key mirror-effect points (see Table 3.2) out of roughly 100 pages of diagnostic. The full power and pertinence of the combination of the ***mirror-effect*** and ***expert opinion*** relies on the intervener-researcher's ability to refer back to fundamental concepts of socio-economic management.

Table 3.2. Selection of Key *Mirror-Effect* Points

Working Conditions
- Changes in top management upset the work atmosphere and worry personnel. This generates rumour and uncertainty, which slows down the work process.

Work Organization
- Agencies feel disempowered by the centralization of the hotline and planning. If agencies fail to operate as profit centres, reactivity is lost with regard to customers.
- The organization chart is not clearly defined. Blurred relations between departments prevent the development of internal customer-supplier type relationships. In addition, the lack of rules for delegation slows the decision-making process.

Communication-Coordination-Cooperation
- Departments are compartmentalized and do not exchange enough information, like operations and marketing. Some departments feel totally isolated, like the command posts or the computer department.

Time Management
- Numerous interruptions at work disrupt *time management*. In particular, telephone calls from technicians or sales people on customer visits are very disruptive.

Integrated Training
- Training is nonexistent for people entrusted with new functions. Similarly, new recruits are not sufficiently trained, both on a technical and marketing level.

Strategic Implementation
- The quality of customer service is not sufficiently developed. The lack of tools to follow up customer relations and poor use of warning indicators, such as the customer complaint ratio, add to this phenomenon.
- Top and middle management question the excessively quantitative and financial basis of the incentive scheme, which tends to maximize the number of end-of-month installations. This phenomenon detracts from the quality of customer service and increases the incidence of drafting predated records.

Source: © ISEO 2001.

Using such tools, the consultant can demonstrate to the client that the analysis reflects actual organizational dysfunctions and how it can serve as the basis for change.

The expression of *non-dit (unvoiced comments)* included 22 points of which the most important are captured in Table 3.3. These unspoken assumptions were well received and accepted in the company. The process of literally "holding a mirror" to the organization triggered the company's desire to change. The clarity of these ideas helped to set a sustainable course for change in a clear and concise way, largely due to their striking and memorable nature. Starting from a series of expert opinion points from the organization (key *mirror-effect* points and unspoken assumptions), the consultant contributed value added by reducing the 1,000 dysfunctions to eleven key points (see Table 3.4).

The fundamental change directions for Générale de Protection were initially expressed in terms of baskets of dysfunctions (see Table 3.5) and

Table 3.3.   Selection of Non-dit (Unvoiced Comments)

- It is astonishing that such a dynamic, innovative and young company should have principles drawn from Taylorism anchored in its genes: division of responsibility, division of labor and a performance approach based on visible *productivity*.

- Some of the company's key operations have had to be centralized for reasons of management security. The failure of past decentralization is probably due to over-*confidence*, lack of *synchronization* and above all, the absence of hierarchical management, or in other words, teamwork.

- Customer relations appear to be insufficiently customized. This customization might be one way to empower teams. In fact, in customer relations, partisan positions and internal strategies between actors lose their interest and impetus, to the benefit of quality of service to which they are committed and on which they are personally judged.

- The authentic and sustainable partnership with the customer is not financial but may be measured in **all relations** with the company, publicized by products and services, in which **everyone** participates through their behavior. One of the company's greatest challenges is to raise each person's awareness of their own role in the sales function.

- It appears that a degree of social peace was paid for using various elements: cars, bonuses, petrol, etc. in response to pressure from various interests. It is therefore quite natural that the social policy was not thought out first and then implemented.

- The steering energy is enormous. Nevertheless, people limit themselves to two basic ultrareactive technologies: telephone and email. One needs to learn to work with more varied technologies: diaries, transformation sheets, task groups, action plans, etc. to better channel energies and adapt them to different circumstances.

- Middle management has a high tolerance for interruptions and disruptions. And yet a little more discipline would not cause any deterioration in the very good state of relations that exists in this company. Everyone should also understand that using time properly is one of the most powerful generators of value added.

- When you enter the premises, you might wonder whether it is a car dealership or an electronic surveillance firm. This may cause some people to lose a sense of the real value of things and the separation of assets.

- Many say that the company is a "marketing company." This is contrary to reality and an honest and serious management analysis. It nevertheless permits justification of the share of earnings taken by those who are lucky enough to be on the right side, i.e. sales. In reality, some people have formed their own company within the company and try to maximise their own profits. The question is to determine whether such an imbalance is strategically sustainable.

Source: © ISEOR 2002.

then translated into a series of mobilizing themes. As illustrated in Table 3.6, these mobilizing themes constituted a long-term transformation strategy and are expressed in terms of constructive objectives.

Using *focus group*, the *dysfunction baskets* (Table 3.5) led to the creation of over 20 multidisciplinary *task groups* in charge of proposing innovative solutions to resolve the dysfunctions.

Table 3.4.   Moving From *Expert Opinion* to *Pivotal Ideas (Idées Forces)*

1. Division of tasks and organization leads to disempowerment.

2. Functions, missions and areas of responsibility sometimes lack precision.

3. The quality of service offered to the customer is insufficient.

4. The relationship with the customer is not sufficiently customized.

5. The rules applying are not sufficiently formalized and maintained.

6. Professional loyalty is not a specific criteria for evaluating people.

7. Confusion between personal and corporate interests.

8. *Working conditions*, both material and psychological, are sometimes poor.

9. The company's human resource policy needs to be set up.

10. Ultrareactive steering is not a culture, it is a regularly maintained cult.

11. Project management lacks method, continuity and structured sharing systems.

Source: © ISEOR 2002.

Table 3.5.   Formulation of *Dysfunction Baskets*

Basket 1: Developing clarity and responsibility in *work organization*

Basket 2: Improving care, quality of customer relations and service to create loyalty

Basket 3: Consolidate and support policies, standards and working rules applying to better support each person's action

Basket 4: Clarify living conditions at work and the human resource policy

Basket 5: Transform the company's steering modes and strategic project management

Source: © ISEOR 2002.

Table 3.6.   *Dysfunction Baskets, Pivotal Ideas* and Mobilizing Themes

Dysfunction Basket (see Table 3.5)	Including Pivotal Ideas (see Table 3.4)	Mobilizing Theme
1	1 and 2	Empowering the organization
2	3 and 4	Quality of customer service and customer loyalty
3	5, 6 and 7	Application of Policies and rules
4	8 and 9	Living conditions at work and human resources
5	10 and 11	Steering mode and strategic project management

Source: © ISEOR 2002

Hidden Costs and the Visible Deficit Link

Although the *horizontal diagnostic* and *expert opinion* had an undeniably stimulating effect on the organization and its members, the *hidden costs* analysis had an even more forceful impact on the change. Theoretically, in view of the link that exists between visible cost-performance and hidden cost-performance, a company that is losing €10 million will have substantial hidden costs: either the company is not producing enough *value-added on variable costs* (*VAVC*) to finance its structure, or it incurs a waste of its *value-added on variable costs* (*VAVC*) (or both at the same time). In the case of Générale de Protection, the *hourly contribution to value-added on variable costs* (*HCVAVC*) was at a perfectly reasonable level (38 €). This amount is in the upper range of averages, thus the company generates good value-added (see chapter 1).

The *vertical diagnostics* on hidden costs covered the company's four business activities and almost one third of its total workforce. The total and per sector results reveal very high hidden costs (see Table 3.7). The intervention conducted with one third of the workforce shows that hidden costs are more or less equal to the amount of the company's loss (€10 million). A simple extrapolation of the average amount of hidden costs per person and per year brings the total amount of the company's potential hidden costs to 33,750,000 Euros for 750 people. These amounts are very high and stand out from the usual averages. The chapter will return to the performance resulting from our intervention, but we can already confirm that over a 2-year period Générale de Protection progressed from a loss of €10 million to a profit of €6 million, a differential of €16 million that equals half its estimated hidden costs.

Credibility and the Power of Inductive Education

The third element in the intervention process involves a constant concern for *inductive education*. Inductive education consists of never tacking

Table 3.7.   *Hidden Costs* per Function and in Total

Sector	Employees Concerned	Total Hidden Cost in the Diagnostic (Euros)	Average Hidden Costs per Person per Year (Euros)
Technical network	44	1,000,000	23,000
Sales administration, sales customer support	39	1,800,000	47,000
Electronic surveillance	73	3,800,000	52,000
Marketing	63	3,200,000	51,000
Totals	219	9,800,000	45,000

Source: © ISEOR 2002

ideas, concepts and tools on existing working methods, using input from the task group (e.g., issues, concerns, objections) to gain genuine buy-in and support. During the intervention process, ISEOR intervener-researchers try to act as mirrors for managers, in essence telling them that "if you treated your colleagues like I'm treating you, what would the result be?" The message is that the intervener not only transfers the content of his or her research, but also a way of being, embodying the three-pronged responsibility of sales person, diplomat, and pedagogue.

This *"constant gentle pressure"* applied by the consultant constitutes a model for managers to develop a style that is more proactive and constructive. This style is based on a number of core principles: acting with kindness, firmness and skill; refusing fruitless discussions and focusing on constructive behavior; refusing to be resigned; and being obsessed about taking decisive and concrete action. At Générale de Protection, over a 3-year period, 90 managers have benefited from 3,200 hours of training in total, whereas internal interveners have followed 1,200 hours. This genuine know-how was rewarded and praised when the client declared at a public conference that the consultants were pleasant but not complaisant. In essence, discipline and the energy to transform are contagious.

Extending the Method to Achieve Socio-Economic Management

The *socio-economic intervention* process tends to produce initial results fairly quickly, which causes many clients to establish the method and extend its foundations to their management. Indeed, socio-economic management involves referring to criteria, value and principles that determine the socio-economic management mode: *synchronized decentralization*, the responsibilizing organization, "everyone a salesperson," shared management control, and so forth. A client can choose to be more or less in line with these reference points. In this way, implementing socio-economic management does not involve dogmatization, but rather the client's choice to progressively change the fundamental characteristics of the management of his or her company.

First Extension (2nd Year of Intervention)

In our experience, the first intervention with a client is usually perceived as a one-time event, especially considering the relative cost and time commitment involved. As firms and their management gradually begin to appreciate both the immediate effects on performance and what still remains to be done to fully implement the socio-economic approach, they are typically eager to follow-up on the actions and ideas raised dur-

Table 3.8.   Objectives for Planned Intervention in the Second Year

Consolidate, develop and maintain the installation of *socio-economic management***:**

- Share top management's *priority action plan (PAP)* with all middle management.
- Train all supervisory personnel in *proximity management*.
- Successfully run the *PNAC* campaign to reinforce the change approach adopted and speed up performance improvement.
- Translate the company's human resource policy into concrete actions in order to implement them.
- Help the company pursue its *internal-external strategic action plan (IESAP)* and create a strategic watch unit.
- Improve the company's profitability by developing *value-added* and reducing *hidden costs*.
- Help the company clarify and formalize its hierarchical and functional *work organization*.
- Formalize the company's marketing know-how and launch a sales personnel day-care center following the sales engineer model.
- Set up a *network of internal interveners* to ensure maintenance of the socio-economic management approach.
- Reinforce the critical mass of change supporters and actors.

Source: © ISEOR 2002.

ing the first phase. While the first year is intense with actions, with concrete steps and outcomes, it is also clear that other changes are simply outlined or run as a pilot program.

Clients that are willing to sign up for a second year of intervention appreciate that socio-economic management is not just about a choice of *tools and processes*. This approach to management is also a way of thinking, an underlying logic that *sustainable performance* demands constant and persevering energy. In the case of Générale de Protection, the desire to provide *implementation* energy is captured in the objectives for the second year of the intervention (see Table 3.8).

Figure 3.3 provides a *chronobiology* of the activity schedule, focusing on the type and frequency of initiatives in the company. One of the recurring observations from our interventions is that during the second year companies often pursue actions in areas that would have been literally unthinkable when the process began (e.g., professionalism, *behavioral management*).

Second Extension (3rd Year of Intervention)

The second extension of the contract resulted from a number of notable outcomes, including very positive financial results, the introduction of *periodically negotiable activity contracts (PNAC)*, and various qualitative

ACTIONS / Month	Sept 02	Oct 02	Nov 02	Dec 02	Jan 03	Feb 03	Mar 03	Apr 03	May 03	June 03
COLLABORATIVE-TRAINING CLUSTERS										
• Implementation of *PAP*	X	X								
• *Vital sales functions*		X		X		X				
• *Socio-economic organization*			X		X		X			
• *Strategic engineering*		X		X		X				
• *Socio-economic management control*			X		X		X			
• *Budgeted actions plans/PNAC*				X		X		X		
PERSONAL ASSISTANCE										
• Clusters: PAP	X	X								
• Clusters: PAB/*PNAC*								X	X	
• Top management	X		X		X					
HUMAN RESOURCES DIAGNOSTIC	X	X	X							
• Human Resource *Project*					X	X	X	X		
• *Strategic vigilance Diagnostic*			X	X	X					
• Top management *focus group*						X	X	X	X	
• *Integrated training plan*						X	X			
• Marketing *integrated training manual*			X		X		X			
• Focus group: creation internal interveners unit		X	X	X	X					
• Training of 7 new internal interveners			X		X		X		X	

Source: © ISEOR 2002.

Figure 3.3.   Projected intervention schedule in the second year.

successes (e.g., new projects, atmosphere, installation of policies and procedures). The third year of intervention was marked by two major directions: (1) the continuing capitalization and transfer of tacit knowledge and (2) concrete work on teamwork interfaces in the company. On this latter point, despite considerable changes in the company's overall *working conditions* and management, problems of *coordination* and *synchronization* can still be problematic and require focused intervention and support (see Table 3.9 and Figure 3.4).

Table 3.9.   Objectives for Planned Intervention in the Third Year

Anchoring the Company in *Sustainable* Improvement and *Development*:

- Ensure maintenance of the socio-economic management method and proximity consulting in terms of organization, management and labor relations, and with top management
- Provide technical help to the manager of the *socio-economic organization (SEO)* department to back up the working of the internal interveners unit and prepare the *Staff Management Integrated Training Manual*
- Reinforce supervisors and personnel so that change and orchestration of practices have become irreversible in the company
- Help the company effectively set up the required *communication-coordination-cooperation systems* for real *teamwork* and strong management
- Continue involving key people in the company in understanding and practicing global shared management

Source: © ISEOR 2003.

ACTIONS / Month	Sept 03	Oct 03	Nov 03	Dec 03	Jan 04	Feb 04	Mar 04	Apr 04	May 04
• *Personal assistance* to CEO • Personal assistance HRD • Personal assistance to SEO department manager • Improvement seminars for *internal interveners* • Personal assistance Administrative and Financial Director • Collaborative-Training "*3C* systems" • Personal assistance "*3C* systems"									
• In-depth training of 5 internal interveners									

Source: © ISEOR 2003.

Figure 3.4.   Projected intervention schedule in the third year.

Third Extension (4th Year of Intervention)

The fourth year focused on strengthening the organization's capacity to evaluate itself and reactivate its performance-improvement loops. The actions included an emphasis on *proximity management* because this aspect remained a weak point in certain parts of the company. After 3

Table 3.10. Objectives for Planned Intervention in the Fourth Year

Continue *Rooting Socio-economic Management* by Developing the Role of the *Socio-economic Organization* department and *Proximity Management*:

- Complete the company's general organization
- Assist top management in implementing a *decentralized organization*
- Share and build on positive local achievements
- Stimulate implementation of *proximity management* in the company by helping middle management of electronic surveillance stations to accept their role of manager
- Continue to integrate and structure regions
- Mobilize *Socio-economic Organization* Department members around *socio-economic innovation actions* in the company
- Continue to involve key people from the company in understanding and practicing global shared management
- Perpetuate the drive for performance and people management through periodic *evaluation* of internal interveners, managers and units from the company.

Source: © ISEOR 2004.

years of intensive collective action, it was necessary to seek out, literally one by one, any lingering holdouts and resistance to the socio-economic approach. Although this typically reflects a residual issue of individual leadership and personal resistance, it must nonetheless be addressed (see Table 3.10).

REFLECTIONS ON THE INTERVENTION: CONFRONTING PROBLEMS AND SUSTAINING CHANGE AT GÉNÉRALE DE PROTECTION

The business-targeted culture of Générale de Protection had resulted in the company's failure to implement a professional, shared management *structure*. The management of the company could be captured by the behavior of its managers and the way they managed the business. When an individual wants to manage effectively and willingly experiments with management models and principles, the opportunity for positive change is much greater compared to those managers who are mired in an authoritarian command and control approach. As Pfeffer (1998) suggests, the failure to effectively manage people is the start of a downward spiral.

An underlying tenet of ISEOR's socio-economic management is that such implicit principles that guide managers should be made explicit and intelligible. A key focus is how managers define performance. Many of the serious operational problems encountered by organizations are borne out of the missing, partial, incomplete or simplistic answers given to this basic challenge.

"When Sales Were Good, Everything Was Good"

One of the difficulties of management is the narrow concept of performance held by many of those involved. In socio-economic management, performance is viewed as multifaceted and dialectic: visible/hidden, *immediate results/creation of potential*, social/economic, qualitative/quantitative/financial, and *effectiveness/efficiency*. This matrix applies to any and all events in the life of the company.

At Générale de Protection, virtually the only performance measurement criterion was the number of contracts sold and the average ticket price; an approach which, ipso facto, excludes any multifaceted view of performance. Generating revenues was a strategy in itself, in fact the only strategy pursued by the company. This single focus was the basis for a number of undesirable effects and the loss of managerial conscience:

- *Denigration of jobs, support services and departments other than those concerned directly with sales.* With sales revenues as the only criterion of performance, only the sales staff can "perform."
- *Failure to take account of the margins made on contracts.* Sales had to be made at any price, because it was sales volume that mattered, not the profitability of what was sold.
- *Low quality service.* Because the *next* customer to be signed was always the most important, those who had already signed merited little attention.
- *Vicious circle of the race to compensate for lost customers.* Because no one was held accountable for lost customers due to poor service, the sales teams became vital in finding new ones to replace those lost.

It's Always "Someone Else's Fault"

Because no one was taking any notice of the difficulties created by the over-emphasis on sales figures, responsibility was diluted as people distanced themselves and the jobs they did from the concept of product and customer service. Générale de Protection had totally lost sight of customer relations. One customer, for example, could have many points of contact, including from the telesales operator who arranged a meeting, the sales person who made the sale, the planning department that arranged the installation, the technician, the administration department (e.g., finance), the control center responsible for taking instructions (e.g., who to notify in the event of an alarm), the technical hotline for operating issues, and the contract renewal sales person at the end of the contract period. No one was identifiable person as the primary point of contact. To complicate matters, the teams, support services and departments were specialized and partitioned from each other. The branches around the country were nothing more than home bases for sales "troops."

The company had approximately 70,000 customers, 7,000 of whom were lost every year—a virtual hemorrhaging of business. Customers were lumped together, torn apart and kept anonymous as they were "processed" through this huge black hole. No one knew which customers had been recruited or what should be done to make sure they stayed with the company. No one knew which customers had left or why they terminated the service.

Management Climate

Instead of creating a framework of formal **resources** and needed systems, Générale de Protection's management seemed to believe in the **effectiveness** of what might be described as an "anthill approach" to managing. In this business model, everyone is constantly on the move, contacting others as the opportunity arises and contributing to the whole by fulfilling a single, precise role. The major difference between an anthill and Générale de Protection, however, is that the anthill seems better organized, with codified activities, an acknowledged need to cooperate, and biologically-programed collective contribution.

At Générale de Protection, people got on so well on the surface that impromptu contact was the main approach to conducting business. A major weakness of the company was the lack of team work, which was replaced by a mass of two-way conversations. This was, in essence, a world of informal hyperreactivity. The combined effect of managing people and business activity in this way tended to create a particular kind of atmosphere or "emulsion," which in turn created a general desire among individuals and teams to perform, especially in terms of sales. A system of challenges and rewards was created, in which benefits and privileges were distributed. What we mean here is that the enterprise sets up challenges based on the fact that free trips and free gifts could be won, which does not help solving the difficult problem of how to properly manage people, but leads people into fleeting, artificial and meaningless entertainment.

The quest to create the right atmosphere is something of an ideology within organizations, many of which naively ignore human nature and the unregulated behavior that such a system is likely to encourage. In the ISEOR approach, getting the best out of people requires voluntary, organized, disciplined action pursued with perseverance, including (1) *people management* by line managers, focusing on behavior, skills, **time management**, **working conditions** and personal **development**; and (2) *business management*, emphasizing the coordination and synchronization of people, the **scheduling**/programming of business activities, quality specification, and quality control. It is clear that informal, occasional, random contact is not sufficient for high quality management of people and business activities. When managers do not pay sufficient attention to their employees, the situation can get out of hand, and it is only human nature that they will be guided by their own self-interest.

There's Always More to Do

The mechanisms, mediated through personal decisions and actions, i.e., observable human activity, by which **value-added** is created and destroyed in organizations pose two inter-connected problems. First, these difficulties can be the product of widespread incompetence, which means that those involved have not learned how to model the effects of what they do. Second, if this is the case, it is more difficult to create the basis for an **active learning approach** designed to involve people in the **creation of value-added**. This is one of the reasons why the principles and mechanisms involved in creating value-added need to be clearly explained and understood in client organizations. It is therefore not surprising that the value created is shared rather chaotically between those involved. The structure within which to negotiate individual contributions/retribution is not sufficiently well developed and forms the basis for many bargaining sessions and conflicted positions. According to the **socio-economic theory of organizations**, each individual harbors a more or less activated economic streak. We have observed that ignorance of their company's economic constraints or lack of awareness vis-à-vis hidden costs renders people heedless claim-makers. Actors then seem to think that the enterprise can literally live on inexhaustible manna from heaven.

At Générale de Protection, managers based their actions on a constant message: "success through money and obvious signs of wealth." The company ended up by buying into a reward structure based on the idea that "if you're a good performer, but you're not happy, I'll upgrade your company car." As a result, many organizational members could not see beyond their own self-interest; in essence, they were taught to behave in this manner. They contacted senior management, often behind the backs of line managers (thus undermining their credibility), to directly negotiate their own personal circumstances and compensation.

Overcoming the Past: The "Dark Side" of Charismatic Leadership

ISEOR's *intervener-researchers* believe that a *socio-economic management intervention* has every chance of success as long as the company's senior management has (1) identified the key *strategic threats* to the company and (2) the will to guide the requisite change. Although many managers may appear to fulfill these conditions, questions linger with respect to the depth of their commitment and involvement.

One of the first problems for managers is securing the commitment of their staff to a structural management method, especially when this approach represents a departure from past behavior. In Générale de Protection, for example, an underlying challenge was to overcome the influence of an organizational "phantom," a now-departed manager whose presence lived on in the company. This individual had literally charmed people through a mixture of fear and hyper-effective admiration. Pushed to the extreme and presented as a management "method," this "charismatic" leadership had, in fact, disrupted team work. Providing guidance through management requires hard work, personal exertion and the ability to win the *cooperation* of others.

The new company manager was faced with the challenge of making others at Générale de Protection understand that managing an organization is much more than managing behavior through the use of a simplistic carrot-and-stick approach. One way of doing this is through the initial *diagnostic of problems*, which creates profound awareness within the company. This diagnostic, which was conducted among the 90 directors and executives of Générale de Protection, revealed over 600 individual problems. This first overall reflection of the negative effects induced by the then current management acted as something of an electric shock. Charismatic leadership becomes distinctly less impressive when you list the huge number of difficulties that this type of management tolerates and, in some cases, cultivates.

Reconstruction and Resistance to Change

Although resistance to change can be harmful to organizations, it can be overcome by embracing the idea of the *necessity for change*. While change is often confronted by various forms of resistance, our intervention experience indicates that harnessing human energy and potential can overcome such resistance. In the Générale de Protection case, from the very beginning, the company manager ceaselessly hammered home the message that unless the company changes quickly it will die. Role modeling the intended behavior he was seeking, the company manager set an excellent example in managing the change, and was a champion of the principles, tools and methods suggested by the intervener-researchers. He was always optimistic and positive, supporting his words with effective action. In essence, the company manager behaved like a results-based leader (Ulrich, Zenger, & Smallwood, 1999).

Although a dominant view of management is based on resistance to change, our observations have shown us that people tend to mimic what they perceive to be the strategy and consistency of their top-level managers. As in many other companies, Générale de Protection's employees did what they were allowed to do, and literally no one spoke negatively about the change or displayed any overt reluctance to change. The clear and visible demands placed on them have had their results. It is important to note that the perceptible determination of the CEO is an essential factor for the involvement of his or her immediate collaborators. Within the enterprise, public statements, demands and stances of the CEO contribute to showing his or her determination.

The company manager understood that the company's recovery would only be successful through a *wide-ranging reconstruction* that would affect every part of the company. Previously a consultant in a major consulting firm, he was able to grasp that change cannot be achieved through half measures, stand-alone projects or mere hype. The company needed to be rebuilt methodically, on the basis of many different actions.

The concept of **reconstruction** differs greatly from that of the **project**. The project concept has become virtually endemic in management practice and theory: project management, management by project, architectural projects, small and major projects, and group projects. The idea of running a project implies a simple achievement of objectives. The concept of reconstruction, in contrast, conveys a much clearer and precise message involving engineering, **coordination**, construction, teamwork, methods, sequence and pace.

Implementing Strategic Change

This section sets out the choices for change adopted by Générale de Protection, choices that were arrived at through the increased awareness that was prompted by the initial diagnostic of the company's problems. A common theme is the intent to reduce the way that work was compartmentalized and reintroduce a sense of responsibility throughout the company.

Respecting the Customer

A lack of *interaction* between company teams had ripple effects on the customer in the form of poor quality service. Générale de Protection succeeded in creating a real cultural revolution by making the transition from the concept of the customer as "prey" to respecting the customer, especially through high quality service. The company chose to organize itself into integrated, autonomous and responsible branches. Each branch was responsible for handling sales and technical support for a particular geographical area and planning its own customer visits. Eventually, each customer was assigned a single point of contact within the branch that was capable of responding to every request and demand at the local level. Regional managers were no longer simply regional sales managers, but were elevated to regional operations managers to ensure that such customer-oriented respect would be sustained.

Creating a Company-Wide Mission

The lack of clarity in employee understanding of the company's mission, and the products and services it supplied, created the opportunity for individual differences and inter-departmental conflicts. In this sense, clarifying a company's mission adds value by creating superordinate goals that cut across jobs and departments. The Générale de Protection company manager decided very quickly to work on defining the mission of the company. The management team then helped refine the company manager's initial suggestions and created a new mission statement:

> To provide European companies with an appropriate and competitively-priced electronic protection service through a combination of remote alarm data management and local, permanent sales and technical support of the highest quality.

Every word of this mission statement is important, because it sends a powerful message to employees about how they should play their part in delivering the company's mission.

Once decided on, the company mission statement was distributed to all employees for their comments. This communication took on an important political dimension as everyone was able to gain a better understanding of the company and how they fitted in to its overall purpose. In this sense, this particular communication performed a structural role, since it helped stabilize employee identities and clarify their responsibilities.

Formalizing and clarifying the mission of the company also provided a strong message that the company was not, as many of them had been led to believe, a "sales" company. Because the mission of the company was not exclusively to make sales, it became easier to change all areas of the company, from remote monitoring control centers, to the technicians and administrative staff. The process of rehabilitation included improving existing premises and equipment, promoting certain staff members, recognizing the contribution made by all organizational members, and introducing a training policy. The company re-balanced its constituent disciplines, reducing the overbearing and at times humiliating power of the sales department in the process. These efforts essentially opened the way for a real spirit of cooperation to develop within the company.

Growing Team Work

Overcentralization results from a fear of losing control over the operational aspects of a company. A basic tenet of the socio-economic approach to management, however, is that it is the nature of control to encourage the very lack of responsibility that leads to the belief in the need for control. A properly led and coordinated team is a team that avoids overdramatizing day-to-day control over its skills.

A distinction was made earlier in the chapter between people management and business management. Générale de Protection had all but abandoned people management and line management teams had little leadership or coordination. Individual activities were managed almost independently of each other. This lack of leadership and coordination gradually led Générale de Protection to trust its people less and less, while making its organization increasingly rigid.

The process of change involved a considerable growth in team work, beginning with *clean-up* and clarifying the line management teams, adopting the principle that no one would be responsible for more than ten staff. The search for solutions to the problems identified in the original diagnostic, together with the initiation of several growth projects, created the foundation for multi-disciplinary team work to develop and contributed to a principle of joint decision making at Générale de Protection.

Changing the Information System and Performance Assessment

Générale de Protection's information system was comprehensive, user-friendly and up-to-date, providing a great deal of data about customers, financial performance, and so forth. However, the potential of this information system was not realized as it was not used to control wasteful or disproportionate expenditures, identify delinquent accounts, or monitor the profitability of the contracts sold.

Within this context, Générale de Protection attempted to make all employees their *own management controller*, a tactic that required a great deal of education about the concept of value-added, as well as the ability to supply everyone with appropriate and relevant data. Générale de Protection approached the problem in two ways. The first was to change certain aspects of its information system; the second was to add new features to improve the specification and control of performance at an early stage. The changes included developing a function that would make the files of "bad payers" (delinquent accounts) flash on the screen, providing the option to withhold assistance from customers who did not pay. Another change was to break the customer portfolio down by brand, so that complete customer management could be devolved to the branch rather than managed by a central department. Additions to the information system consisted of introducing an authorization procedure for expenditures. The variable salary systems that caused a number of undesirable consequences were revised. A profitability calculation was also introduced so that the company could decide whether or not to provide certain services on the basis of their forecast profitability.

Structuring and Coordinating Support and Operational Staff

The process of *synchronized decentralization* is based on reinforcing the status and role of guides (territories), ensuring greater autonomy and *integration* of business activity parameters, and the developing a high level of consistency throughout the organization. On one level, Générale de Protection was a fairly stereotypical organization:

- Operational staff (sales people, technicians, remote alarm monitoring control centers) focused strictly on the operational aspects of their jobs, without concerning themselves with the jobs done by others.
- Support departments were responsible for carrying out those tasks that operational staff either refused to do, did badly or even completed in ways that threatened the security offered by the company. All such tasks were centralized.

Générale de Protection decided to create a core work force with fewer, but better equipped and trained, support specialists and strengthen the operational regions and branches. In this way, the operational staff had, wherever possible, every resource they required to carry out their work in a decentralized way, backed up by the power to make decisions locally. Support managers played a "triple role" on behalf of senior management: defining rules, standards, procedures and policies; distributing information and training; and evaluating practices and capitalizing on the most successful.

Results of the Consultancy Intervention

The results of the Générale de Protection project can be broken down into three major categories of performance indicators: (1) quality of service and *productivity*; (3) financial performance (2) *social performance* (sustainable workforce satisfaction).

Quality of Service and Productivity

As an example of the changes made, Générale de Protection was able to reduce operating faults and improve remedial maintenance, allowing the company to refocus its technicians on preventive maintenance. In the space of 18 months, the number of remedial maintenance jobs fell by 20%, while the total number of maintenance jobs rose by 29%. This policy change resulted in improved customer satisfaction and paved the way for contract renewals.

The reduction in faults has also meant fewer calls to the technical hotline and, as a result, improved hotline response, which rose from 92% to 96% in 1 year. The company introduced a policy of dismantling and reconditioning equipment on customers' premises. Accordingly, equipment costs as a percentage of total installation costs fell from nearly 16% to just 4%. Finally, an interactive voice server was installed, which enabled technicians to carry out their own installation tests and emergency repairs, guided purely by a synthesized voice. Their productivity has therefore increased, since they no longer need to wait for an operator to become available.

Workforce Retention and Social Performance

Générale de Protection introduced an induction course and on-going training plan for its sales people. This initiative has resulted in a sales force where 80% of the staff had, at the end of 2003, been with the company for over 6 months, compared with 55% at the start of the year. In 2 years, the high turnover previously seen in this group of employees fell by over 60%. The company's overall staff turnover fell by 50% in the same period.

Along similar lines, the company's absentee rate fell below 5%. In 2003, salaries rose by an average of 2.5%, and Générale de Protection also introduced an individual target-related bonus of 3.3%, thus giving employees a potential average rise of 5.8%. A profit sharing agreement was also planned for the following year (2004).

Financial Performance

ISEOR began work with Générale de Protection in September 2001, at a time when the company had just published a $10 million loss for the previous year. By the end of 2001, the firm's financial position improved, reducing the loss to $3 million. In 2002, Générale de Protection made a profit of nearly $1.5 million and the estimated figure for 2003 was almost $4 million.

Although the mainstream wisdom suggests that a struggling company cannot survive without drastic reengineering, our experience indicates that an energetic and methodical integrated quality approach can deliver spectacular and rapid results through: (1) improving the organization and operation of the company; (2) improving people management and business activity management; and (3) improving strategy implementation and information support systems. In Générale de Protection's case, the company's cash surplus after investment is $1 million, or 8% of sales revenues. This contrasts starkly with the fact that in 2000 and 2001, shareholders had to inject $15 million in cash to ensure the company's *survival*.

CONCLUSION

The introduction of socio-economic management consists of contributing tools, methods and management principles. This approach to **management engineering** may be compared with installing a company-wide networked information technology (IT) system. In this analogy, the IT system's terminals are organizational members with energy and intelligence, and the connections are lively and progressive teams. The organization is a living system, not a machine, but what manager would take the **risk** of installing a company-wide information system without the support of specialists? Too many system approaches identify the decision-making system, the operating system and the information system, but virtually ignore the consequences that these systems have on the **interaction** between people at work.

The socio-economic approach can be conceptualized as a *set of reasoned approaches, implemented and monitored to achieve a purpose.* An underlying problem is that the actions undertaken by companies are not always methodical. Changing ways of doing things and accepting the need to look at problems in different ways is, in itself, a basis for changing results. The lessons to be drawn from our consultancy work are that **releasing energy,** whether in terms of leadership, motivation or accepting the necessity for change, is not enough unless it is accompanied by a focused management method, capable of simultaneously stimulating, channeling and evaluating how well those involved apply those energies.

NOTE

1.   This chapter draws on the work of Henri Savall (notably Savall 1974, 1975, 2003a, 2003b; Savall & Zardet, 2005; Savall, Zardet & Bonnet, 2000) and

work done by the author at ISEOR (Bonnet & Cristallini, 2003; Christallini, 1995, 2006).

REFERENCES

Bonnet, M., & Cristallini, V. (2003). Enhancing the efficiency of networks in an urban area through socio-economic interventions. *Journal of Organizational Change Management, 16*(1), 72-82.

Cristallini, V. (1995). *Contribution de l'énergie des acteurs au management et à la transformation des organisations. Cas d'entreprises et d'organisations.* Unpublished doctoral dissertation, University of Lyon, France.

Cristallini, V. (2006). *Le rôle de responsable hiérarchique: Du malentendu original au concept d'énergie de transformation.* French National Authorization to Supervise Research, University of Lyon 3, France.

Pfeffer, J. (1998). *The human equation: Building profits by putting people first.* Boston: Harvard Business School Press.

Richerme, B. (2003). Amélioration des performances et mutations profondes dans une jeune entreprise de télésurveillance [Performance improvement and in-depth transformation in a young remote-surveillance company]. In ISEOR (Ed.) *L'université citoyenne* (pp. 303-318). Paris: Economica.

Savall, H. (1974, 1975). *Enrichir le travail humain dans les entreprises et les organisations* [Work and people: An economic evaluation of job enrichment]. Paris: Dunod.

Savall, H. (2003a). An updated presentation of the socio-economic management model. *Journal of Organizational Change Management, 16*(1), 33-48.

Savall, H. (2003b). International dissemination of the socio-economic model. *Journal of Organizational Change Management, 16*(1), 107-115.

Savall, H., & Zardet, V. (1987). *Maîtriser les coûts et les performances cachés: Le contrat d'activité périodiquement négociable* [Mastering hidden costs and performances: The periodically negotiable activity contract]. Paris: Economica

Savall, H., & Zardet, V. (1992). *Le nouveau contrôle de gestion: Méthode des coûts-performances cachés* [New management control: The hidden cost-performance method]. Paris: Éditions Comptables Malesherbes-Eyrolles.

Savall, H., & Zardet, V. (1995). *Ingénierie stratégique du roseau, souple et enracinée* [Strategic engineering of the reed, flexible and rooted]. Paris: Economica.

Savall, H., & Zardet, V. (2004). *Recherche en sciences de gestion: Approche qualimétrique. Observer l'objet complexe.* Unpublished English translation: *Research in management sciences: The qualimetric approach. Observing the complex object.* Paris: Economica.

Savall, H., & Zardet, V. (2005). *Tétranormalisation: Défis et dynamiques* [Competitive challenges and dynamics of tetra-normalization]. Paris: Economica.

Savall, H., Zardet, V., & Bonnet, M. (2000). *Releasing the untapped potential of enterprises through socio-economic management.* Geneva, Switzerland: International Labor Office-ISEOR.

Ulrich, D., Zenger, J., & Smallwood, N. (1999). *Results-based leadership.* Boston: Harvard Business School Press.

CHAPTER 4

ORCHESTRATING COMPATIBILITY BETWEEN ART AND MANAGEMENT

Socio-Economic Intervention in a National Opera House

Philippe Benollet

The chapter details an ***intervention-research project*** (Savall & Zardet, 2004) conducted in a national opera house, exploring the possibility of getting artists and managers to work together to improve organizational "performance" within the context of a cultural institution. "Performance" in this sense refers to the ***effectiveness*** and ***efficiency*** of resource allocation in relation to the results achieved, that is, making sure that critical resources produce the desired, intended effects from the vantage point of the artists, the audience, and the larger community. The intervention (see Savall, 1974, 1975, 2003a, 2003b; Savall & Zardet, 1987, 1992, 1995; Savall, Zardet & Bonnet, 2000)does not dwell on the quality of deployment in terms of allocation of means (e.g., "quality" of artistic policy, rationality of investments) and type of resources (private, public), although the quality factor may affect both the ability to capture new ***resources*** and, naturally, the quality of artistic creation.

Socio-Economic Intervention in Organizations: The Intervener-Researcher and the SEAM Approach to Organizational Analysis, pp. 99–122

THE NATIONAL OPERA HOUSE INTERVENTION

The national opera house is a nonprofit organization (see Table 4.1), endowed with a 27.5 million euro budget, 80% of which is derived from public subsidies (as is the case for many French cultural institutions). These subsidies are granted by a variety of authorities such as the town council, and the departmental and regional councils and government.

These authorities contract a 5-year agreement with the opera house that sets various objectives for the institution in terms of attendance, programming, artistic creation and access to performances by young audiences. In return for achieving these objectives, the level of subsidies is guaranteed for five years. In addition to this agreement, the cultural com-

Table 4.1.   The National Opera House (2000)

BACKGROUND & FINANCES

- *Status*: Association whose board of directors is chaired by the town mayor, under government contract *Total receipts*: 28.7 million Euros (1999/2000 season)
- Own receipts (ticket receipts + theater rental receipts + sale of audiovisual products: 4.9 million Euros (or 17% of receipts)
- *Operating and investment subsidies*: 23.8 million Euros (or 83%). Subsidized: Government (10%); region and department (10 %); town council (80%)
- *Indirect financing*: In addition to investment and operating subsidies allocated by the town council, indirect financing by payment of salaries for municipal staff (over 8.7 million Euros)—see workforce below.
- *Traditional cost structure*: 80% overheads (mostly for salaries); 20% allocated to artistic production

PRODUCTIONS

- *Ratio*: [(Production budget)/(Annual budget including salaries covered by the town council)] × 100 = 26.5%
- *Number of performances* [Lyric total (i.e. town + national and international tours)] = 90
- *World creation and new productions* = 9

WORKFORCE

- *Global workforce*: 300 people full-time equivalent (ETP)
- *Workforce*: 341 (109 association-private; 232 municipal-public). The municipal-public workforce is paid by the town council [77% of the overall full-time equivalent workforce: 121 artists (musicians, choir members, dancers); 111 technicians (set, costumes, scenery production departments)]
- *Orchestra*: 61 musicians; 27 choir members; 14 lyric artists; 95 children's choir; ballet including 31 dancers
- *Support & Logistics*: 21 stagehands; 8 costume assistants; 6 dressers

CAPACITY & INSTALLATION

- *Capacity*: 1,200 seats (large theater) + 250 seats (amphitheater)
- *Set design* shop: 1,000 square meters
- *Institution* (except set design shop): fully renovated including stage machinery and lighting

Source: © ISEOR 2000.

pany benefits from the "National Opera" label awarded by the Ministry of Culture. Five cities in France currently benefit from national opera status: Lyon, Strasbourg, Paris (Bastille and Garnier), Bordeaux and Marseille.

The opera house studied in this intervention has a permanent staff of nearly 350 salaried employees (Durel & Bergeot, 2002). Two thirds of the staff is are servants, from the Territorial Civil Service or workers from the town council and the remaining one third are under private contract hired by the association. In addition to this permanent staff, there are approximately 100 additional temporary employees, including people from the performing arts sector, artists and technicians. The opera house is equipped with full production facilities, including technical (e.g., ability to produce its own costumes, accessories and sets) and artistic (e.g., all the artistic trades required for a performance) capabilities.

The Intervention

The intervention was conducted from 2000-2003 and the project was an extension of other *socio-economic management* model experiments run by ISEOR in cultural institutions, including work with a television channel (1980-82) and a ballet company (1989). The intervention was carried out in three phases: (1) a *Horivert process* focusing on the set design shop (December, 1999–September, 2000); (2) an *extension phase* consisting of intensive management training for the institution's top executives and directors (March–July, 2001); and (3) an in-depth *evaluation* of results, focusing specifically on the scope of the design shop and the technical management (April–July, 2003). The schedule for the first two phases is shown in Figure 4.1.

The Horivert Process

The first phase of the *project* was the *Horivert process* in the set design shop (December 1999 to September 2000). Established on approximately 1,000 square meters, the set design workshop involved the manager and his assistant, the head of the research unit, the head of the design section (who supervises a minimum of 4 carpenter-designers and 2 locksmiths), and the head of the painting section (who supervises a minimum of 4 painter-designers). A dozen casual workers spread between the two sections eventually complete this *structure*.

The role of the workshop was to ensure the construction of sets within a specified time frame, according to a threefold process:

- Collecting the scenery set-up plans and assembly books (two-dimensional plans of the marking and putting-up of each set, axometric projection of complex units, list of the units with their mark-

INTERVENTION SCHEDULE - OPERA HOUSE											
MONTH / **INTERVENTION ACTIONS**	Dec. 1999	Jan. 2000	Feb. 2000	Mar. 2000	Apr. 2000	May 2000	June 2000	July 2000	Aug. 2000	Sept. 2000	Oct. 2000
STEERING GROUP		1		2			3			4	
COLLABORATIVE TRAINING of the Board of Directors and the Management team of the Sets Design Workshop (6 sessions -2.5 hours each)	1	2		3		4	5	6			
PERSONAL ASSISTANCE in implementation for the Board of Directors and Management (4 sessions-1 hour each per person)		1	2		3					4	
HORIZONTAL DIAGNOSTIC Board of Directors and Management		▮	▮								
FOCUS GROUP Board of Directors (4 sessions)		▮	▮	▮	1	2	3			4	
IN-DEPTH *VERTICAL DIAGNOSTIC* of the Sets Design Workshop					▮	▮					
FOCUS GROUP Sets Design Workshop							1	2	3	4	
TASK GROUPS piloted by Management and including non-management personnel					X	X	X			X	

POSTINTERVENTION IN-DEPTH TRAINING ACTION	March 2001		April 2001		May 2001		June 2001		July 2001		Aug. 2001
STEERING GROUP (3 sessions - 1 hour each)	1				2				3		
COLLABORATIVE TRAINING - 2 Groups (4 sessions-2 hours each)	1				2		3		4		
PERSONAL ASSISTANCE - 4 Sub-Groups (3 sessions - 2.5 hours each)			1				2		3		
FOCUS GROUP (4 sessions - 3 hours each)			1		2		3		4		
TASK GROUPS			X		X		X		X		

Source: © ISEOR 1999.

Figure 4.1.   The *Horivert intervention* and management training schedule.

ing and numbering systems, where units are fastened to the flies, weight of the units). The assembly books are sometimes cobuilt with the workshop's research unit;

- Studying the feasibility and the possible adjustments with the different interlocutors of the opera;
- Actual assembly.

The process comes within the scope of a more general schedule that includes upstream (e.g., an overall survey, detailed estimates, presentation of models and projects, technical studies) and downstream (e.g., partial then complete assembly of the scenery for rehearsals) activities. As a rough estimate, this process represents a 9-month schedule, including 3 months for scenery build-up.

Table 4.2. *Horivert Intervention* Contractual Obligations

LIST OF OBJECTIVES
Horivert Intervention Agreement
(December 1999)

The company wishes to undertake a revitalization and clean-up approach for the set design shop in order to optimize its budgetary and human resources to achieve its internal missions.

This intervention agreement falls within the framework of a global objective to implement a strategic and organizational project to mobilize top management, middle management and set design shop personnel.

From this principal objective, the intervention process, focused on implementing change, locating and mobilizing sources of hidden potential, meets the following objectives:

1. Conducting change procedures to improve overall efficiency in the set design shop
2. Point out the organization's dysfunctions and hidden costs so as to improve cost control and optimize budgetary and human resources in the set design shop
3. Make the organization of the set design shop more efficient while preserving its professional diversity
4. Transfer ***management tools*** to top and middle management to improve steering and implementation of the company's strategic project
5. Introduce practical tools and methods to support the workshop's internal vocation within the company by providing it with priority action plans
6. Ensure the deployment and permanence of ***socio-economic management tools*** and systems with top and middle management and workshop personnel

Additional Management Training Agreement *(February 2001)*

Besides knowledge transfer, the additional training will:

1. Develop the managerial capacity of middle management by providing it with methods and tools to improve steering of the company's ***strategic implementation***
2. Reinforce the quality of management by developing areas of responsibility
3. Create management unity through the use of tools and a shared vocabulary
4. Equip each manager with diagnostic, project and evaluation tools for department ***functioning***
5. Develop reflexes and know-how in terms of ***synchronization*** and ***communication***

Source: Agreements between the National Opera and ISEOR.

The global objective, which was jointly defined by the institution's top management and ISEOR, was to "permit ***implementation*** of a strategic and organizational project that would mobilize top management, middle-management and staff in order to enhance the workshop's effectiveness and efficiency." This broad objective was broken down into a series of specific operational objectives, which are summarized in Table 4.2. Listed by ***method-products*** and ***service-products***, these objectives are accompanied by the schedule and ***intervention architecture*** as well as a financial appendix, which together constitute the terms of the ***intervention agreement***.

The choice of the set design shop as the initial intervention point must be viewed from two partly interrelated concerns. The set design shop was experiencing serious *functioning* problems and a number of *strategic threats* that threatened its very existence. In 1999, the workshop had a budget of 153,000 Euros, and key tasks that fell within its function and *competence* was subcontracted out for 92,000 Euros. This situation became a source of bitter contention within the town council (the Opera's trustee). Outsourcing set construction, seen by certain trustees as a means of cutting the cost burden on the town council, was linked to broader concerns about outsourcing of activities that were "peripheral" to the opera's core business. Remember that the concept of hidden cost-performance ratio challenges this outsourcing logic, portrayed as the fast track to improving performance simply by cutting costs. The complex governance of the opera with its two-fold characteristics, that is, its hybrid legal status (multiple financial partners [state, region, county, city], dual status of the personnel from both the private and public sectors) and the nature of its objectives (artistic and economic) demands an intervention architecture that is characterized operationally by a number of concerns.

The notion of *complex governance* is useful in analyzing this situation, enabling for instance the operational management of the intervention's architecture (see Figure 4.2). This complexity appears in its hybrid legal status, plurality of financial partners (e.g., government, region, department, town council), the dual status of staff (e.g., civil servants on temporary assignment, private sector), and the dual nature of the objectives which are both artistic and economic. Thus *horizontal action* and intervention are necessary to simultaneously involve (1) the director general, in charge of strategy and artistic programming, and the administrative and financial director, in charge of budgets and administrative management, in the *horizontal diagnostic* and project and in the steering group meeting; and the (2) town council, through its management controller in charge of artistic and cultural organizations, in the same segments. Structured in this way, the work practices encourage a certain unity in decision making, create converging points of view, make artistic and management decisions consistent, and involve the opera house's principal partner in decisions and their respective activity requirements.

The *vertical action* in the intervention is characterized by a single sector—the set design shop. Although *socio-economic intervention* typically involves at least two sectors for tactical and operational reasons (e.g., avoiding distortion and obstruction), in the opera house it was not conventionally possible to intervene in two sectors. The main reason for the limited intervention appears to be based on top management's fear of pushing an institution not yet used to management requirements, as is still the case with many artistic and cultural institutions. "*Capillary attrac-*

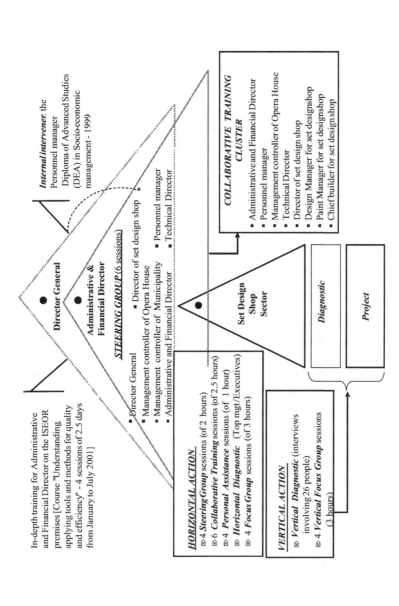

In-depth training for Administrative and Financial Director on the ISEOR premises [Course "Understanding applying tools and methods for quality and efficiency" - 4 sessions of 2.5 days from January to July 2001]

Internal intervener: the Personnel manager Diploma of Advanced Studies (DEA) in Socio-economic management - 1999

Director General

Administrative & Financial Director

STEERING GROUP (6 sessions)
- Director General
- Management controller of Opera House
- Management controller of Municipality
- Administrative and Financial Director
- Director of set design shop
- Personnel manager
- Technical Director

COLLABORATIVE TRAINING CLUSTER
- Administrative and Financial Director
- Personnel manager
- Management controller of Opera House
- Technical Director
- Director of set design shop
- Design Manager for set designshop
- Paint Manager for set designshop
- Chief builder for set designshop

Set Design Shop Sector

Diagnostic

Project

HORIZONTAL ACTION
- 4 *Steering Group* sessions (of 2 hours)
- 6 *Collaborative Training* sessions (of 2.5 hours)
- 4 *Personal Assistance* sessions (of 1 hour)
- *Horizontal Diagnostic* (Top mgt./Executives)
- 4 *Focus Group* sessions (of 3 hours)

VERTICAL ACTION
- *Vertical Diagnostic* (interviews involving 26 people)
- 4 *Vertical Focus Group* sessions (3 hours)

Source: © ISEOR 1999.

Figure 4.2. *Horivert architecture* in the *initial set-up phase.*

tion," the socio-economic model's measured extension to the rest of the organization, thus takes place here, not through several vertical diagnoses but by additional management training, described in the extension phase of the intervention.

The initial *Horivert process* intervention represented roughly 540 hours by the intervention team, which was made up of three people (the ISEOR director, the author of the chapter, and a junior consultant). This work volume breaks down according to the segments presented in Figure 4.1: 10% for the steering group (preparation and running), 19% for *collaborative-training* and *personal assistance* on socio-economic management and control tools, 19% for horizontal diagnostic, 11% for the *horizontal project*, 24% for *vertical diagnostic*, and 17% for the *vertical project*.

Additional Management Training and Evaluation

As stated previously, the "*Capillary attraction*" of the socio-economic model was done by additional training for the institution's top managers and executives. The contractual objectives of this training, which are reflected in Figure 4.3, were intended to provide top and middle management with the requisite tools to implement the strategy. This training was organized over the period from March to July, 2001. Although titled "training" in the intervention contract, this *extension phase* included a number of different segments, the pace and orchestration of which are shown in the timetable in Figure 4.1 and the architecture in Figure 4.2.

The action led by the ISEOR research team in this phase corresponds to an intervention volume of 240 hours, including a cooperation-training segment on management and quality tools, and work with steering groups, *focus-groups* and *task groups*. As part of the training, an underlying goal was to trigger an initial awareness of possible improvement actions (with the steering groups), refining these actions and their implementation with the focus and task groups. It is important to note that ISEOR does not engage in any interventions that are limited to training in the traditional sense. The performance concept is not just about "pushing" players by making them aware of *dysfunctions* and gaps; it also involves setting up systems to facilitate improvement actions and guiding them through the appropriate assessment meetings and indicators.[1]

The vast majority of the training sessions and work with the steering, work and *focus groups* in this extension phase was handled by an *internal intervener*, with methodological assistance from ISEOR intervener-researchers. The internal intervener was the institution's staff manager, who was also studying for a diploma of advanced studies (DEA) in *socio-economic management* in its professional training program. The internal intervener was thus familiar with the theory underlying the socio-eco-

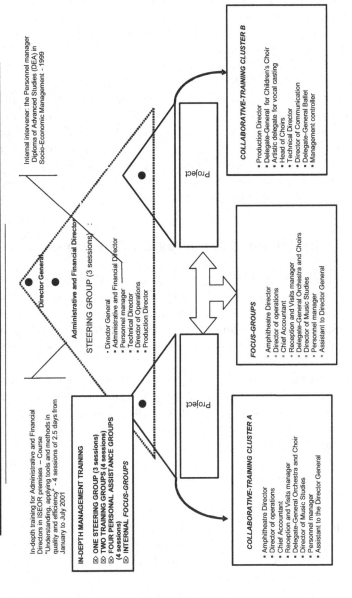

Source: © ISEOR 2001.

Figure 4.3. **Architecture** of the additional in-depth management training action in the **extension phase**.

nomic model. Furthermore, at the time of the Horivert intervention, the internal intervener has already conducted vertical diagnostic interviews (qualitative and quantitative) and worked with the ISEOR team in running vertical focus groups.

As an illustration of how organizational networking can contribute to the "*capillary attraction*" of socio-economic management within an institution, the staff manager who served as an internal intervener during the intervention introduced ISEOR to the opera house. At the end of 1999, he presented the socio-economic method to the organization's top management, leading to a series of contacts between the opera house and ISEOR senior management, and the first intervention agreement in December 1999. It is also worth noting that the administrative and financial director, acting as deputy director general of the opera house, attended specific additional management training in ISEOR premises for four 2½ day sessions (January-July 2001).

INTERVENING IN CULTURAL INSTITUTIONS

This section focuses on the diagnosed dysfunctions and loss of potential that they generate, highlighting the *effectiveness* of the socio-economic model in intervening in cultural institutions as much as other organizations. In this instance, the concept of *hidden costs* appears to be a productive way of modelling and orchestrating compatibility between art and management.

This compatibility may form the basis for signs of convergence between the artistic and managerial logic, or, to put it another way, the aesthetic and the utilitarian[2] (Evrard & Chiapelo, 2004). Management is opening up to rationality, new ways of thinking and acting that were hitherto completely unfamiliar (e.g., Midler, 1993; Thévenet, 1992), and at the same time the "romantic" view of art and artists is being re-examined by perceiving the creation pf art as the *implementation* of a process (Becker, 1983; Wolff, 1981).

In examining the actors' statements, questions could be raised as to whether the socio-economic grid reading for *dysfunctions* (in six domains, with the database of key ideas in the *SEAMES*® *expert system*) distorted the portrayal of the actual situation. Based on our analysis, it appears that this did not occur. All key statements voiced by the institution's actors could be processed, without any manipulation, distortion or change on the part of the consultants. When the diagnostic was returned, in the context of the *mirror-effect*, the actors fully recognized their institution and its dysfunctions, and they were aware of and subscribed to the need to act.

A similar question could be raised as to whether cultural institutions are sufficiently different from other organizations that the model might leave out other major dysfunctions. Again, based on our analysis, it also appears that this did not occur. All stages of the work, from consolidating the input and placing quotes in the *dysfunction basket*, to the *mirror-effect priority* and *non-dit (unvoiced comments)* and key ideas, were approved by the institution's actors. The *dysfunction baskets*, which form the operative resolution unit in the *project phase*, as usual in the socio-economic method, are placed at a sufficiently deep level that the causes of problems are established as a foundation for improvement actions. A causal analysis of these baskets, going back to the *root causes of dysfunctions*, brings us back to the traditional socio-economic concepts of cleaning-up, *synchronization* or even the level of *HISOFIS* problems (Humanly Integrated and Stimulating Operational and Functional Information System; for a clarifying definition, see the *SEAM*-related terminology in the end of volume glossary).

As an illustration of these dynamics, Figures 4.4 and 4.5 present the key ideas and "baskets" from the *vertical* and *horizontal diagnostics*. The evaluation column, in the latter *figure* corresponds to the quotation the company gives on the degree of effectiveness of the proposed solutions to their problems.

The socio-economic model is fully applicable to research-intervention in cultural institutions. There are, of course, specific constraints that are traditionally recognized as being part of these organizations—their project-based organization, the unpredictable nature of artistic production, the product's uniqueness, and the limited impact of experience and automation. However, these characteristics are all the more comprehensible when examined through the *hidden cost-performance theory*, as the latter re-examines three traditional concepts in management teaching: production function theory, economies of scale, and the effects of automation.[3] The creation of value is traditionally analyzed as dependent on capital and labor factors. The underlying hypothesis in this theory is that capital ranks high in the factors accounting for value creation. This underestimates the pre-eminent role of actors in creating the value, distorting our comprehension of cultural company phenomena and how art, creativity, innovation and renewal interact to support their projects.[4] The cultural sector is an activity of "prototypes,"[5] so the notion of economies of scale is barely relevant in seeking economic *efficiency* in this type of activity. In view of the nature of cultural companies' activities, automation is hardly feasible; on the other hand, improving processes is a very promising way to *progress*.

The qualitative expression of problems and work practice covers almost all latent calculated cost-performance ratios. The ensuing results

DYSFUNCTION *BASKETS*

P1 Develop planning–scheduling tools for set construction and implement more effective steering tools

P2 Improve working conditions through regular cleaning of physical structures

P3 Develop capacities for anticipating and negotiating through synchronized practices and more stimulating 3C* systems between the construction workshop, the company and the artistic supervision

P4 Clarify work organisation and develop procedures for concerted delegation and integrated staff training

KEY IDEAS BASED ON EXPERT OPINION

- IF 2 – Lack of steering for set design–production activities
- IF 3 – Not enough reliable management indicators and lack of visibility on realised and projected activities
- IF 8 – Absence of formalised tools for set construction planning and time management

- IF 6 – Partial discrepancy between workshop's physical means and construction and safety objectives

- IF 4 – Lack of synchronization between workshop, company and artistic supervision
- IF 7 – Lack of operational information flow between construction workshop and company
- IF 10 – Decline in workshop's negotiation capacities and company's different decisionmaking levels with artistic supervision

- IF 1 – Lack of definition or cleaning up of working rules and procedures
- IF 5 – Lack of clarity in division of responsibilities in construction workshop
- IF 9 – Lack of concerted delegation of work

* 3C: Communication-Coordination-Cooperation

Source: © ISEOR 2000.

Figure 4.4. *Dysfunction baskets* and *key ideas* from *vertical diagnostic*.

Dysfunction Baskets

Key Ideas	Improvement actions	Assessment of achievements
IF 1 – Lack of planning, scheduling and steering on design-production times for performances	· Definition of typical planning for performances per season (pilot: JYB)	■
IF 7 – Absence of intermediate objective and lack of priority management		■
IF 2 – Lack of precision in specifications and decline of professional discussion between artistic supervisor and technical departments	· Definition of standard content of specifications for users and set construction (pilot: PMU)	■
	· Definition and implementation of a system of meetings and internal synchronization to monitor set construction (pilot: JMP)	
	· Definition and implementation of standard contract between Opera house and the artistic supervisor (pilot: JMP)	■
	· See Vertical project	
IF 11 – Physical work environment in workshop partially unsuitable	· Implementation of scheduling sheet per operation tool (FOO) (pilot: JYB)	☐
IF 13 – Lack of formalized tools for steering activities	· Set up of strategic piloting indicators (pilot: JYB)	◢
	Actions/Tools not released	

Box 1: Controlling design - production flows for performances by developing the notion of internal contract

Assessment scale for fulfillment rate of progress actions:

■ Actions/Tools implemented

◢ Actions/Tools in advanced state of production or to be approved

☐ Actions/Tools prepared, started

Source: © ISEOR 2000.

Figure 4.5. **Key ideas** from the horizontal project—*dysfunction* basket 1.

that have been validated, which will be discussed in the next section, correspond to a considerable reduction in hidden cost-performance ratios that is comparable to results achieved in other sectors of activity, in other research-interventions conducted by ISEOR.

Hidden Costs in the Opera House

The ability to calculate *hidden costs* in an artistic or cultural company is an important managerial prerogative, especially in light of the problems set out in the introduction. Because the operative system for calculating hidden costs includes qualitative (e.g., actors' statements), quantitative (e.g., expression by frequency and volume of *regulation* facts) and financial performance data, hidden costs can be calculated. As illustrated in Figure 4.6, in this case the intervention revealed hidden costs amounting to over 640,000 Euros. Of course, the set design shop operates as a cost center and it bears certain costs without necessarily generating any dysfunctions. This is why, although the workshop's budget is only 153,000 Euros, the total amount of hidden costs calculated in this sector is 4 times higher. This amount represents 12% of the lyric production budget and 8% of the opera house' total production budget. In terms of the number of people (full-time equivalent) in the set design shop, this equals a hidden cost of 34,000 Euros per person per year. These figures provide an idea of the minimum resources diverted from normal (*orthofunctional*) use, in particular what is needed to ensure a high quality of artistic activity.

Changes Observed and Resource Savings

To evaluate the pertinence and penetration of improvement actions implemented in 2000 and 2001, an *evaluation* of results was conducted from April to July 2003 in the set design workshop and its technical department. This evaluation represents an investigation of nearly 120 hours, run by a consultant from ISEOR. Through personal interviews, we were able to identify improvement actions and their qualitative, quantitative and financial impact. The assessment of financial impact is made up of two parts: one expressed in terms of the reduction of hidden costs, and the other in budgetary terms.

The results obtained during the intervention in 2000 and 2001, as well as those identified in 2003, were largely the result of the *Horivert* architecture's flexibility and adaptability to the special features of different types of organization. The work systems that were implemented in the opera house were modified to take advantage of the institution's underly-

	EXCESS SALARIES	OVERTIME	OVER-CONSUMPTION	NON PRODUCTION	NON CREATION OF POTENTIAL	RISKS	TOTAL
ABSENTEEISM (1)	€10 100	N.E.	N.E.	€4 500	N.E.	N.E.	€14 600
WORK ACCIDENTS	See absenteeism	See absenteeism	See absenteeism	See absenteeism	See absenteeism	N.E.	See absenteeism
PERSONNEL TURNOVER	N.E.	N.E.	N.E.	N.E.	N.E.	N.E.	N.E. (2)
NON-QUALITY	N.E.	€228 000	€31 100	N.E.	N.E.	N.E.	€259 100
DIRECT PRODUCTIVITY GAPS	N.E.	€310 000	€52 100	€10 200	N.E.	N.E.	€372 300
TOTAL	€10 100	€538 000	€83 200	€14 700	N.E.	N.E.	€646 000

Number of people: 19; or 34 000 Euros per person and per year on average – Ratio indicative because a part of hidden costs is generated outside the workshop

(1) In the absence of an indicator specific to this workshop, these amounts are calculated by taking global opera house costs, 300 people full time equivalent, by workshop workforce

(2) N.E. No evaluation given the time allowed for the study

Source: © ISEOR 2000.

Figure 4.6. *Evaluation of **hidden costs** in the set design shop.

ing elements, including its political and strategic situation, its legal status, and the nature of its activity. By getting key actors to work together, based on focused indicators as part of a measured process, socio-economic intervention provides the foundation for transformational outcomes.

As an illustration of this work, Figure 4.7 is an extract of the vertical project evaluation. The complete evaluation, which was done in October 2000, shows a reduction in *hidden costs* of over 305,000 Euros, or almost 50% of the initial amount of hidden costs diagnosed. This figure is an indication of the opera house actors' ability to define and implement improvement actions. It is useful to note that a typical *socio-economic project* leads to an approximate 30% reduction of hidden costs over a period of one year, which further underscores the applicability of this type of intervention in cultural institutions.

Figure 4.8, which is based on the evaluation conducted from April to July 2003, shows that from nearly 214,000 Euros' worth of hidden costs, over 99,000 Euros were reduced. This saving, expressed in time, was partially reinvested in capacity to reintegrate the subcontracted workload and proportionally increase the value added generated by the set design shop with the same staffing level. The consequent release of a budget equivalent of 91,500 Euros was redeployed in the artistic and creative domain.

These examples illustrate the domino effect that can occur when resources are redeployed in ways that create value for a variety of stakeholders, including the quality of service for the audience, artistic quality for staff, and the effectiveness of resources allocated by taxpayers via public subsidies (which are fed by local taxes). These savings come from change actions that either helped to resolve costly dysfunctions or identified a less costly mode of *regulation*. Some examples of these changes are presented in Figure 4.9.

Through evaluation interviews conducted in April 2003, the changes observed during the 2000-2001 period were updated. This analysis revealed a certain degree of continuity in management insight and behavior that were developed during the *Horivert intervention* and subsequent management training initiative. As an outcome example, the cost and volume of activities subcontracted to external companies by the set design shop fell from 92,000 Euros in 1999 to no activity by 2003. We think it likely that the rather inconsequential loss of "good" management reflexes is the result of innovative *intervention architecture*, capillary attraction systems and structuring tools implemented during the *socio-economic intervention*.

We believe that this durability is due to the pervasive nature of the intervention architecture (from the director general and the administrative and financial manager to carpenters, locksmiths and painters), to the capillary systems (e.g., proper extension of the model is obtained through

CONVERSION OF HIDDEN COSTS INTO VALUE-ADDED

Quality Level	QUALITATIVE	QUANTITATIVE	FINANCIAL
Services to customers	✔1 –Improvement in planning-scheduling and steering of productions 2 - Development of synchronized practices between workshop, company and artistic supervisor 3 – Improvement in operational information flow between construction workshop and company	• 1 Not Assessed • 2 Reduction of 2/3 of initial amount Valued [€ 50 000] • 3 Reduction of 2/3 of initial amount Valued [€ 79 000]	• 1 Not Assessed • 2 Saving: € 33 000 • 3 Saving: € 52 000
Working	✔1 - Reduction in alterations made to construction plans and scenery ✔2 - Decrease in stock shortages of equipment 3 - Development of piloting indicators and better visibility on work done ✔4 - Clarification of work rules and procedures in workshop ✔5 - Improvement of work organization and division of responsibilities 6 - Better balance between physical means of workshop and construction and safety objectives	• 1 Not Assessed • 2 Not Assessed • 3 Reduction of 2/3 of initial amount Valued [€ 192 000] • 4 Reduction of 1/3 of initial amount Valued [€ 50 000] • 5 Reduction of 1/3 of initial amount Valued [€ 117 000] • 6 Reduction of 1/3 of initial amount Valued [€ 146 000]	• 1 Not Assessed • 2 Not Assessed • 3 Saving: € 29 000 • 4 Saving: € 16 000 • 5 Saving: € 39 000 • 6 Saving: € 48 000
Management	✔1 – Development of negotiating capacity of different levels of the company with artistic supervisor ✔2 – Better communication of middle management of workshop with on and off workers	• 1 Not Assessed • 2 Not Assessed	• 1 Not Assessed • 2 Not Assessed
			€ 217 000 Total

Source: © ISEOR 2000.

Figure 4.7. *Qualitative, quantitative and financial* effects of the *socio-economic intervention* in the opera house's set design shop (October 2000).

Initial dysfunction	Adjustment at the outset	Amount	Adjustment on arrival	Conversion into Value-Added
Interruption at work	3 technicians from design dept are interrupted 5 times a day by the production staff. Each interruption last 15 minutes	€ 10 300	By better staff information (meeting, resolution sheets, posting, physical relays) half the interruptions eliminated	€ 5 100
Modification of scenery at moment of assembly (interpretation errors)	10 extra hours for 10 people are required to make adjustments on assembly	€ 51 000	By regular project reviews (including visit by supervisor), and orchestrated specifications (mock up 1/33), saving of 10 hours in construction and 20 hours in stage design per production	€ 15 300
Time lost spent looking for equipment and tools	12 people spend 10 minutes per day on trips to look for tools	€ 28 000	By individual allocation of tools, by better control of loaned tools (signing loan list) and collective (cupboard) 2/3 of time originally lost was recovered	€ 18 600
Supply shortage	2 technicians return 2 times a week to suppliers. Each trip takes 1 hour	€ 18 300	By better identification of technical specifications desired (see mock up 1/33, installation and cut 1/50) supplies are better defined and managed. Trips are still made once a week for "specials" (e.g. foam rubber for floors)	€ 9 100
Alterations to scenery during construction	For each production, modifications require 10% extra work for 12 woodworker–locksmith technicians	€ 104 000	Half the wastage has been recovered as a result of better circulation of information, definition of needs and specifications (see mock up, installation). The remaining half is considered as fixed.	€ 52 000

Source: © ISEOR 2003.

Figure 4.8.   Examples of reductions in **hidden costs** and **conversion into value-added** (March 2003).

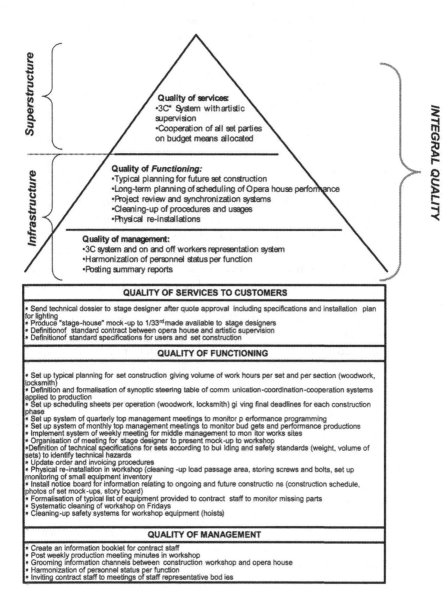

* 3 C : communication-coordination-cooperation

Source: © ISEOR 2000.

Figure 4.9.   Examples of changes observed in the opera house.

in-depth training in management and its associated tools; see Figure 4.5), and implemented structuring tools and *game rules* (e.g., weekly management meetings, disposing of a theatre-box model, regularly displaying *progress* reports).

Enhancing Cooperation and Coordination

One way of looking at improvement actions is through the synoptic table of coordination–cooperation (see Figure 4.10). This tool illustrates the *HISOFIS concept* showing the importance of stimulating information tools for actors to help increase effectiveness and efficiency. The long-term planning of *scheduling* for the opera house "orchestrates" the management of the artistic creation process with the necessary compliance with "final deadlines," a core issue for all artistic and cultural companies. It is worth noting that the *development* and implementation of this particular tool benefited from the personal assistance of the opera house's director general, who was in charge of the artistic domain. His acceptance reflects a cultural "revolution" of sorts, especially in view of the conventional wisdom that the artistic spirit would not accept tighter restrictions and organizational methods.

The tool shown in Figure 4.10 reflects an action orchestration tool for meeting "final deadlines" and it emerged from the work that was jointly organized in *horizontal and vertical projects*. As regard to its relevance for the set design workshop, its budget of 153,000 Euros was found to have roughly 640,000 Euros in hidden costs that were calculated in the *vertical diagnostic*. Such analysis provides a relatively clear picture of well-known flow problems, where the last actor has to regulate certain dysfunctions generated upstream in order to meet production deadlines. Orchestrating and steering the design-construction-rehearsal process in this way then allows workshop personnel to focus on other issues, such as the required/available capacity ratio, physical workflows in the workshop, or managing and informing on and off workers. In essence, the designing-constructing-rehearsing process enabled workshop participants to have a clear sense of potential problems and prevent a number of dysfunctions (e.g., lack of cooperation, lack of warning signals, lack of information). A result was a better balance between the workload and the working capacity of the workshop (the ratio between the two determines the time to be allotted).

This tool also illustrates the *principle of synchronization* as a factor for improving performance. In this case, improving the synchronization between actors and their activities helped save time in preparing performance and production costs by not having to rework sets or costumes, and enhancing the ability to adapt to changes in artistic creation. Similarly, the stress of the final weeks before a performance and rehearsals was

	Deadline	Decisive stages	3C mechanism	People and departments	Alert procedure (GM Intervention)	Theatre documents → project manager	Project manager's documents → Theatre	Interdepartmental documents
Design phase	15 months	MO commitment.**		D.G/ M.O/ Production		Engagement letter		
	-14 months	Signature Contract		D.G/ M.O/ Production		Technical dossier		
	-14 to -10 months	Artistic Project Design	Project review	D.G/ Tech. Dir. / M.O.			Pre-mock-up Story board	
	-10 months	Mock-up returned		D.G./Tech. & Workshop Dir./M.O	If late delivery mock-up>1 mth.		Mock-up1/33 Set up and cut1/50	Photos mock-up story → workshop
	+3.4 days	Specifications: Installation plans	Budget approval meeting	Dir. Tech / Prod. Mgr				Workshops, Tech. Dept heads
	+3 wks	1st quote		Tech. Dir./Workshops/Dept.heads	If budget dispute			Workshops, Tech. Dept heads
Construction phase	-9 months	Quote appr oved		Tech. Dir./ Set des./ Workshop Manager		Layout adjusted→ deco lighting		Layout→ Tech. Dept. heads
	+1 week	Mock-up presented to workshops		Workshops/ Set Des./ Prod manager				
	-9 mois → 1ère répétition	Construct.	Monthly meeting – Construt. and set.	Technical Dir. / Workshops/ Set des./ Prod Assistant		Detailed minutes of each meeting	Samples (A4), Colour chart,material samples, sketches	Detailed minutes→ heads of dept. and Tech. Dir.

*3C: Communication/coordination/cooperation   ** M.O.: Artistic Supervision

Source: © ISEOR 2000.

Figure 4.10.   Long-term planning schedule (new management tool) for an opera performance.

largely eliminated. Satisfaction of artists and audience members was increased, while hidden costs decreased.

This approach, which was developed in a *focus group*, highlights senior management's understanding of the efficiency of the source/application of funds ratio. As discussed earlier in the chapter, the Director General (an artist also in charge of the artistic domain), played a key role in implementing this tool. He was the first to use it in his negotiations with top artists, thus strongly promoting its use throughout the rest of the organization.

CONCLUSION

The socio-economic model experiment in a national opera house shows that such performance can be improved, as reducing resource losses in favor of higher *value-added* creates an economic stability compatible with the artistic and service qualities offered to audiences. This involves seeing such reductions in resource losses as an opportunity to offer artists greater means, all other things being equal. Therefore, the challenge is indeed to make the manager and artist's work complementary. Overall, the managerial and organizational rigor that was introduced in the opera house by socio-economic management was well received, including long-term planning of performance scheduling, a core concern for the profession.

The resulting management rigor complements the great artistic rigour that drives artists in their own profession, which is, perhaps, an underlying reason for this acceptance. By listing the dysfunctions, remedial actions and ensuing results, this chapter has attempted to illustrate how the socio-economic model can be helpful in solidifying the interdependence between art and management. By allowing better allocation of resources to jobs, artists benefit from socio-economic management by the freedom they receive to fully exercise their profession. For his part, the manager finds it easier to strike a balance between cultural imperatives and economic needs. Yet, although this analysis tends to negate "the discursive opposition of art and the economy" (Chiapello, 1998), such movement does not come about "naturally." The body of knowledge on *socio-economic theory of organizations* (Benollet, 1996; Savall & Zardet, 2005), along with its intervention theory, systems, methods and tools, can help to bring out a desire for convergence by allowing actors to discover for themselves and at their own pace the benefits of change. The scientific approach contained in the socio-economic theory of organization approaches management as being an art, in the sense that any activity, when all is said and done, is both an art and a science, and as a result anyone today who shapes his life and world is an artist.[6]

NOTES

1. The dynamic improvement actions involve both push and pull forces. This concept for achieving and steering performance is presented in chapter 1 of this volume.

2. On this point see chapter 1 of this book. The range limit for the descriptive field of socio-economic management, which integrates these different dimensions, includes the aesthetic emotion as a value and a resource.

3. For a more in-depth presentation of these elements, see chapter 1 in this volume and Savall and Zardet (1999).

4. Savall and Zardet (1999) note that econometric models (e.g., Malinvaud, Carré, & Dubois, 1972) have shown that capital and labor only partially (45%) account for value creation. The theory of hidden cost-performance ratios involves precisely the analysis of the unexplained 55%.

5. The term "prototype" activity is found in the work of Evrard and Chiapello (2004).
6. For further discussion of this issue, see Ingold (1985, 2001).

REFERENCES

Becker, H. S. (1983). Monde de l'art et types sociaux [The world of art and social types]. *Sociologie du Travail, 4*, 404-417.

Benollet, P. (1996). *Principe de régulation sociale: Le rapport structure-comportemental dialectique-recherche expérimentale.* Unpublished doctoral dissertation, University of Lyon, France.

Chiapello, E. (1998). *Artistes versus managers: Le management culturel face à la critique artiste* [Artists versus managers: Cultural management confronted with artistic criticism]. Paris: Métailé.

Durel, A., & Bergeot, V. (2002). Démarches innovantes à l'Opéra National de Lyon [Innovative management approach at the National Opera of Lyon]. In ISEOR (Ed.), *Le management des entreprises culturelles* (pp. 87-100). Paris: Economica.

Evrard, Y., & Chiapello, È. (2004). *Le management des entreprises artistiques et culturelles* [Management of artistic and cultural enterprises]. Paris: Economica.

Ingold, F. P. (1985). Jeder keinKünstler [Everybody can't be an artist]. *Neue Rundschau, 2*, 5-24.

Ingold, F. P. (2001). Auteurs et managers. Organiser et créer, un parallèle [Authors and managers: Organizing and creating in parallel]. *Passages-Passagen. Culture et Management. Histoires de couples. Magazine culturel Suisses, 31*, 18-22.

Malinvaud, E., Carré, J. J., & Dubois, P. (1972). *La croissance française. Un essai d'analyse économique causale de l'après-guerre* [French economic growth: An essay on causal economic analysis]. Paris: Seuil.

Midler, C. (1993). *L'auto qui n'existait pas: Management des projets et transformation de l'entreprise* [The car that didn't exist: Project management and company transformation]. Paris: InterEditions.

Savall, H. (1974, 1975). *Enrichir le travail humain dans les entreprises et les organisations* [Work and people: An economic evaluation of job enrichment]. Paris: Dunod.

Savall, H. (2003a). An updated presentation of the socio-economic management model. *Journal of Organizational Change Management, 16*(1), 33-48.

Savall, H. (2003b). International dissemination of the socio-economic model. *Journal of Organizational Change Management, 16*(1), 107-115.

Savall, H., & Zardet, V. (1987). *Maîtriser les coûts et les performances cachés: Le contrat d'activité périodiquement négociable* [Mastering hidden costs and performances: The periodically negotiable activity contract]. Paris: Economica

Savall, H., & Zardet, V. (1992). *Le nouveau contrôle de gestion: Méthode des coûts-performances cachés* [New management control: The hidden cost-performance method]. Paris: Éditions Comptables Malesherbes-Eyrolles.

Savall, H., & Zardet, V. (1995). *Ingénierie stratégique du roseau, souple et enracinée* [Strategic engineering of the reed, flexible and rooted]. Paris: Economica.

Savall, H., & Zardet, V. (1999). *La décision managériale multidimensionnelle comme fondements des sciences de gestion* [Contribution to the collective work in honor of Jacques Lebraty]. Nice, France: Editions IAE.

Savall, H., & Zardet, V. (2004). *Recherche en sciences de gestion: Approche qualimétrique. Observer l'objet complexe.* Unpublished English translation: *Research in management sciences: The qualimetric approach. Observing the complex object.* Paris: Economica.

Savall, H., & Zardet, V. (2005). *Tétranormalisation: Défis et dynamiques* [Competitive challenges and dynamics of tetra-normalization]. Paris: Economica.

Savall, H., Zardet, V., & Bonnet, M. (2000). *Releasing the untapped potential of enterprises through socio-economic management.* Geneva, Switzerland: International Labor Office-ISEOR.

Thévenet, M. (1992). *Impliquer les personnes dans l'entreprise* [Involving people in the enterprise]. Paris: Liaisons.

Wolff, J. (1981). *The social production of art.* London: Macmillan Education.

CHAPTER 5

FROM MANAGERIAL TREND TO PERMANENT CHANGE

Intervening in the Public Sector

Olivier Voyant

The powerful mobilization of trade unions and their rank and file membership in an array of sectors appears to be aimed at challenging the image of an expensive, deficit-prone and overprivileged public service. At a deeper level, this "battle of opinion" against those who govern appears to be fighting off the heralded "death penalty" for public services in favor of private enterprise. As if by sleight of hand, we are asked to believe that public service will disappear and that the private sector will emerge as the white knight overcoming all society's problems. This picture seems to ignore the fact that the public sector has no need to be ashamed of its efficiency and efficacy in comparison with the private sector (and vice versa). It would also be a mistake to believe that the public service does not have the capacity and means to evolve in order to guarantee its *survival*, **development**, and long-term effectiveness.

On the basis of this principle, the chapter illustrates the ability of a public service organization to adapt and master change (Voyant, 1997).

Socio-Economic Intervention in Organizations: The Intervener-Researcher and the SEAM Approach to Organizational Analysis, pp. 123–145
Copyright © 2007 by Information Age Publishing

After presenting its system of governance, which is viewed as quite complex, the discussion focuses on time as a critical resource and a determining factor in any strategy for change. The analysis then explores signs of genuine social advances and renewed economic demands, together with the formation of a team of *internal actors*, which is a necessary (albeit insufficient) prerequisite for *sustainable* development of organizational methods and tools (Savall, 1974, 1975, 2003a, 2003b; Savall & Zardet, 1987, 1992, 1995, 2004, 2005; Savall, Zardet & Bonnet, 2000). Finally, so as to avoid lumping together results and communication processes, the chapter concludes with a discussion of the policy choices made when communicating the results of change. By bolstering actors' desire for change, these initial advances clearly show that it is possible for a public service organization to readily influence a considerable part of its future.

The organization Le FOREM, directed by Jean-Pierre Méan "administrateur général," which is the focus of this chapter, employs 3,300 people and is in charge of public service for employment and training in the Walloon region (Belgium). Its mission consists of helping people to better express their vocational plans, acquire more qualifications and find jobs; helping companies recruit and train staff; and coordinating and supporting all those involved in the employment market in the Walloon area (see Bossens, 2006; Méan & Ombelets, 2003). In 2002, the organization set up a new structure, better adapted to its missions, organizing its activities into three complementary entities coordinated by senior management:

- The *Consultancy* entity, which offers general advice to job seekers and human resources to companies;
- The *Training* entity, which acts as a provider of training for companies and individuals;
- The *Support* entity, which provides support services (e.g., human, material and financial resources) for the Consultancy and Training entities.

GOVERNANCE COMPLEXITY IN THE PUBLIC SECTOR

It may seem peculiar to describe a system of governance as complex given that complexity is inherent in the concept of governance. This choice reflects a desire to compare systems of governance, with an emphasis on the level of threat faced by the organization. Although governance systems often make one think of companies in difficulty, we in fact consider that they are all in difficulty, the only difference being that some of them are aware of this reality while others are not. The latter category is particularly interesting because ignorance often engenders inertia and consequently a natural opposition to change. In this context, the threat intensifies and the system of governance becomes that much more complex. Therefore, since "public service" is commonly associated with "inertia," especially in comparison to the private sector, it is reasonable to assume that the public sector has specific features compared to the private sector and that the said features are immutable.

Special Features of the Public Sector

Numerous books, articles and studies use the term "public" as a sort of flag implying there is something specific in the field observed, as is often the case when we refer to "public management." Yet, based on numerous change management cases compiled by ISEOR intervener-researchers, on one level it does *not* appear that there are necessarily managerial issues and problems that are unique to the public sector. The types of *dysfunctions* and *hidden costs* discussed throughout the book, for example, show considerable convergence between the public and private sectors.

Following in-depth observation of public service organizations, however, we do believe that there are some specific features that differentiate public and private management, which relate to strategy. In the private sector, relations with the environment are governed by the well-known "law of the market." While the public sector is also affected by this law, there is another, less visible and more informal law that comes into play—the "policy law" or "law of politics," which has highly complex ramifications. Since we launched our intervention-research in this case in 2001, for example, our client organization has had to deal with two elections and a new minister at the head of the supervising administration. During change management phases, such cyclical elements generate a centrifugal force that encourages mental or even physical avoidance among actors. While it is true that actors' avoidance is a recurrent phenomenon, in such situations it is typically intensified. In this particular case, the *interaction* between political actors and the organization's actors is further shaped by a high level of unemployment in the region. As such, the public service organization's social responsibility has become highly politicized, to the point where unsatisfactory qualitative and quantitative results could precipitate questions about the organization's very legitimacy. Consequently, political programs and politicians themselves are directly concerned with the efficiency, efficacy and consistency of any actions undertaken by the organization.

A second specificity lies in the composition of the organization's board of directors. Referred to as the "management committee," its members are not only politicians and staff from the organization but also employer and trade union representatives. The ISEOR code of practice prevents us from discussing the state of relations between these members. However, we feel it should be pointed out that the political-institutional logic at work in this committee is often at odds with the managerial logic of the organization. The reporting of data provides a good example. For external purposes, reporting is composed of a patchwork of indicators (given the mixed nature of the parties involved) that tend to show a "forced" link of efficiency and efficacy between the political-institutional line and the organization's acts. Internally, the concern is to build a structure for steering people and activities. The result is a divide aggravated by the *progress* achieved within the organization. On the one hand, managerial logic is based on an increasingly convergent *project* (trust is growing); on the other, political-institutional logic is based on a project that evolves more slowly (control vis-à-vis the organization tends to become more wary).

Thus, there do appear to be some legitimate differences across the public and private sectors. However, they do not necessarily engender inertia; on the contrary, they result in perceived threats, which (1) act as an incentive for progress at the internal level and (2) create the need to develop a capacity for the positive contagion of the ***external environment***. More specifically, leaders have to display political courage internally and political caution externally. On this point, our experience has taught us that advances are only of value as long as they do not bring into question the mandate assigned to leaders by their constituents, that is, by politicians and institutional representatives.

Time as a Strategic Resource

In many cases, the past has much to teach us and we should heed it with respect and modesty. But if thought of in terms of a committed *resources* liability, where any break with the past is seen as a rift, the past then becomes a formidable threat. In this case, certain actors' collective memory was like a virus that propagated over the years. The resulting dependency was that they could not perceive the need to change or to pursue it with any reasonably efficient forms of action. This phenomenon only reinforces the complex system of governance and should therefore be taken into account. As such, it is useful to utilize time as a lever, in essence transforming it into a strategic resource for relaxing constraints. As an Asian proverb states, "even with nine wives, we cannot make a child in one month." There is no point in endlessly increasing the means devoted to change if the time dimension is not taken into account.

A Soft Approach to Change

The present case represents over 1 million kilometres traveled, 200 hotel nights and 10,000 person-hours split between 10 intervener-researchers. As part of the *project*, over 900 people were interviewed, 7,000 dysfunctions were identified, summarized and processed, 300 people were involved in training sessions, and 25 million Euros in *hidden costs* were partially converted to *value-added*.[1] In addition, the field notes were captured in roughly 15,000 post-it notes, 300 ring-binders and 200 writing pads. The work was completed between 2001 and 2004.

An underlying question, of course, is why spend four years in an intervention-research involving over 3,000 people? The answer, which is illustrated in Figure 5.1, can be summarized in three points:

1. Change literally fills all spaces (all hierarchical levels, all sectors) over an annual period: considered as "hard," it is rejected by certain actors and accepted by others who will reject it once the action has ended;

2. Change "benefits" from a focused *evaluation* of the resistance to change and penetration of those areas, which happens over a long

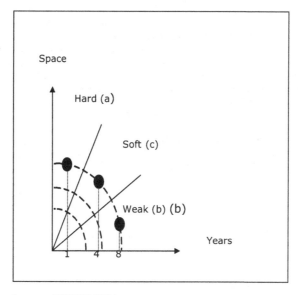

Source: ©ISEOR 2004.

Figure 5.1.   Types of change.

period: if considered as "weak," the change will not modify practices among actors; and

3. Change operates as a gradual transformation, using time as an element for "softness": actors have the right to apply "brakes" to change and such "softness" can be helpful in finding points of agreement on the *progress* to be realized.

Similar to a plane coming in to land, time can facilitate change touching down gently on the organization. The effects of this type of change are infinitely better than the so-called "hard" or "weak" approaches, particularly with regard to a total transformation of professional behavior.

As an example, Figure 5.2 reflects possibilities in a meeting in which everyone is asked to analyze a particular subject. Boxes "1a" and "1b" describe two inefficient situations. In Box 1a those present spend a great deal of time on analyses, to the point where a "battle of wits" emerges as the importance of words replaces the importance of action. All too often, this box concerns the top level of the hierarchy. Box 1b more generally concerns the "bottom level." The actors rarely meet and cannot devote much time to analysis, and as such, the analyses are too superficial to

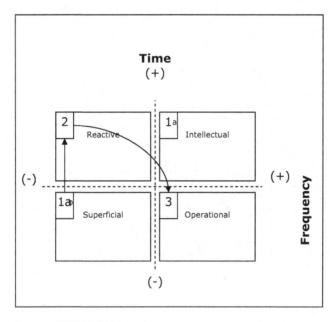

Source: ©ISEOR 2004.

Figure 5.2.   Types of analysis.

prompt a move to action. In Box 2, analyses are seriously undertaken, but the main drawback is their reactive nature. At ISEOR, the preference is for Box 3, where the depth of the analysis is less costly due to its regularity and more effective because it bridges the gap between the organization's "head" and its "legs." However, it often takes a realistic time period to develop this type of orientation. To apply this observation to the present case, 2001 and 2002 were Box 1 years, 2003 was a Box 2 year, and ultimately 2004 transitioned to Box 3. In comparison to intervention-research missions in companies of similar size and with similar problems, the public service organization's *development* was extremely positive, which further underscores the need to take time into account when undertaking change.

Determination for Deep-Rooted Change

Any change, even if brought about gradually over time, runs the risks of the "surface effect," whereby new practices fail to become rooted. One of the symptoms signaling this problem is the dichotomy actors experience between "change activities" and "professional activities": contrary to what consultants may desire, organizational members often feel that when they make a change they are not doing their job (and vice versa). Of course, such schizophrenia tends to diminish over time, at a rate that varies according to the actors and their needs. However, without any other input, this effect can remain rooted in people's minds.

One solution consists of raising the actors' awareness regarding the notion of intangible investments. The organization's legacy in that respect is a series of short-lived trends. Like shooting stars, new methods come and go, leaving behind them social and economic waste. One such example was the opportunity presented to the public service organization with the arrival of budget restrictions at the end 2002. The organization was no longer able to invest on several fronts and therefore made some arbitrary cutbacks. A result was a marked insistence on introducing *socio-economic management* and a reduction in or even abandonment of other potentially redundant, inefficient intangible investments.

Another contribution came from the intensity of the change initiative, which can be linked to the growing awareness among the actors of the extent of the organization's inertia, and the strong determination of the director and that of his main partners in the governance system. This is how socio-economic management transcended the stage of being a mere trend and has over the years become a new mode of management tending to become lasting practices.

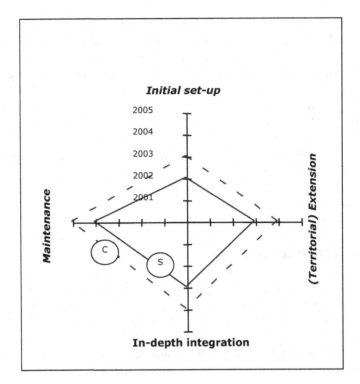

Source: ©ISEOR 2004.

Figure 5.3.   Phases of the *intervention-research*.

Finally, it is important to note the effect of the method used for insti-
gating the change as one reason for its deep-rooted nature. As illustrated
in Figure 5.3, the intervention was based on four major phases in the
socio-economic intervention life cycle as the organization "discovered"
socio-economic management.

- *Initial set-up phase*: This process involves building, strengthening
  and installing common work methods to coordinate all functions
  and specialties. For large companies, intervention is focused on priority areas involving senior management and some vertical areas.
  In the case of the public service organization, we ran an initial set-
  up on the Support entity (workforce of about 400 people) in 2002,
  another on the Consultancy entity (about 1,500 people) in 2003.
  These entities are respectively indicated in Figure 5.3 by the letters
  "S" and "C."

- *Extension phase*: Continuation of the initial installation phase, focusing on areas that were not touched by the common work methods and socio-economic management perspective.

- *Improvement phase*: The purpose of this phase is to run specific actions to consolidate the installation and extension work. In the present case, this phase was launched at the same time as the extension phase (in 2003 for the Support entity, in 2004 for the Consultancy entity). This type of phased approach (over time) can actually reduce the implementation time of socio-economic management: while some people are discovering socio-economic management, others are running in-depth improvement sessions. Regarding the latter, they can be classified into two groups: (1) those helping to improve the tools initially by creating closer links with other activities and (2) those implementing diagnoses and projects on transversal activities (e.g., management control, strategic intelligence, marketing). In terms of the "*time management* self-analysis" tool, the satellites help to go beyond a simple analysis context and revitalize time management practices and reflexes in a concrete manner (see Figure 5.4). By bringing greater depth and strength to the

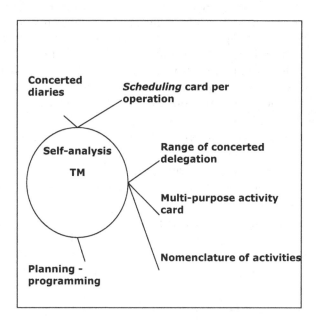

Source: ©ISEOR 2004.

Figure 5.4.   Satellite *time management* tools.

action, this phase helps socio-economic management to "put down stronger roots."

- *Maintenance phase*: As with equipment, intangible methods and tools also suffer from the effects of time in the form of either erosion or evaporation. Thus, they require maintenance. This phase, which was initially very demanding (conducted in the Support entity in 2004), is gradually becoming a minor intervention taking the form of a long-term partnership.[2]

Gradually Constructing a Partnership Together

The term "partnership" is regularly used to describe the relations that unite various different actors. Nevertheless, there is a notable difference between temporary and long-term partnerships. Moving from temporary to long-term partnerships requires a combined effort whereby gradual and modulated use of the time resource is essential.

In this case, the intermediary indicators for this joint construction took the form of a series of *intervention agreements*:

- *Intervention Agreement No. 1 (2001)*: Contrary to ISEOR's usual practice, we agreed to intervene in the organization in a limited area (senior management) and theme (strategy). The reasons for this exception were the history of the organization, the number and quality of our contacts with internal and external members of the organization, and an urgent need for rapid implementation due to strong pressure from the Minister in charge, who recommended an ISEOR intervention.
- *Intervention Agreement No. 2 (2002)*: After initial implementation with senior management, the client organization's Support entity was considered as a priority "target" for introducing socio-economic management. At this stage, the intervention was framed on the *HORIVERT model*.
- *Intervention Agreement No. 3 (2003)*: The Consultancy entity took over from the Support entity for the introduction of socio-economic management. This initial introduction phase in the Consultancy entity was followed up with extension and improvement phases in the Support entity.
- *Intervention Agreement No. 4 (2004)*: This phase of the project involved extension and improvement tasks in the Consultancy entity, maintenance tasks in the Support entity, and additional work with senior management to reinforce the organization's cred-

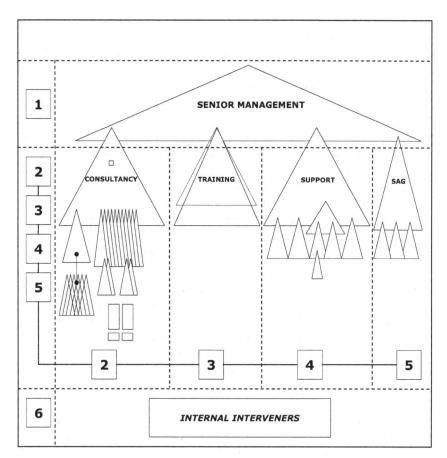

Source: ©ISEOR 2004.

Figure 5.5.   Overall *intervention architecture* (4th year: 2004).

ibility with the local government. The full range of the tasks in this phase is presented in Figure 5.5, which also shows *intervention clusters*. For educational reasons, the diagram is segmented into 6 areas: (1) senior management, (2) the Consultancy entity, (3) the Training entity, (4) the Support entity, (5) an offshoot of the previous entity called "GAS" (General Administration Services), and (6) "*internal-interveners*."

To illustrate what was done during the intervention, the work carried out on each area included:

- *Area 1*: **Collaborative training** and **personal assistance** on improvement (**IESAP, PAP,** satellite tools); a **IESAP/PAP** seminar; an audit of **communication-coordination-cooperation** systems; three focus groups on consolidating **socio-economic management control,** the **periodically negotiable activity contract (PNAC)** charter, and socio-economic management certification.

- *Area 2* (23 clusters): Collaborative training and personal assistance for improving socio-economic **management tools,** creating and implementing regional **internal-external strategic action plan (IESAP),** a steering group for implementing the horizontal project and strategic intelligence (services provided during the course of 2003); two transversal diagnoses and projects on the "recruitment consultancy" and "partnership" business.

- *Area 3* (comprising one limited cluster and one extended cluster): Collaborative training and **personal assistance** for improving socio-economic management tools.

- *Area 4* (totalling 8 clusters): Collaborative training and personal assistance for improvement and maintenance of socio-economic **management tools** in marketing (one diagnostic and one **socio-economic marketing** project), finance (one diagnostic and one vertical project on the financial resources function), management control (continuing socio-economic management control project), and the human and material resource sectors (two **socio-economic evaluations**).

- *Area 5* (totalling 4 clusters): One project and personal assistance on socio-economic management tools; one diagnostic and a vertical project on the "international relations" function.

- *Area 6*: Advanced training on acquiring socio-economic management methods and tools organized at ISEOR Headquarters (involving several companies), a focus group for improving the internal-intervener network, collaborative training and customized training on the internal-intervener profession, and synchronization and steering sessions on the intervention-research with the organization's socio-economic management pilots.

Over this 4-year period, the team of ISEOR intervener-researchers completed a total of over 10,000 hours of activities whereas the initial partnership (i.e., the first agreement) provided for only 350 hours. What happened? Although there are doubtless many reasons, one of the most interesting reasons was the actors' satisfaction with the work accomplished, at both the team and individual levels. In essence, success came from establishing a true partnership and working on multiple levels, initially focusing on the requisite improvement actions, and then moving to the practical implementation of desired actions.

The collaborative nature of the project can be seen in the work-sharing involved in drawing up the agreements. To prepare the first year's agreement (2001), for example, some of the client organization's senior and middle managers, after hearing about ISEOR's achievements in Belgium, came to ISEOR Headquarters in Lyon to meet one of our partner companies. After this visit, several meetings were organized in Lyon to outline a framework for the initial intervention. For subsequent agreements, consolidation of the partnership took roughly six months (from July to December). Although this extensive period helped to achieve greater maturity for the annual project, it also precipitated a degree of tension between fulfilling the ongoing partnership and preparing future interventions. As an example, during discussions on the partnership for the second year of intervention (2002) the managers assured us they were ready to set up a *Horivert process* but then balked at initiating diagnoses and *hidden costs* analyses Despite these tensions, there was a perceptible shift in the attitude of the organization's actors toward the intervener-researchers over time, especially during the subsequent agreements (2003 and 2004), which can be described as a shift from a customer-supplier relationship to one of long-term partner. This new relationship, which was apparent in the action-plan *negotiation* stages, was also noticeable at the implementation stages (see Table 5.1 for a breakdown of all actions over the 4-year period).

Drawing on the information in Table 5.1, there are four points that illustrate the joint construction of the partnership. The first is the virtual balance over the four-year period between efforts focusing on "policy decisions" (25% of the time), "process" (33%) and "tools" (29%). This is an important point because in the long term, some companies tend to neglect the "process" part, querying the equilibrium that makes this approach so effective. The second point is the sharp increase in actions based on "political decisions" between years 1 and 2 (50 days) and years 3 and 4 (301 days). In our experience, strong "intrusion" in this direction generally indicates a mark of trust. Similarly, there was a sharp increase in the amount of work done with internal-interveners: 159 days in years 3 and 4 compared to 23 days in the first two years. This shift reflects the organization's determination to appropriate this approach in such a way that socio-economic management can survive and evolve after our team's departure. Finally, the fourth point concerns the balance in the number of people trained at ISEOR Headquarters: 24 internal-interveners and 23 relay managers; 24 people trained over the first two periods compared to 23 people during periods 3 and 4. This considerable investment, given

Table 5.1. Summary of Actions (in Days) During the 4-Year Intervention

		2001 (Year 1)				2002 (Year 2)				2003 (Year 3)				2004 (Year 4)			
	1418	46				302				534				536			
		D	S	C	F	D	S	C	F	D	S	C	F	D	S	C	F
		44	2	0	0	33	229	17	0	136	64	217	28	209	122	111	24
Policy-decisions axis (351 days, or about 25%)																	
Steering	58	7	—	—	—	13	—	—	—	22	—	—	—	16	—	—	—
Str vigilance	58	—	—	—	—	—	—	—	—	—	—	—	—	58	—	—	—
Mgt Control	88	—	—	—	—	—	30	—	—	—	25	—	—	14	19	—	—
Marketing	66	—	—	—	—	—	—	—	—	—	—	28	—	38	—	—	—
PNAC chart	14	—	—	—	—	—	—	—	—	—	—	—	—	14	—	—	—
Certification	14	—	—	—	—	—	—	—	—	—	—	—	—	14	—	—	—
Audit tools	53	—	—	—	—	—	—	—	—	25	—	—	—	28	—	—	—
Improvement-Process axis (475 days, or about 33%)																	
Evaluation	53	—	—	—	—	—	—	—	—	—	—	—	—	—	53	—	—
Implement.	27	—	—	—	—	—	—	—	—	14	—	—	—	—	13	—	
Diag-Project	395	25	—	—	—	—	153	—	—	—	—	133	—	27	—	57	—
Management-Tools axis (410 days, or about 29%)																	
Maintenan.	105	—	—	—	—	—	—	—	—	—	9	84	—	—	12	—	—
In-depth int.	208	—	—	—	—	—	—	—	—	89	16	—	—	38	—	41	24
Set-up/ext.	97	12	2	—	—	20	46	17	—	—	—	—	—	—	—	—	—

Set-up *internal-interveners* team (182 days, or about 13%)

		2001	2002	2003	2004
Int. Interv.	182	0	23	89	70

In-depth training of *internal-interveners* and key managers in Lyon (number of people)

	47	12				12				13				10			
		D	S	C	F	D	S	C	F	D	S	C	F	D	S	C	F
		0	9	1	2	0	10	1	1	2	5	4	2	1	3	2	4
Int. Interv.	24	—	9	1	—	—	10	—	—	—	2	2	—	—	—	—	—
Managers	23	—	—	—	2	—	—	1	1	2	3	2	2	1	3	2	4

Source: ©ISEOR 2004.
Note: D = senior management; S = Support entity; C = Consultancy entity; F = Training entity.

the organization's geographic distance, shows that senior management wanted *their* people to be responsible for the approach.

SIGNS OF GENUINE PROGRESS IN SOCIAL PERFORMANCE

The elements of a soft approach to change and the rooting process presented in this chapter, as well as the joint partnership effort, have produced numerous socio-economic advances. Although we cannot provide an exhaustive list, the following sections highlight those that constitute the principal strengths, in essence creating a foundation for the survival and long-term development of socio-economic management.

A radical transformation in professional behavior is one of the main features of evolution among the organization's actors. To identify the root causes of this change, it is useful to focus on the client organization's new *communication-coordination-cooperation* systems (the *gears*), how they help to deploy strategy (contents) and the effects that their efficacy produces on delegation at various levels of responsibility and qualification.

Revitalized Communication-Coordination-Cooperation Systems

In accordance with the principles of intervention and research, our team found itself interacting with the organization's actors over a four-year period. On the "process" aspect, the team conducted interviews with over 900 people. These interviews, conducted during *horizontal diagnoses* (3 in all), *socio-economic marketing diagnoses* (2), strategic intelligence diagnoses (2) and *vertical diagnoses* (9), help to identify over 7,000 *dysfunctions*, which were categorized, analysed, summarized and processed in focus groups. These elements enhanced the "tools" aspect, on which we held about 500 collaborative training sessions with the help of internal-interveners and performed over 2,000 individual and group personal assistance actions with the direct involvement of over 300 client organization executives, senior and middle managers and experts.

Of course, these contacts brought out the traditional resistance that are typically observed. In the client organization's case, the most common resistance was a collective lack of discipline among the actors in collaborative training sessions: absence, lateness, coming and going in the middle of a cluster, reading mail, and, naturally, using mobile phones. To mitigate the effects of these problems, various *socio-economic intervention techniques* were utilized:

- Methodological reminders to explain that a consultant cannot run a session in the absence of the pilot. This rule stems from the fact that while consultants are qualified to present methodological materials, they are unable to effectively take the place of an organization's management to convey "content" on the organization's policy or how to adapt socio-economic management methods to the firm's specific context.

- *Internal-interveners* ran "remedial" sessions to promote the *integration* of an individual into an already tightly-knit team and to help absentees to remain (or become) actively involved in the change process.

- Launching an *intervention-research* involves setting up "biodegradable" *collaborative training clusters*: when a set of services is complete, the *cluster* can theoretically be dissolved. In the client organization's case, both the formation and composition of these clusters were regularly called into question. As a result, the clusters ended up reflecting the organization's structure and gradually became a "living" organization chart—the distribution of responsibilities is not immutable, but varies according to the organization's needs.

These actions had a twofold impact: a reduction in physical and mental *absenteeism*, and a determination to multiply meeting points. The Support entity set up a management staff in 2003, accompanied by a seminar in 2004; senior management organized a strategy seminar in 2003 and 2004, bringing together for the first time approximately 60 people from different clusters. Such examples show that actors use and develop collaborative training clusters to revitalize their own communication-coordination-cooperation practices. It was undoubtedly no coincidence that senior management wanted the consultants to run an audit of *communication-coordination-cooperation systems* in all the entities for 2004. Such efforts to improve the way clusters function naturally prompt people to raise questions about how communication-coordination-cooperation systems are functioning.

Concrete Deployment of the Organization's Strategic Plan

Initially, collaborative training clusters constitute a type of methodological container. Once they have been assimilated into the *communication-coordination-cooperation systems*, the question of their content arises in order to sustain a permanent collective process. In this context, the *IESAP* and *priority action plan (PAP)* tools are useful. As with the *PNAC*, these tools represent a recurrent work theme as they maintain collective work between and within teams and promote concrete deployment of the organization's strategic plan. As illustrated by the strategy seminar discussed earlier, in 2003 and 2004 preparation for the seminar mobilized members of the senior management, support, consultancy and training clusters for several months. Setting up this ritual helped to meet an ambitious objective: mobilizing the firm's 60 top managers so that they may in turn mobilize all the organization's actors to implement the strategy. Inevitably, this strategic deployment involves clusters and *communication-coordination-cooperation systems*, where roles are allocated between actors, the necessary arbitration between daily activities and strategic activities takes place, and specific actions, such as entering PAP actions into diaries, can occur.

While the work on *IESAP* and *PAP* contributes to periodic exchange systems and rolling out the strategy, it also has a permanent impact on all activities. When actors work on planning and scheduling priority actions for the *PAP*, they develop reflexes for planning and *scheduling* activities, with two positive consequences: better team efficiency and efficacy, and an increase in delegation.

INCREASED DELEGATION FOR HANDS-ON MANAGEMENT

Setting up the basis for solid management (actions), as opposed to "hot-air management" (talk), transforms managerial responsibility into a driving force, a participative movement, thereby reducing certain harmful centralizing effects. In this respect, the development of the "senior management" cluster is highly significant. In the public service organization case, for a thee-year period all diagnostic tasks were first submitted to the "senior management" cluster before being presented to the other sectors. Besides longer delays in returning diagnoses, these presentations reduced the cluster's rate of progress on other issues. Beginning in 2004, this necessary centralization was replaced by *synchronized decentralization*, where diagnoses were directly presented to the appropriate sectors with a copy sent to the CEO and a summary report presented to the "senior management" cluster. The benefits of this approach coincide with those of the *Horivert model*: a decentralization thrust enables middle management to take on the role of policy transmission, while leading employees and activities. Furthermore, this decentralization thrust is most efficient when it is synchronized. When such is the case, the thrust follows an "elevator" movement: top-down information eventually returns from the bottom-up, notably to obtain successive validations.

RENEWED ECONOMIC RESPONSIBILITY THROUGH STRATEGIC PATIENCE: CONCEPTS AND TOOLS FOR IN-DEPTH IMPROVEMENT

Developing and strengthening delegation not only involves improving *social performance* by developing accountability at all levels and fostering meaningful work, it also includes improving economic management. A sense of economic responsibility was acquired gently and gradually, yet with steady progression. The terms "gently" and "gradually" refer to Figure 5.1 of this chapter. Economic requirements were understood and accepted because they were transmitted gently, without harshness or laxness. Furthermore, these requirements proved to be stimulating, to a large extent because they were constant. This constant progression can be compared to the staircase principle: each individual step requires little effort, making it possible to maintain "constant effort." Every step takes the actor higher, which enables progression. The first signs date from 2002, with requests to justify work hours. In 2003, the issue of staff department *productivity* was raised and there were numerous calls to cut back on strategic dispersal. By 2004, the organization moved toward *socio-economic management control*.

Socio-Economic Management Control

In times of budget cutbacks, it may seem quite natural to make economic decisions a top priority. The problem is that such decisions relate back to ideological principles that actors will not easily abandon merely on the grounds of budget cuts. It is therefore necessary to persuade people and make sure that demands slowly increase.

In the present case, a critical first step in persuasion and building demand concerned the management control specialists. Between 2002 and 2004, three goals were set: design a *socio-economic management control architecture*, effectively implement it, and support its adoption and activation within the organization. At this stage, our observations, which reflect the outcomes of previous experiences, suggest that the main opponents of the introduction of socio-economic management control are the specialist management controllers who must take charge of it and adopt new management control concepts and models. This mandate is often perceived as a literal change in profession, arousing inevitable resistance to change.

With the passing of time, however, resistance weakened and brainstorming sessions gave way to action. These discussion sessions were an indispensable prelude to reducing misunderstandings, fostering the emergence of transformed opinions, and then carrying out actions in collaboration with non-specialists in management control.

In parallel, another step entailed calculating the **hidden costs** of controlling dysfunctions discovered in diagnostic interviews, with the indispensable **cooperation** of managers and support from certain management controllers. If this cooperation during the diagnostic phase, regarding the configuration of quantitative and financial indicators, is not viewed as a methodological "obligation," it provides significant credibility and reinforcement for the results. Indeed, the corollary of this double control, carried out by both the hierarchy and the specialists, was a significant diminution of opposition during the diagnostic.

The third step involved bringing together specialist management controllers and operational staff. Over a six-month period in 2003, a number of senior management cluster sessions tackled the subject of socio-economic management control: presenting concepts, sharing roles between operational and specialist staff, and presenting **economic balances** through vertical project pilots. Two policy aims marked the end of 2003: (1) the need to use economy balances as a decision-making instrument; and (2) the challenge of making 2004 the year of socio-economic management control. Given all the work achieved during this period, it appears that another level of operation has been reached. First, increased awareness of management control and socio-economic management control concepts and tools was achieved in all collaborative training clusters. Second, in their day-to-day activities, actors have acquired the reflex of using economy balances to assist in making and assessing decisions. Finally, the **HCMVC (Hourly contribution margin on variable costs)** indicator was used to work out the economic value of the workload generated by new projects, a process that is very helpful when negotiating new outside funding.

While considerable progress has been achieved, there is still much to do, which accounts for the CEO's new aim for 2005. As he noted, "2005 will be the year of socio-economic management control and *this* time we will ensure we have the necessary resources." Although the term "management control" is often perceived as synonymous with cost-cutting, within the context of socio-economic management it is also about increasing the number of products with more value added. In this respect, the contribution of strategic intelligence is pivotal.

Strategic Intelligence

If management control suffers from ideological handicaps, the same is true for *strategic intelligence*. Although managers are typically quite willing to use it as an instrument for input (e.g., for **IESAP** and **PAP**), they hesitate to make it a proactive instrument for positive and determined action on the firm's external environment. In the public service organization, after presenting and explaining the links between internal and **external environments**, the Senior Management Cluster reached a consensus on the need to transform organizational members into *sales people*. The initial ideological assumption in public service, however, was that there was nothing to "sell"—selling is about acquiring money, and money implies a form of "theft." In 2003, this notion was challenged. Since the budget was the result of the ability of all company staff to negotiate with the authorities, they ultimately adopted the slogan "we're all salespeople." This view point (awareness), although still quite new, was expressed in the 2004 agreement in the form of a diagnostic, a project and a strategic intelligence seminar involving political and institutional actors from the organization's governing circles. The goal, which was now in sight, was twofold: (1) to make the organization's **IESAP** an input instrument for the management contract[3] and (2) to create awareness among external actors of the progress achieved by the organization, involving them in order to obtain greater consistency between set objectives and the resources the relevant Ministry allocates to achieve them.

Periodically Negotiating the Activity Contract (PNAC)

From a methodological point of view, **PNAC** is a team management tool. On a practical level, this tool also tends to arouse heated debate on the financial bonus issue and raises some difficult questions: (1) Does recognition of actors necessarily mean additional pay? (2) In a public service organization, can bonuses be awarded on individual objectives? The gentle pressure applied to certain members of the Board of Directors and Management by the director at large to set up PNAC constitutes a positive response to the first question. For the second question, practices common in other organizations indicate that "public service" is not incompatible with awarding bonuses according to individual and team performance appraisal criteria.

The main difficulties in introducing the tool, in fact, appear to be institutional. The possibility of financially rewarding actors' efforts must be negotiated, while ensuring that this is indeed a redistribution of benefits rather than another budget allocation. In this respect, the work with institutions will doubtless greatly help to introduce the tool as an instrument of ongoing motivation and a way of sharing the benefits of improved economic performance among all the parties involved: users, authorities and the organization's staff.

SUSTAINED DEVELOPMENT OF METHODS AND TOOLS: THE INTERNAL-INTERVENER TEAM

To support senior management's efforts in setting up and institutionalizing these different techniques, early successes must be supported and intensified to ensure that they can be replicated over time. In this respect, the formation of a team of internal "consultants" is highly significant.

Growing Efficacy and Efficiency

Creating an internal-intervener cell is similar to growing a new organ inside the firm's body, which is subject to attacks and a series of legitimate queries: (1) What legitimacy do internal-interveners have? (2) Is it a spy cell for senior-level management? (3) What did they do to get appointed? Contrary to initial appearances, these stated reservations can be helpful in developing the network as they oblige internal-interveners to practise modesty and patience. In the public service organization, the early years (2001, 2002) were a learning period for these individuals, in essence serving as assistants to the external consultants. It was only in 2003 that they began to take charge of collaborative training services and personal assistance, and in 2004 they undertook vertical diagnoses. While this development reflects both efficacy and efficiency on the part of the internal-intervener team, considerable progress still remains to be made on a number of points: technical control of services, evaluating the cost of change actions, preparing work for the years ahead, teamwork, and steering change leadership.

The Gradual Takeover of Copiloting Action

As a prerequisite for the independent steering of change, copiloting actions have been perfected since the initiation of the intervention-research program. Over the period 2001–2002, the organization needed a liaison contact in the ISEOR to draw up progress reports on services. As of 2003, the points of contact have been multiplied. To manage everyday affairs, there were weekly telephone meetings between the organization's action pilots and ISEOR intervener-researchers. As a way of crafting an objective view of change dynamics and what it would take to lead the change, the internal and external consultants held an off-site meeting to share their experience and explore possibilities.

Once again, the initial customer-supplier type of relationship was transformed into a genuine partnership. Resisting pressure, avoiding traps, and guaranteeing proper conduct of services and timetables have all become common concerns, managed through constant cooperation.

CONCLUSION

At the start of this intervention, ISEOR's goal was to make the public service organization a "showcase" for socio-economic management. The results of this four-year intervention are highly encouraging. Admittedly, the management system that was put in place is still searching for the right balance between social and economic performance. Tools like the *competency grid* and the *strategic piloting indicators* must still be further developed. The mechanism whereby actions are deployed vertically is still too fragile and prone to being challenged. The achievements of the internal-interveners are also well short of achieving true autonomy in managing change.

Ultimately, the dynamics of organizational improvement should be viewed within the lens of continuous improvement. Be that as it may, this interim report and progress to date suggests that public services can draw on internal resources to evolve, facilitating their long-term *survival*, development and success. Let us hope that other public sector organizations seize the opportunity and use this successful experience as an inspiration for their own evolution and development.

NOTES

1. This hidden cost amount had been found in areas representing over 800 people, which represents an average of over €30,000 hidden costs per person, per year.

2. On this point, we wish to point out two exemplary partnership experiences respectively lasting 20 and 10 years. The first concerns a group "processing food products" (Viennese pastries, frozen cakes, continental toast) Brioche Pasquier (see chapter 2); the second involved an industrial group, Technord, whose CEO is Michel Foucart, in state-of-the-art technologies (automation, industrial software, networks) with revenues of over €40 million for a workforce of 300 people.

3. The local government and the organization's management committee signed a management contract defining priorities, objectives and directions to be pursued by the CEO and his teams. The current contract covered the period from 2001-2005. It positioned the organization in a context of combined management of the job market inferred by ratification of Agreement No. 181 of the International Labour Organization.

REFERENCES

Bossens, C. (2006). Une démarche d'installation du contrôle de gestion socio-économique: Le cas du FOREM [Socio-economic management control set-up procedure: The case of FOREM]. In ISEOR (Ed.), *Le management du développement des territoires* (pp. 183-194). Paris: Economica.

Méan, J. P., & Ombelets, N. (2003). Le management socio-économique dans une institution publique belge: Le FOREM [Socio-economic management in a Belgian public institution: The FOREM]. In ISEOR (Ed.) *L'université citoyenne* (pp. 270-290). Paris: Economica.

Napoli, B., & Simar, P. (2005). Positionnement stratégique, accroissement de l'efficience et de la qualité de service à l'usager. Organisme de l'emploi et de la formation professionnelle (Belgium) [Strategic positioning, increasing efficiency and quality of user services: An employment and training organism]. In ISEOR (Ed.) *Enjeux et performances des établissements sociaux: Des defies surmontables?* (pp. 145-164). Paris: Economica.

Savall, H. (1974, 1975). *Enrichir le travail humain dans les entreprises et les organisations* [Work and people: An economic evaluation of job enrichment]. Paris: Dunod.

Savall, H. (2003a). An updated presentation of the socio-economic management model. *Journal of Organizational Change Management, 16*(1), 33-48.

Savall, H. (2003b). International dissemination of the socio-economic model. *Journal of Organizational Change Management, 16*(1), 107-115.

Savall, H., & Zardet, V. (1987). *Maîtriser les coûts et les performances cachés: Le contrat d'activité périodiquement négociable* [Mastering hidden costs and performances: The periodically negotiable activity contract]. Paris: Economica

Savall, H., & Zardet, V. (1992). *Le nouveau contrôle de gestion: Méthode des coûts-performances cachés* [New management control: The hidden cost-performance method]. Paris: Éditions Comptables Malesherbes-Eyrolles.

Savall, H., & Zardet, V. (1995). *Ingénierie stratégique du roseau, souple et enracinée* [Strategic engineering of the reed, flexible and rooted]. Paris: Economica.

Savall, H., & Zardet, V. (2004). *Recherche en sciences de gestion: Approche qualimétrique. Observer l'objet complexe.* Unpublished English translation: *Research in management sciences: The qualimetric approach. Observing the complex object.* Paris: Economica.

Savall, H., & Zardet, V. (2005). *Tétranormalisation: Défis et dynamiques* [Competitive challenges and dynamics of tetra-normalization]. Paris: Economica.

Savall, H., Zardet, V., & Bonnet, M. (2000). *Releasing the untapped potential of enterprises through socio-economic management.* Geneva, Switzerland: International Labor Office-ISEOR.

Voyant, O. (1997). *Contribution à l'élaboration d'un système de veille stratégique intégré pour les PME-PMI* [Contribution to the development of an integrated strategic alert system for small and medium-size businesses and industries]. Unpublished doctoral dissertation, University of Lyon, France.

CHAPTER 6

SEAM AND NONPROFIT SPORTS CLUBS

Activating Intraorganizational Negotiation as Piloting Levers

Miguel Delattre

Of the top three world events open to the public, the top two—the Olympic Games and the World Football Cup—emanate directly from nonprofit clubs and associations. These organizations represent a fertile field for management scientists and practitioners. Indeed, given the important role that this part of the nonprofit sector plays in our society, its contribution warrants reconsideration as a factor of societal well-being. Clubs and associations are a vital part of our daily life. We all come into contract—sooner or later, more or less durably, directly or indirectly—with nonprofit associations, whether during recreational activities (sports or cultural pursuits), in the course of preparing for professional life, as part of our community and societal involvement (social advocacy, unions) or in connection with quality-of-living concerns (health, environment).

Socio-Economic Intervention in Organizations: The Intervener-Researcher and the SEAM Approach to Organizational Analysis, pp. 147–170
Copyright © 2007 by Information Age Publishing

In France, between 1990 and 2004, there were approximately 800,000 to 900,000 clubs and associations, including 170,000 sports associations and 160,000 associations dedicated to culture, tourism and recreation (Archambault, 1996). By 2006, the number of these clubs and associations exceeded 1 million, which represents a 10% increase over the past decade. These are businesses and not merely a marginal organizational phenomenon. In-depth knowledge of this sector was initiated in France during the 1990s by a study that was part of the John Hopkins Program for International Classification of Nonprofit Organizations (ICNPO),[1] which classifies the nonmarket "third-party organization" sector by primary areas of economic activity, that is, the goods and services produced (see Salamon & Anheier, 1992). The actual number of French associations is difficult to establish, due to the fact that the data gathering system is under construction (Camus, 2006). In France today, approximately 70,000 associations are created annually, that is, 190 new associations per day. Knowledge of this phenomenon resulted in the creation of the French Ministry of Associations. Institutional recognition of the important factor that associations represent led to setting up a nation-wide framework starting in 2002 (national survey, Febvre & Muller, 2004) to promote the *development* of associations.

The John Hopkins Program distinguishes 12 activity areas, which are in turn further subdivided by the French "third-party" sector classification into 124 activities (Archambault, 1996). Currently, French associations are grouped into three major programs: sports (24.5%), culture-tourism-leisure (23%) and social/health (16.5%). In the 1990s, club and association revenues were estimated to reach approximately €15 billion, based on data collected from federations (Cheroutre, 1993; Padieu, 1990). According to an assessment of expenditures during the early 1990s, revenues for the nonprofit sector were estimated between €33-35 billion. In comparison, revenues for the water, gas and electricity distribution sector, as well as revenues for the textile, clothing and leather industrial sector, were €33 billion for the same year (Tchernonog, 1994a, 1994b).

The evolution of association and club financing in recent years reflects an increasing commitment on the part of public institutions. Funding rose from €15 to €25 billion, that is, from 44% to 54% of the total budget. The French donate approximately €1.9 billion annually to associations. They donate €35 million per year per million inhabitants, that is, less than the United Kingdom (€140 million), Switzerland (€90 million), Holland (€80 million) and Germany (€60 million), but more than Spain (€20 million). The associations with the highest budgets belong to the health and social sector, with average budgets of approximately €200,000. They also have the highest payroll. The sectors of culture-leisure and sports have lower budgets (€20,000 to €30,000 on the average).

The nationwide survey conducted in 2002 by the National Institute for Statistics and Economic Studies (INSEE) counted 12 million volunteers, 3.5 million of whom with permanent status, since they devote at least 2 hours a week to an association. Four out of five associations operate exclusively with volunteers. Twenty million people over the age of 14 are members of an association, among which one out of four is a teenager. The 2002 INSEE survey on associations constituted an important step because it made it possible to become better acquainted with official volunteer workers in France, specifically those working in associations and clubs (Febvre & Muller, 2004; Prouteau & Wolff, 2004).

Clubs and associations account also for a significant number of salaried and volunteer jobs. In 1996, paid jobs[2] in associations employed roughly 800,000 full-time people (Mescheriakoff, Frangi, & Kdhir, 2001)—the equivalent of the entire transport sector workforce, or twice the total workforce of the two largest companies in France's private sector, *Alcatel* and the *Générale des Eaux*. It should also be remembered that full-time equivalent paid jobs in clubs and associations represent 4.5% of all paid jobs in France.

From 1981–1991 employment stagnated or regressed in most economic sectors, but increased by 40% in associations. In 1994, clubs and associations employed 1,300,000 people, a surge probably due to State grants and assistance packages for job creation. In 1992, 119,000 associations had at least one paid employee, 26,000 associations had more than 10 employees, and 413 associations had more than 200 employees (Alix, et al., 1993). Twenty percent of the 700,000 associations registered and operating in France[3] have employees. Volunteer work[4] in associations is estimated at 1,079 million hours (i.e., 138 million working days of 7.8 hours; see Halba & Le Net, 1997). Between 1988 and 2002 the number of days worked by volunteers virtually tripled. This progression can be attributed to funding from the State to encourage the creation of employment. In 2006, 1.6 million salaried employees were on the payroll. The health and social sector remains the main "employer," with 560,000 wage earners, that is, the equivalent of 380,000 full-time jobs. The cultural and sports sectors amount to the equivalent of 85,000 full-time jobs. Finally, it seems important to underline the fact that 70% of all association jobs are occupied by women.

The economic weight of non-profit institutions in France in 2002 represented a gross wage package of approximately €26 billion (i.e., more workers than the transport sector and as many as the construction sectors), with a *value-added* more than €55 billion (i.e., 2.5% of the GDP) and an overall budget of approximately €60 billion (Savall & Zardet, 2004).

This chapter is based on the case study of *intervention-research* in a sports club. Associations account for a broad spectrum of jobs, ranging from exclusively full-time paid personnel (mostly commercial businesses) to exclusively volunteer workers. A good example of the latter is the sports association that is the focus of this study—an organization producing sports services relying mostly on volunteers. This case provides an example of *socio-economic intervention* (Savall, 1974, 1975, 2003a, 2003b; Savall & Zardet, 1987, 1992, 1995; Savall, Zardet, & Bonnet, 2000) in an organization where effective and efficient governance is not defined by *economic performance* in the conventional sense but as an essential condition for the growth—and very *survival*—of the organization.

In this type of organization, the *producer-consumer-citizen* triangle (Savall,1979; Savall & Zardet, 2005) is crucial. Volunteer members are first and foremost citizens belonging to an organization, while at the same time involved in the quality of the services provided. In essence, they are simultaneously producers and consumers. They are as sensitive to the quality of a sports service (consumer status) as they are to the conditions of working life inside their organization (producer status). These three poles constitute both explicative variables as well as action levers for negotiating improved *piloting* of the organization (Delattre, 2000).

INTERVENING IN A SPORTS ORGANIZATION: THE CASE OF A SOCCER CLUB

The intervention was carried as part of the author's research work (Delattre, 1993, 1998) in the ISEOR doctoral program. The sports club that served as the intervention-research site is characterized by a strong volunteer component. Paid employees do not have a major role in the determining and deploying strategy. This type of organization, which is characteristic of over 80% of French associations, typifies one of the upstream stages in the trend towards better defined organizational *structures*. Most of these organizations lack *visibility* in terms of their social usefulness as perceived by outsiders and also in terms of the *effectiveness* and *efficiency* of their internal working procedures.

Intervention Structure and Objectives

The case concerns an association that has been a member of the French Football Federation (which has over 2 million registered players) since 1925. The association provides services to football clubs in the Drôme and the Ardèche departements (a French geographical district between state and county). Its mission is "to promote, organize and develop all activities related to soccer." Although its services are only accessible to members, the association is recognized by the government and, as soccer is very poplar with the general public, also concerns the "outside world." The budget of the association is approximately €366,000 per year. Covering two departments, the association has 26,000 member players. The organization of this mission is divided between the headquarters (HQ) staff and the 266 clubs (see Figure 6.1). The HQ operates with 130 people, four of whom are paid—two secretaries and two technical advisers per department, dispatched to the club from the Ministry for Youth and Sports (a legal obligation).

The association operates with elected representatives. The sovereign body is the Club Members Assembly (266 clubs). The assembly mandates a 26-member executive committee for 4 years to run the organization under the responsibility of its president. The executive committee is composed of 19 committee chairpersons (these basic activity management units include rules, sports, promotion and technical aspects), a secretary, treasurer, technical advisers, and vice-presidents (who represent geographical sectors) as well as representatives and partners designated by

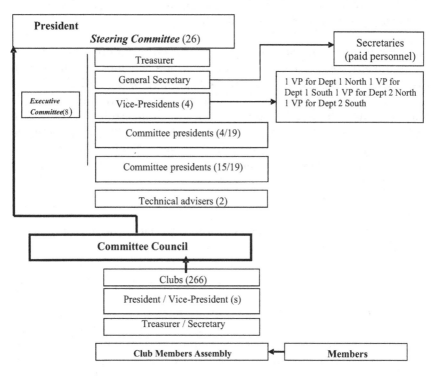

Source: © ISEOR 1993.

Figure 6.1.   Organization chart of the committee.

the statutes. An eight-member steering committee meets every week for day-to-day operational issues and to prepare the monthly meetings of the *plenary executive committee*. Decisions are made by a majority vote. All the clubs are structured in the same way. All plenary executive committee members are proposed by the clubs and allocated to the various committees. Some have several missions and act simultaneously as vice-presidents and committee chairpersons. As one of the respondents in the study noted, "There are no well-defined criteria to select the people who sit on the district executive committee, quite typical in nonprofit organizations."

The credibility of the members of the steering committee is vital when conflicts of interests emerge in activities. An underlying problem, however, is that steering committee members are co-opted rather than elected, which reduces their credibility and emphasizes a political dimension. Clubs lobby for people whose implicit mandate is to defend the interest of "their" club. According to one of the respondents:

> many people would be well qualified to be on the steering committee but there are obstacles to their election. In fact, steering committee membership depends on personalities and not abilities.

The executive committee's request for an intervention reveals doubts about the *efficiency* of the way the Association's activities were managed. Prime concerns that were voiced at the beginning of the project reflected a lack of commitment of volunteers and the lack of *effectiveness* of myopic management without a long-term vision. Overcumbersome working procedures further undermined the association's dealings with the outside world and led to a significant drop in subsidies, a fall in active participation by members, discontent at the grass roots level, and a failure to attract sponsors.

The Intervention

The *intervention agreement* was designed to clarify the role of the various parties in the change process. Since the decision-making process is elective, the roll-out had to reflect existing rhythms and systems to prevent allegations that our intervention was only an alibi for the fact that in-house members were making a play for power. Indeed, delegation of power to representative authorities within the organization was limited to the operational management of sports activities related to the organization of matches. In this type of intervention, only the general assembly has the sovereign power to ratify decisions: the outcome of voting on decisions in the clubs procures a status of "internal law." In this context, the decision to carry out an intervention is made through the following decision-making process: the intervention framework, approved by a committee of the executive board, was added to the agenda of the plenary executive board to validate its addition to the agenda of the general assembly. The piloting of the intervention was adapted to existing rhythms and **structures** to avoid suspicion that **internal actors** were "capturing power" under the pretext of the external intervention. As part of the acclimatization to the **intervention architecture**, the **collaborative training** services were not discussed beforehand.

Indeed, during presentations of intervention protocol, objectives were formulated in terms of the constraints involved in cooperative training, especially the participants' lack of availability to manage the activity, as well as to carry out the training sessions (time devoted to the activity was taken away from time with their families). Another objection was related to the fact that many participants were retired or mature professionals, unwilling to adhere to a framework they considered excessively scholastic and demanding. To avoid sacrificing contributions made possible by a methodological approach, an intervener's participation in executive committee and board of director meetings, with the right to speak up, permitted injecting forceful messages about the intervention and delivering a presentation on the basic **management tools**.

The main consultant was granted the right to speak at the steering and executive committee meetings, making it possible to raise key points about the intervention and to present the basic tools that would be employed: the **self-analysis of time grid**, the **competency grid** (CG), the **strategic piloting logbook** (PLB), and the **priority action plan** (PAP). The intervention strategy reflected a sense of *methodological opportunism*. The introduction of the messages was not conditioned by a planned agenda but intuitively based on topics introduced by the participants. One of the limits of this approach is that participants required a longer time period to understand, assimilate and adopt the messages, that is, the **socio-economic management** principles.

The **diagnostic phase** of the intervention (see Figure 6.2) involved 67 interviews, largely due to the diversity and degree of representativity of the participants (e.g., position in the organization, mission, soccer club league level, geographical localization). Twenty-eight people (22%) of the HQ staff and 26 clubs (19% of all member clubs) were interviewed. In addition, 13 interviews were held with external partners including the League (which federates the district committees), institutional partners (who influence association business by the construction of amenities and the promotion of sports policies), the local media and sponsors. The interviews with external parties were undertaken (1) to assess the extent to which the association's internal problems were creating difficulties with key stakeholders and (2) as a basis for the development of aggressive action plans.

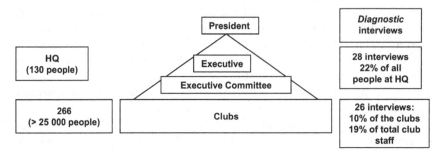

Source: © ISEOR 1993.

Figure 6.2.   *Intervention architecture.*

Chronology	1993	1994	1995	1996	1997	1998	>
Negotiation of the intervention	X						
Presentation of the intervention to the Council	X						
Diagnostic	X						
Analysis of the diagnostic and reporting	XX						
Assistance provided to the Executive and Steering Committees	▓▓	▓▓	▓▓	▓▓	▓▓	▓▓	
Focus group		X	▓▓	▓▓	▓▓	▓▓	
Assessment report to the General Council		X	X	X	X		

Legend:

▓▓	100% involvement and piloting of the intervention process
▓▓	Progressive retreat (1): support focused on project piloting
▓▓	Progressive retreat (2): role of assistance in the change process

Source: © ISEOR 1997.

Figure 6.3.   Chronology of the *SEAM intervention*.

Since the *chronobiological rhythm* of the organization was rather slow and, due to the procedures laid down in the statutes, a full year was necessary for the general council to ratify decisions about global operations. Since the chronobiological rhythm of the organization was rather slow, a full year was necessary for the general council to ratify decision about global operations. As summarized in Figure 6.3, the intervention lasted 6 years, with a progressive withdrawal over the last three years. The diagnostic report to all committee members and the creation of a *focus group* system were ratified at an annual general meeting.

SEAM Implementation

It took 3 months to prepare the *diagnostic*. Participants mainly highlighted the "lack of visibility" about organization missions and the association's failure to express its requirements in terms of jobs and the skills required to mobilize members and organize production. As one of the respondents noted, "people who come here are linked by soccer, everyone has his or her individual reasons but the mission seems to have no common objectives."

The topics which provoked the most dissatisfaction (see Table 6.1) were the implementation of strategies (project organization control), communications between participants, and *work organization* (job repartition). This observation confirms the initial assumption that volunteers are more sensitive to the conditions of membership and to the values and goals of the project than to the procedures and conditions by and in which the mission is executed (e.g., *working conditions*, time management, *integrated training*). Work organization was only the third most important topic—which plays down the hypothesis that volunteers are indifferent to the methods through which they contribute to the association.

To qualify these initial results, the materials were examined using what Savall (1986) refers to as "spectral analysis," examining the data in the context of epistemological reflection on research methodology and the *quality of scientific data* in management *sciences*. *Spectral analysis* designates the study and analysis of phenomena orbiting around a situation in which *interactions* occur (see Plane, 1994). In a research process, emphasis is placed by Savall and Zardet (1986) on the role of the field and the dialectic between visible and hidden phenomena, the research materials and

Table 6.1.   Distribution of *Key Ideas* Spontaneously Expressed by Interviewed Members

Dysfunction Topics		*Rank*
Strategic implementation	32%	1
Communications-coordination-cooperation	22%	2
Work organization	16%	3
Time management	14%	4
Working conditions	9%	5
Integrated training	7%	6
Total	100%	

Source: ©ISEOR 1993.

their citation. This posture has led to the principle of **HISOFISGENESIS**, which consists of inducing impacts and behavior stimulation in members through the intervention itself. This approach provides meaningful information about the phenomena observed "by and for action."

Spectral Analysis of the Participants' Behavior Patterns

In this case, the spectral analysis consisted of highlighting complementary points of view about the operation of the organization from different perspectives—from the vantage point of the individual and the collective positioning of the participants. Every individual possesses an energy dimension beyond the conventional mechanistic views of a work force. Associating participants with active units (Perroux, 1975) makes it possible to relativize this limitation and formulate a hypothesis of the positive impact of human energy on organizational outcomes. In this representation, both the organization (collective dimension) and the member (individual dimension) have an effect on their environment. Both levels—collective and individual—can mobilize *resources* (resource management/ways and means management), to deploy them to achieve objectives (*synchronization*/ flow organization), to pursue an activity in a relatively permanent context (Hierarchy/Compatibility with strategy), and to negotiate and mutually adjust the conditions of *implementation* (Strategic Scope/Variations in Vision).

Figure 6.4 shows the rankings for the expression of "collective" and "individual" priorities and a "mapping" of these rankings according to whether participants were leaders (P1) or members (P2). Overall, the collective expression dimension is focused on project deployment and the deficiencies and/or appropriateness of necessary means design.

The individual dimension is characterized by dissatisfaction with the way activities are implemented as well as the means available to organizational members. This indicates that the collective image shared by participants shows strong dispersion of mental representations of the organization. By improving image clearness according to the category which is described, for the collective dimension, a strong differential overlap between the various populations which represent the dispersion of the collective image shared by the actors is observed. This is due to the "watertight" separation between the different levels in the association—the steering and executive committees and the clubs. The individual dimension shows overlapping *cohesion* (with no difference in expression). This pattern is not surprising given the way the association operates, as the structure is identical on every level; any action only needs to be approved by equals.

Examining *dysfunctions* using this "dimensional" filter made it possible to refine the diagnostic to identify the *root causes* that the project was

COLLECTIVE DIMENSION	Ranking		INDIVIDUAL DIMENSION
Resource management	4	2	Ways and Means
Synchronization	2	4	Flow Organization
Hierarchy	3	1	Compatibility with strategy
Strategic Scope	1	3	Variations in Vision

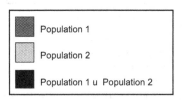

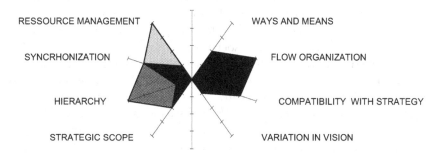

Source: © ISEOR 1997.

Figure 6.4.   Expression of *key ideas* by participants ranked by their collective and individual perception.

designed to address. It highlighted strong expectations for the deployment of a strategy for the Association in a context of an atrophied decision-making process, on both political and operational levels (cross-referencing collective and individual dimensions). The characteristics of the control system (collective operation) led to the adoption of the **HORIVERT** grid approach to the change process (see Figure 6.5). The horizontal dimension reflected the level of the leaders in contact with

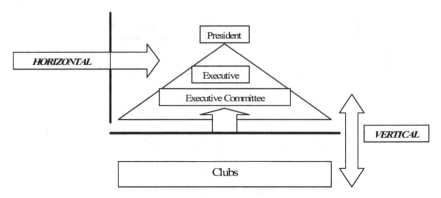

Source: © ISEOR 1993.

Figure 6.5.   Activating the **HORIVERT process**.

public authorities and the vertical dimension reflected the interfaces between the various levels in the association's hierarchy. **HORIVERT** analysis permits the confrontation of different group logic systems (HQ logic and club logic) and to bring to the surface the **scope of negotiations** between the members in a powerfully stimulating context.

The diagnostic makes it possible for the participants to change their positions in relation to the Association through the emergence of a strengthened collective intelligibility (the collective image that is used as the work platform following the **mirror-effect**). The diagnostic reports were used to clarify the "**game rules**" by which the committee agreed to operate. Game rules were defined in discussions between the various groups of Association members and also by negotiations between work enjoyment and **productivity** (sports activity) to facilitate the **development** of tools and procedures for operational needs. In essence, these **game rules** are defined through a dialectic relationship among the different categories of actors: each category becomes aware of its impact on the activities of the others.

As part of the diagnostic report, executive committee members made commitments to clubs that they understood must be honored. The entire executive committee expressed satisfaction with the outcome of the diagnostic, which focused on their vague feeling of having a "lack of purpose." The "small problems" each committee member encountered were smoothed out in a discrete, informal manner that was invisible to other members. For example, following the presentation of the diagnostic, zones of spontaneous dialogue appeared, outside the institutional framework, where the new operating modes were discussed. Thus, over a drink or at the stadiums, volunteers exchanged their personal "recipes" for organizing certain club activities or managing relations with the various commissions at the head-office (championship *scheduling* management, transportation organization for the children, interpretation of sports *regulations*). These actions affected the clubs both directly and indirectly through compromises reached between members of the different Committees. However, the collective problems still had to be resolved (e.g., decompartmentalization of the organization, development of internal communications, quality improvement of the sports activities offered).

Club members were on the defensive. During the discussions, they also gave commitments, in particular after *misunderstandings* were cleared up, such as, misunderstandings about certain security regulations on the playing field, about the obligation for clubs to train referees or the amounts of certain financial penalties for cases referred to a commission. Regarding security, club managers also bore legal responsibility and failure to respect certain norms can incur serious human and financial consequences in the event of an accident: friendly practice of sports should never sacrifice the security of players. Without referees, there can be no championships, hence the obligation for every club to train referees with certified competency; this is the very condition for the survival of sports activities. A club is fined if it does not conform to this training obligation. Conversely, it can benefit from subsidies (redistributed penalties) to alleviate the cost of its training program. Finally, the commission in charge of supervising the interpretation and respect of *regulations* justified their decision to establish and raise financial penalties in cases where clubs contested the outcome of a match based on an article in the regulations. The objective was not to financially penalize clubs, but to foster the spirit of sportsmanship and to limit procedural abuse; a match is won, first and foremost, on the playfield. It was the incremental increase in the number of disputed cases, while the volume of those with a positive outcome remained constant, which led to the commission's decision. Many club presidents were surprised, even astonished, by the change in committee members' attitudes. They also discovered that they were not wholly to blame for the problem and that the committee was not infallible.

Club presidents also felt the report was constructive and helped them to realize that other clubs were facing the same problems. But there were many unsatisfied complaints that still needed to be resolved and doubts lingered about the committee's ability to honor its commitments. For example, the committee's proposal to organize delocalized cooperative meetings with clubs by geographical sector to foster more personal discussion, thus contributing to the reflection on work-session topics of general interest.

Reconstructing the dialogue gave *visibility* to all expectations and stakes to be negotiated. For instance, a strong desire was expressed by head-office managers to become better acquainted with the work of people in the field. Among committee-member stakes is mastering the situation: conducting a structured dialogue while avoiding the pitfalls of partisan quarrels relative to particular situations. Points to be negotiated stimulate collective energy (the will and the need for discussion). The ensuing discussions, and the points they bring to light, in turn reveal new elements to be clarified or areas in which a potential for *progress* exists. For example, reinforcing club participation by increasing involvement in work-sessions charged with making proposals on specific topics such as "fair play" or the development of girls' and women's soccer teams.

The protagonists awaited the second step with mixed feelings. The diagnostic report spread through the district, particularly through gossip at matches and training sessions and arguments ensued between "believers" and "nonbelievers" who thought the whole procedure was a "smokescreen." The tool-axis of the process was activated as well. Discourse was accompanied with formal action, for example, up-stream display of posters that spelled out the game rules of discussion and decision frameworks and work methods, as well as writing up and circulating documents on the organization's strategy. These instruments had a structuring effect, making it possible to formalize and capitalize on improvement actions to strengthen the operation of the executive committee, which will be further discussed later in the chapter.

Results and Assessment

The *intervention-research* in the association comprised the first three phases of the trouble-shooting procedure: diagnostic, project and implementation. The "toolbox" used for implementation was based on *SEAM* methodology, but with extra efforts made to ensure that members understood and adopted the game rules as their own. The main priority—political and strategic decision making—was more complicated due to the collective process, and required a high level of involvement by the *intervener-researcher* since it involved monthly participation at the Executive Committee meetings and attendance of the general and advisory meetings throughout the research program period.

The presentation of the results of the diagnostic to the executive committee and subsequently to the annual general meeting led to the decision to set up a focus group. It also led to a commitment to track the change process on an ongoing basis and put "progress" on the agenda as a recurring item. The main constraint in conducting this intervention was the wide geographical dispersion of actors (they were spread across two counties, one of which is very mountainous and difficult to access). This dispersion of the people involved caused prolonged time delays in the implementation of improvement actions. As an example, certain executive committee members traveled 150 kilometers to the head-office to attend meetings and/or carry out their activities. In this context, it did not seem reasonable to multiply the number of trips required to hold committee meetings. The intervention-research was integrated, as much as possible, into existing frameworks (commission meetings, core and plenary executive committee meetings) in order to maximize the dissemination of messages and information concerning, in the first stage, the diagnostic, and then the advancement of implementation work.

Figure 6.6 presents the scaled-down project framework in comparison with the socio-economic analysis framework used in organizations with closer-knit structures. In this instance, the problem-solving process required the involvement of top management in the context of collective decision-making. The figure illustrates the fact that the scaled-down framework adopted in this instance does not conflict with an overall framework set-up. Typically, four structures are involved in *socio-economic projects*—the *project leader,* the *steering committee (SC),* the *plenary executive committee (EC),* and *specific task group*. In the current case, the collective operating mode of the Association required a different solution-finding framework. As envisaged by the *socio-economic intervention* method, the project system must not place the executive committee in an uncomfortable position vis-à-vis the clubs and the league because it is the legitimate decision-making authority. Many executive committee members refused to participate in the project, saying it was "a waste of time." In reality, executive actors took refuge behind an institutional alibi to avoid individual exposure in a system that privileges collective decision-

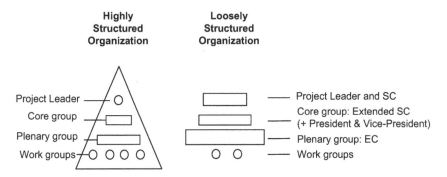

Figure 6.6.   Scaled-down *focus group* framework for seeking solutions to key *dysfunctions*.

making. The selected option was to integrate this operational specificity into the general process of the socio-economic *focus group*.

As a result, the focus group was set up in the following way:

1. *Piloting* leadership was modified. In this intervention, the project piloting function took the form of a volunteer group within the executive committee (referred to as the *steering committee* during the intervention). It was composed of eight actors including the *intervener-researcher* and the secretary general of the executive committee. The steering committee's role was to exploit the diagnostic, proposing axes of work to be validated by the executive committee, referred to during the intervention as the *plenary steering committee* (core group in the methodology).

2. The *core group*, called the *plenary steering committee*, met every other month. It was composed of the members of the executive committee who were not involved in the steering committee: the president and vice-president of the association. The purpose of their participation was to guarantee that steering committee proposals were not disconnected from policies supported by executive committee members. This method involved the president, but because he was an elected member he did not want to lose his margin of political maneuvering through exposed involvement. His participation was never fully acquired, but continuously negotiated throughout the investigation.

3. The *plenary group* comprised all the executive committee members, which facilitated the validation of the proposals made by the steering committee and provided participants for the specific task group.

4. The *task group* set up by the steering committee was led by one of its members, to ensure continuity in the search for solutions. This level of involvement is part of the socio-economic approach.

Progress in the Way the Association Operates

An ad hoc working group, comprised of the president and executive committee members and led by the consultant, formalized action guidelines, using the *priority action plan (PAP)* tool derived from the diagnostic to drive implementation (*proportional directivity*). Four **strategic axes** were proposed to promote improvements in the way the executive committee operated:

• *Axis 1*: To develop internal **communication-coordination-cooperation**.

- *Axis 2*: To develop external **communication-coordination-coopera-tion**.
- *Axis 3*: To define and implement a sports policy.
- *Axis 4*: To improve the quality of services provided and the internal operation of the committee.

Socio-economic methodology for producing project-action themes was implemented in formalizing the association's strategic axes. The diagnostic was reprocessed in the following manner: (1) **key ideas** and their corresponding **fieldnote quotes** were synthesized into **idées forces (pivotal ideas)**; and (2) in a second phase of synthesis, idées forces (pivotal ideas) were grouped together to form the four strategic axes.

CONCLUSION: REFLECTIONS ON THE PROJECT PHASE AND IMPLEMENTATION PHASE

The breakdown and deployment of actions throughout the Association, by translating them into objectives at successively lower levels, was not a success. The explanation is twofold. First, the organization's structures were too feeble at the connection level among its components. Second, knowledge of the instrument was insufficient due to a lack of actor training, which can be attributed to an intervention choice, that of integrating training sessions into existing meetings.

Despite these difficulties, the **priority action plan** was distributed to executive committee members before being published in the liaison newsletter sent out to clubs. The action plan provided the opportunity for individuals and groups to acquire personal awareness, because readers could position their own improvement action plans within the structure of the document. Implementation of the action plans improved the **cooperation** between the HQ and the clubs, which was a main priority of the intervention. The decision to strengthen the interface with the clubs was adopted, in particular, in the context of a committee mission to support club activities with the purpose of prolonging their existence. Emphasis was placed on improving the:

- management of relationships between the clubs and departmental and regional authorities;
- assistance due to the complexity of the three levels of regulation (federation, league, committee) both for the implementation of the sports activity and for the participation of clubs in the committee;
- control of financial losses due to failures to respect rules and regulations; and
- workload and the time investment by club secretaries and directors.

Table 6.2.   The Eight Headings in the Guide for Club Secretaries

Chapter 1:   The club secretary's job (mission, tasks and relationships assumed)
Chapter 2:   The day-to-day and week-to-week operations to be carried out
Chapter 3:   License management (players, directors, rules, regulations and examples)
Chapter 4:   Referees (licenses, rules, regulations and examples)
Chapter 5:   Game management committee (rules, regulations and examples)
Chapter 6:   Club operation (club meetings)
Chapter 7:   Committee meeting procedures
Chapter 8:   General information

© ISEOR 1997.

The executive committee secretary-general authored a "Guide for Club Secretaries," which was distributed to executive committee and subcommittee members for amendments and corrective comments, and to encourage additions to the topics covered before it was published. The guide, which emerged as a binder containing 70 files organized under eight headings (see Table 6.2), attempted to provide, in the simplest possible way, answers to questions frequently asked by club secretaries.

The loose-leaf binder format makes the guide easy to update with articles published in the executive Committee newsletter. The guide contains explanations of the rules and regulations, "checklists" for various events and activities, calendars, examples to ensure the homogeneous denomination of the various liaison documents between the club and the committee, explanatory charts (logic diagrams, flow charts), and advice on organization. One file, for example, explains the purpose and utility of the committee liaison newsletter, including such detailed information as: (1) what should be read in the law gazette, (2) what information is essential, (3) how to file information for use later reference, (4) how to communicate the information to the individuals concerned, and (5) how to file the law gazette. The Guide for Club Secretaries is intended to clarify and stabilize Association "know-how" and shape the way members act. It also contributes to structuring and institutionalizing the frame of reference used for committee procedures.

The *diagnostic* created a commitment to modifying the *game rules* for relationships between the participants, with a goal to enhance *productivity*. This led to the introduction of *socio-economic management tools*, such as *priority action plans* and other techniques particular to this type of association (such as Guide for Club Secretaries), which helped to reduce club management *charges* and lower membership fees. The creation of sports activity *management tools* was negotiated with all membership levels. They concerned methods designed to define the way the executive committee members and the general council (the plenary body) made decisions about sports activities.

Modifications to the general council were necessary to strengthen adhesion to the collective project and to improve the participation of the clubs. The diagnostic had highlighted that the general council as a governing body was viewed unfavorably by all the association members. Clubs expressed major dissatisfaction with its organization (e.g., the system of fines for absent members, excessively long meetings) and decision-making methods (e.g., the lack of dialogue, the impression that decisions were taken before the meetings, the difficulty of being allowed to speak). Executive committee members said they felt far removed from the clubs due to a breakdown in the nature and quality of dialogue and discussions.

Success in preparing a "new approach" was achieved through participation by the clubs in the committee via sector meetings (delocalized and chaired by a vice-president) and the consultative assembly. The first "new approach" general council meeting lasted only two hours—setting a new record for brevity—and the reception that followed it was, according to the participants, the "most friendly ever," a clear demonstration of changes within the executive committee, the result of a permanent system of *cooperation* spread over time, structured in space and focused by activity. The organization of the "new approach" general assembly was the concrete outcome of the committee's modified consultation modes. Committee actors were more involved in this framework because different actors' concerns and constraints were taken into consideration, and because it reinforced the role played by committee members vis-à-vis the clubs due to the implementation of structured consultation at the different levels. In the end, the general assembly embodied a *permanent consultation framework* (regular meetings), spatially structured (meetings by geographical sector) and focused on the activity (taking into account the constraints and concerns of parties involved).

Executive committee members were now obliged to make personal commitments reflecting the concerns and policies wanted by association members. The new structure, which is based on consultation at every level, strengthens the power of executive committee members in their relationships with the clubs.

The Consultative Assembly

The role of the consultative assembly is to prepare the annual general meeting (AGM); The consultative assembly is made up of delegates from clubs and executive committee members. It examines motions proposed by the clubs to amend rules and modify game organization and operations in general. Before the intervention, the consultative assembly was

Table 6.3.   The First New Consultative Assembly Agenda

   I. Fair play
  II. Payment of referees by the committee and the return of game reports
 III. Improvements in refereeing
 IV. Women's soccer
  V. The committee party
 VI. Participation of reserve teams in the Giraud Cup and the creation of a league club cup
VII. Other expectations:
    a.   Age categories
    b.   *Alignment* of financial sanctions with penalties applied in the leagues
    c.   Proposals for subscription rates
    d.   Budget estimate for the coming year.

© ISEOR 1998.

the place where the executive committee filtered complaints from the clubs before the AGM. Following the diagnostic, it was decided to modify the procedures used by this preparatory assembly to make it a place for debate and reflection on executive committee activities.

The introduction of local meetings between the clubs and an executive committee member was intended to focus the executive committee agenda on a set of specific topics. Following the report from the *focus group*, the consultative assembly meeting became a three-stage event comprising an introductory plenary phase, a system of task group per topic, with representatives from the district and the clubs, and a system of reports to the consultative assembly followed by the presentation of the findings to the AGM. As an illustration, Table 6.3 summarizes the topics of the first consultative assembly.

The introduction of sector meetings for the negotiated settlement of project issues reflected the decision to organize proximity meetings with the clubs before the consultative assembly. A geographical sector was allotted to each executive committee member for meetings with clubs, to facilitate the process of submitting topics the executive committee wanted to address and reporting on the comments, observations and proposals made by the clubs. These reports were consolidated by the secretary-general and published in the liaison newsletter.

Containing Lingering Resistance

As some executive committee members refused to participate in meetings with clubs, pairing operations were organized and courses in meeting leadership were provided voluntarily by a professional corporate trainer.

It proved so difficult to motivate members for the introduction of delocalized meetings that the executive committee feared it was sullying its credibility with clubs by this attempt to improve operations. As these meetings sought to improve the interface between local and central levels by obliging participants to take positions and defend them, duplicity obviously became more difficult.

The operation of the executive committee changed. After the sector meetings, the participants formed pairs representing the interests of the clubs and the executive committee no longer considered "clubs" as an abstract notion which hindered its work. Proposals for the inclusion of topics for commissions—for example dealing with violence—were agreed upon in discussion (particularly since the district president found himself in court twice over a period of a few weeks as the representative of the association). The framework was complete: it was bottom-up and top-down (between the clubs and the committee) and transversal (among various members at each level thanks to formal reports and informal debriefings after matches).

Concluding Thoughts

The discussion in this section of the chapter reflects the *progress* negotiated in the way the executive committee operates and how it has impacted the entire association. Using the *assessment of achievements* tool, the intervention illustrates that it is possible to modify the behavior of association members without promoting a priori the *decisive role of the official hierarchy* as the vector in the introduction of organizational change (see Delattre, 2002).

We are aware that the objectives were formally achieved because of the participation of the official hierarchy. However, the official hierarchy was not the driving force which impelled the changes. The implementation process for operation improvements was far more diffuse, anchored in explanation process of the piloting framework that reinforced the bargaining power at the club level.

The nonprofit sector is a major economic force (Salamon et al., 1999). Total spending amount to $1.22 trillion, with the equivalent of 31 million fulltime workers, representing 6.8% of the workforce (not including agriculture) or 19.7 million fulltime employees and 11.3 million fulltime volunteers. Contrary to common belief, it is services and not philanthropy that provide the main portion of organization sector financing, followed closely by public aid. This funding varies widely from one geographical area to another. Volunteer donation of time accounts for only 28% of the sector's revenue. Nonprofit organizations have been a major source of job-creation over the past years. Between 1990 and 1995, in a sample of eight countries, while employment in the classic economy increased by 8.1% the figure went up to 24.4% in the nonprofit sector.

What is at stake in improving overall piloting of associations and clubs is improving their mastery of resource consumption for their internal operation, thus enabling them to reinforce their capacity for sustained *survival*. Our research leads us to believe that discourse in associations on professionalism seems to overshadow the analysis of professional behavior development among actors. The principal instruments for action and canalization of actor energies introduced by socio-economic analysis make it possible to go beyond appearances in organizational situations and to deal both with the visible and the hidden aspects of a common project, which is submitted to stakeholder and third-party judgment that evaluates the quality of the production. Thus, initiating an engineered process, sustained by techniques, tools and frameworks deployed according to an overall *intervention architecture* that respects the intelligence of the organization's actors (actor, producer, citizen) contributes to improving the overall operation of structures that constitute an active framework of social and societal *cohesion*.

NOTES

1. The International Classification of Non Profit Organizations is an international program (13 countries) that compares capitalistic companies with voluntary associations. Its main objective is to demonstrate the social and economic contribution that the nonprofit, voluntary sector makes at national and international levels.

2. Data about employment in associations needs to be handled with care because "employment" has several meanings: paid or unpaid, and expressed by notions of equivalence such as "full-time equivalence" (FTE).

3. The total number of associations in 1990 was estimated by INSEE to be 700,000, approximately 2,000 of which were classified as foundations.

4. Estimates of volunteer work in France were developed in a study on the role of the third sector in France. Two questionnaires were used: (1) a sample of 2,000 people over 18, representative of the French population in 1991 and repeated in 1994; and (2) a sample of 2,300 from associations out of a population of 15,000.

REFERENCES

Alix, N., Baguet, R., Baruch, M. O., Bourel, E., Bruneau, D., Cheroutre, M. T., et al. (1993). Associations, nouveaux espaces, nouveaux enjeux [Associations, new spaces and new stakes]. *La Tribune Fonda, 99/100*, 17-18.

Archambault, E. (1996). *Le secteur sans but lucratif: Associations et fondations en France* [The non-profit sector: Clubs and foundations in France. Paris: Economica.

Camus, B. (2006, March). *Projets et programmes de la statistique publique française* [Projects and programs of French Public Statistics]. Paper presented at the 20th Colloquium of the Association pour le Développement de la Documentation sur l'Economie Sociale, Paris.

Cheroutre, M. -T. (1993) *Exercice et développement de la vie associative dans le cadre de la loi du 1er juillet 1901* [Exercise and development of the life of associations under the framework of the July 1st 1901 Law]. Report presented by the Conseil Economique et Social. Paris: Journal Officiel.

Delattre M. (1998). *Contribution à l'élaboration d'une mode de pilotage de l'organisation à forte composante bénévole—Cas d'expérimentations* [Contribution to the development of a model for piloting organizations composed of a majority of volunteers: Cases of experimentation]. Unpublished doctoral thesis, University of Lyon, France.

Delattre, M. (2000). *Le pilotage de l'organisation vivante? Bilan et perspectives spectrales en sciences de gestion* [Piloting a living organization: Spectral appraisal and perspectives in management sciences]. Paper presented at the ASAC-IFSAM World Congress, University of Quebec, Montreal.

Delattre, M. (2002). Professionnalisme et bénévolat: alibi de l'amateurisme ou crise des discours? *Revue des Etudes Coopératives, Mutualistes et Associatives, 283,* 53-66.

Febvre, M., & Muller, L. (2004). *La vie associative en 2002: 12 millions de bénévoles* [The life of associations in 2002: 12 million volunteers]. Report by the National Institute for Statistics and Economic Studies (INSEE Première publication No. 946). Paris: INSEE Printing Office.

Halba, B. & Le Net, M. (1997). Bénévolat et volontariat dans la vie économique, sociale et politique [Volunteer work in economic, social and political life] [Special issue]. *Les études de la Documentation Française, 5055.*

Mescheriakoff, A. -S., Frangi, M., & Kdhir, M. (2001). *Droits des associations* [Legislation of associations]. Paris: Presse Universitaire de France.

Padieu, C. (1990). *Statistiques de l'économie sociale, constat et propositions* [Statistics of the social economy: Summary and proposals]. Report presented to the Minister of State for Social Economy, Paris.

Perroux, F. (1975). *Unités actives et mathématiques nouvelles : Révision de la théorie de l'équilibre économique général* [Active units and new mathematics: Revision of the theory of general economic equilibrium]. Paris: Dunod.

Plane, J. -M. (1994). *Contribution de l'intervention en management au développement de l'entreprise—cas d'expérimentations* [Contribution of management intervention to development of the firm: Cases of experimentation]. Unpublished doctoral dissertation, University of Lyon, France.

Prouteau, L., & Wolff, F. C. (2004). Le travail bénévole : Un essai de quantification et de valorisation [Volunteer work : An attempt at quantification and valorization]. *Economie et statistique, 373*: 33-59.

Salamon, L., & Anheier, H. (1992). *Toward an understanding of the International Nonprofit Sector: The John Hopkins Comparative Nonprofit Sector Project.* Baltimore: Institute for Policy Studies, John Hopkins University.

Salamon, L., Anheier, H., List, R., Toepler, S., Sokolowki, S. W., & Associates. 1999. *Global civil society: Dimensions of the nonprofit sector.* Baltimore: The John Hopkins Center for Civil Society Studies.

Savall, H. (1974, 1975). *Enrichir le travail humain dans les entreprises et les organisations [Work and people: An economic evaluation of job enrichment].* Paris: Dunod.

Savall, H. (1979). *Reconstruire l'entreprise : Analyse socio-économique des conditions de travail* [Reconstructing the enterprise: Socio-economic analysis of working conditions]. Paris: Dunod.

Savall, H. (1986). Le contrôle de qualité des informations émises par les acteurs des organizations [Quality control of information emitted by organizational actors]. *Méthodologie fondamentale de la recherche en gestion*, ISEOR-FNEGE Conference Proceedings funded by the Fondation Nationale pour l'Enseignement de la Gestion, Lyon, France.

Savall, H. (2003a). An updated presentation of the socio-economic management model. *Journal of Organizational Change Management, 16*(1), 33-48.

Savall, H. (2003b). International dissemination of the socio-economic model. *Journal of Organizational Change Management, 16*(1), 107-115.

Savall, H., & Zardet, V. (1987). *Maîtriser les coûts et les performances cachés: Le contrat d'activité périodiquement négociable* [Mastering hidden costs and performances: The periodically negotiable activity contract]. Paris: Economica

Savall, H., & Zardet, V. (1992). *Le nouveau contrôle de gestion: Méthode des coûts-performances cachés* [New management control: The hidden cost-performance method]. Paris: Éditions Comptables Malesherbes-Eyrolles.

Savall, H., & Zardet, V. (1995). *Ingénierie stratégique du roseau, souple et enracinée* [Strategic engineering of the reed, flexible and rooted]. Paris: Economica.

Savall, H., & Zardet, V. (2004). *Recherche en sciences de gestion: Approche qualimétrique. Observer l'objet complexe.* Unpublished English translation: *Research in management sciences: The qualimetric approach. Observing the complex object.* Paris: Economica.

Savall, H., & Zardet, V. (2005). *Tétranormalisation: Défis et dynamiques* [Competitive Challenges and Dynamics of Tetra-Normalization]. Paris: Economica.

Savall, H., Zardet, V., & Bonnet, M. (2000). *Releasing the untapped potential of enterprises through socio-economic management.* Geneva, Switzerland: International Labor Office-ISEOR.

Tchernonog, V. (1994a). *Le poids économique du secteur associatif* [The economic weight of the nonprofit sector]. Paris: Laboratoire d'Economie Sociale, Centre National de la Recherche Scientifique.

Tchernonog, V. (1994b). Une dichotomie marquée du secteur associatif [A major dichotomy in the nonprofit sector]. *Revue des Etudes Coopératives Mutualistes et Associatives, 253/254*(3), 118-146.

CHAPTER 7

REBUILDING THE IDENTITY OF CHAMBERS OF COMMERCE AND INDUSTRY

Reinforcing Legitimacy and Effectiveness[1]

Nathalie Krief

In recent years, the mission of French chambers of commerce and industry (CCI) has become nebulous. To maintain their former positions as driving forces in local economic affairs, they have had to introduce strategic and organizational change, focused on rebuilding their internal identity and enhancing the nature and quality of the services they provide to ensure that the external world finds legitimate reasons to use them in a competitive environment.

This chapter uses *intervention-research* (Savall & Zardet, 2004) structural and economic analysis techniques to examine the applications of the strategies adopted by two chambers of commerce and industry in France, illustrating how these complex-governance organizations manage their approach to change and the characteristics of change management in this type of establishment, positioned halfway between public service and private enterprise.

Socio-Economic Intervention in Organizations: The Intervener-Researcher and the SEAM Approach to Organizational Analysis, pp. 171–195

THE MISSION OF FRENCH CHAMBERS OF COMMERCE

French chambers of commerce are public services. They have a three-tier structure: the national assembly (Assemblée des Chambres Françaises de Commerce et d'Industrie, ACFCI), 20 regional chambers of commerce (Chambres Régionales de Commerce et d'Industrie, CRCI), and 159 local chambers (Chambres de Commerce et d'Industrie, CCI). This network extends beyond France with 80 French chambers of commerce around the world.

French chambers of commerce have two categories of management staff: unpaid company directors and personnel elected by the businesses and industries in each circumscription (approximately 4,500 people) and a paid staff running the services provided to this electorate (approximately 26,000 employees). Chambers of commerce and industry are funded by the IATP tax (a supplementary tax to business tax), the government, loans, and income from the sales of products and services.

A Network of Corporate Services

The core business of chambers of commerce is to assist and advise commercial, industrial and service companies. Chambers also run training courses for students, apprentices and trainees in technical, business and engineering colleges and schools and manage locally-implanted national *infrastructures* such as airports, ports, road complexes, warehouses and logistics parks, congress centres, exhibition parks, and bridges). Chambers of commerce and industry handle approximately 60% of company creation, acquisition and transmission formalities in France.

Local Chambers as the "Anchor Point" of the Network

Local chambers are government-owned establishments and employees have the status of civil servants. The initial law, by which they were created (1898), defines a CCI as the intermediary between "commercial and industrial interests in their circumscription and government authorities."[2] This law also empowers chambers for two other missions:

- to promote the interests of local companies and participation in the elaboration of national policies; and
- to provide an interface between private companies and national and regional administrative authorities.

Chambers have many fields of intervention, including addressing the *development* of business activities, equipment and economic development of their circumscriptions, the environment, transport, tourism, training and employment, urban planning, national and international trade, workplace safety, and exports. They assist local businesses in several ways, including corporate consultancy, training and support services for economic activity in local areas, and the *development* of zones for specific business activities. The activities of all chambers fall into three categories:

- *support to companies*, the main economic-development tool, which accounts for an average 38% of chamber expenditures;
- *training* (e.g., schools for apprentices, higher schools of management, engineering schools), which accounts for an average 29% of chamber expenditures; and
- *management of public equipment* (e.g., airports, ports, road complexes, warehouses and logistics parks, congress centers, exhibition centers), which accounts for the remaining 33%.

Management Structure: Complex Governance

There are four particular operating characteristics that make chambers of commerce and industry highly complex governance systems: politics, their "two-headed" structure, hybrid missions, and the diversity of their areas of expertise and authority.

Politics

Chambers are politically managed organizations staffed with civil servants. Each chamber of commerce is run by a board of directors elected for five years, representing the general assembly (the sovereign decision-making body), associate members empowered to contribute to deliberations (nominated by the "Préfet"), and technical advisers co-opted by the general assembly. This political organization means that the rate of change of chambers often reflects the general assembly's 5-year election cycle—a handicap to the extent that the economic rhythm of the organization is by nature different from its political rhythm. This dynamic leads to "stop/go" approach to management since no strategic changes are made during election campaign periods (Charreaux, 1997; Gilbert & Thoening, 1993; Larcon & Ritter, 1979; Laufer & Ramanant-soa, 1982).

A "Two-Headed" Structure

Chambers of commerce and industry are "two-headed" organizations, composed of unpaid, elected officials and paid full-time staff. Elected officials, representing the general assembly, determine the chamber's external policy and the ways and means required for its implementation. The political management structure includes a president and an executive committee, which assists the president in setting the chamber's policy and the operating strategies for services in contact with the outside world. The president represents the chamber in dealings with public authorities, is legally responsible for all acts and guarantees the execution of decisions taken by the general assembly. The employees of the chamber assist the elected members in their missions and implement, under the responsibility of the general manager, the strategy defined by the elected officials. The general manager assists the president of the chamber and the elected members in the performance of their missions, orchestrates the implementation of the actions decided by the general assembly, and manages all the services provided by the chamber.

The two-headed nature of the chamber confers a duality, even antagonism, between the policies decided upon by the elected representatives and the realities of execution by the civil servants concerned (see Figure 7.1). From a strategic and managerial point of view, this dual level of management and objectives makes governance of the chambers particularly com-

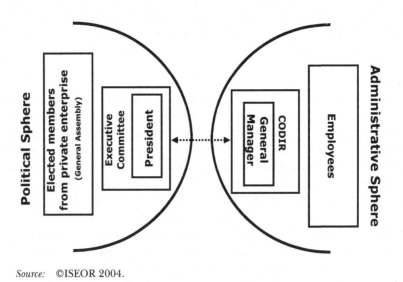

Source: ©ISEOR 2004.

Figure 7.1.   Chambers of commerce "two-headed" structure.

plex as elected members like to make speeches reflecting political choices. This "democratic process" slows project implementation significantly and can become seriously penalizing when the system of governance is disturbed by conflicts between the two "heads" (elected vs. paid officials) and at times even within the same "head" (e.g., between elected officials).

Hybrid Missions

Chambers of commerce and industry are hybrids, existing halfway between civil service and private enterprise (Puaux, 1998). They are public organizations in the way they operate (e.g., in their human resource management policies), despite the fact that they are run by industrialists, merchants and service providers from private enterprises. Moreover, some of their activities are commercial and are sold to generate income, while other actions are not (e.g., public utility management, which is financed by taxation and government funding). It is sometimes difficult to find a harmonious balance between these missions, serving the general public as well as the more capitalistic directives from the boards of directors who are less dedicated to public service.

Diversity of Expertise and Authority

The diversity of areas of expertise and authority increases the complexity of each chamber's problems. Given the degree of specialization that private companies demand, it is difficult to operate and position chamber responsibilities between running airports, ports, training courses, and business schools and traditional chamber services. The fundamental question focuses on whether all these activities should (or should not) have a common denominator. A similar issue concerns the advisability of respecting the technical and operational individuality of all the different missions handled by each chamber of commerce versus the need to seek a common identity that federates the efforts of all chamber actors. The risk of making the wrong decision is that chamber customers may be convinced that it serves no clear purpose at all.

Another difficulty comes from inside the chamber itself, since the people concerned always assert the specificity of their services. "Profit-making" services are often assumed to be "better" than "non-profit-making" activities. Similarly, people in "profit centres" seem to believe they compensate for those in "cost centers." This attitude, which is highly unlikely to promote internal cohesion and solidarity, exacerbates squabbles over power and territory.

Challenges Facing Chambers of Commerce and Industry

In France, the dawn of the third millennium ushered in decentralization laws, which opened the doors to competition from local authorities to operate certain areas that had been the province of chambers (e.g., ports and airports, promote economic development, provide vocational training). As a result, local authorities have developed significant human and financial resources to penetrate markets traditionally reserved for the chambers, sidelining their local economic development missions. Regional Authorities have also extended their prerogatives in the sphere of teaching and training, in particular in adult education, a field to which the chambers devote between 25 and 30% of their budget. In essence, the chambers' sustained development strategy is now positioned in an unprotected, increasingly competitive battleground. They must learn new trades and advance through a fluid, unstable and highly competitive minefield. This war is a fight between different emanations of government authority. To win, both identity and perceived legitimacy have become the weapons of mass attraction.

Many chambers believed that the decentralization policy was just part of the political ebbs and flows of normal cycles and did no do anything to introduce change, waiting for the next elections and the inevitable "counter-order"—which has not come. Chambers now have to rush to catch up on several years of neglect and adopt new strategic positions and develop new solutions contributing *value-added* that meet the demand for functions that are not available elsewhere.

Currently, chambers, and more generally the network they form, perceive decentralization as a medium-term *strategic threat* for a number of reasons (Savall & Zardet, 2005). Chambers have never succeeded in modernizing their missions, their financing and their two-headed management system. A significant number of reports criticizing the performances of chambers of commerce and industry since the 1990s (e.g., CNPF, CGPME, the Ministry for the Economy, the General Inspectorate of Finances and the Economic and Social Council; Eggrickx, 1999)[3] have all questioned their organizational methods and recommended in-depth readjustment to the economic environment.

The changing context has created a crisis situation, prompting the need for emergency action by most chambers. They realize they have to find a reason for continuing to exist by answering a series of key questions: What are they for? What do they exist to do? Have they sufficiently adapted to current economic conditions? How should they define their strategic position, motivate their staff with new objectives, and recognize and reward performance?

It was within this general context that two chambers of commerce and Industry chose to adopt a process of in-depth change using ISEOR's *socio-economic management* methods (Savall, 1974, 1975, 2003a; Savall & Zardet, 1987; Savall, Zardet, & Bonnet, 2000).

A COMPARATIVE INTERVENTION STUDY

Table 7.1 briefly presents the two chambers (on which the chapter is based) at the start of the *socio-economic intervention* in each. Both chambers are "relayed" geographically across their "circumscriptions." Chamber 1 is in an urban community and Chamber 2 is in an area referred to as a "department," something between a county and a state.[4] This relay system complicates the coherence of every chamber operation, since each satellite asserts a degree of independence and positions itself in terms of its own "territory" rather than that of the chamber as a whole.

The local enterprises that chambers assist can be viewed as increasingly demanding "shareholders," especially since part of the funding (20 to 40% of the budget) is based on a tax paid only by local businesses who

**Table 7.1.   A Comparative Overview of the
Two Chambers at the Beginning of the Intervention**

	Chamber 1 (May 1999)	Chamber 2 (January 2002)
General mission	To help commercial, industrial and service companies to set up and grow in France and abroad by targeted, collective actions and personal assistance and advice.	Represent companies in dealings with public authorities, drive and promote the economic fabric of the area, support the creation, development and transmission of companies, by *personal assistance* and collective actions.
Paid staff	Approximately 500 full-time staff	Approximately 250 full-time staff
Elected representatives	48 Elected officials	36 elected officials
Budget	80 million €	25 million €
Departments and directorates	International relations and trade foreign/training and vocational training/urban projects/economic impetus and corporate support/ territorial trade and action/airport/business school/ economic information, studies and development/human resources/administration and finances/communication	Support to companies (industry, trade, services)/ airport/port commercial/ training/administration and finances/human resources/ communication/quality and futurology
Territorial space	38,000 companies (urban community)	20,000 companies (*département* geographic areas)

Source: ©ISEOR 2004.

expect value for their money. As a result, a major recent change is that these local enterprises now audit expenditures and use cost benefit ratios to evaluate the performance of the services that chambers render to companies in the area.

When these two chambers decided to adopt the *socio-economic intervention method* they were starting from a strategy of aggressive conquest rather than retrenchment. Their policy was to improve the *visibility*, transparency, measurement and control of the costs of their activities. At the precise moment when the socio-economic model (Savall, 2003b) was adopted, both chambers were in a pre-election period, which readily complicated their operations and planning efforts.

A Realization of the Need for Change

Chamber 1 decided to initiate socio-economic change in 1999. The main objective, which was introduced gradually until early 2001, was to implement a strategic and organizational project to mobilize all employees in new projects and optimize use of the chamber's resources. The approach was initiated by the changing economic context that the general manager of the chamber described in the following way:[5]

> The purpose of chambers of commerce and industry is increasingly questioned by the world of business and public authorities. Many national reports and audits show that chambers are badly perceived, positioned and understood. To rectify this situation, it is now essential to clarify the missions and the actions of the chamber and to strike a transparent balance between the means deployed and the performances achieved. This requirement is all the stronger that the pressure for full auditing of how the IATP tax levied on companies is spent increases continuously.
>
> Chambers must not only be recognized for their capacity to make efficient use of public funds, but by the quality of their actions in the service of companies and their ambitions for the regions they serve. To this end, we must be able to demonstrate real capacities for reactivity and anticipation and by improving our use of our human and budgetary *resources*. For these reasons, the Chamber intends to deploy a strategic and organizational project harnessing all our staff and our resources by implementing projects which meet new expectations from our constituents and optimize the resources of the Chamber. We must become a driving, effective, force in promoting local companies and our constituency.

Chamber 2 (Chambre de Commerce et de l'Industrie du Morbihan) initiated the *socio-economic intervention* in 2002, focusing on the main objective of "federating the chamber's actors through shared tools and a common management project." This desire for change was propelled by several difficult years, particularly infused with financial difficulties. At the end of the 1990s, a financial recovery plan was adopted to improve the chamber's profitability and reduce its working capital requirement, which was structurally overdrawn for several years. Despite the efforts made by every member of the chamber's staff, the effects of this plan remained unsuccessful, largely due to a reduction in state funding and a continuous increase in overhead. This endemic financial loss led to a shared awareness among the board of directors and elected officials alike: inadequate *management tools* made it impossible to obtain reliable and coherent provisional figures. This situation led the chamber to embark on a new venture—the "search for a million Euros." The objective was clear and required external support in a structured approach addressing both social and *economic performance*.

Michel de Trogoff, the general manager expressed the ambition of the chamber as the challenge of "learning cost reduction and *creation of potential*" (de Trogoff, 2003). Strategic implementation was expressed in seven major action plans: answerability of everyone responsible for the chamber; improvement in skills and formalizing know-how; development of teamwork and communication among elected officials, employees, services and hierarchy to facilitate the creation of *value-added*; strengthening of investment and project cost control, particularly for infrastructures, to eliminate over-runs; development of the *vital sales function* in all services; promotion of *strategic vigilance* to improve decision quality with the introduction of a reliable and regular reporting system; and reinforcement of the link between strategy and human resources by introducing an integrated *socio-economic strategy* (Krief, 1999; Savall & Zardet, 1995).

The goals of both chambers addressed the *creation of value-added* and the *creation of potential* for their *external environment*, plus the control and coherence of its internal working methods.

The Construction of the Socio-Economic Intervention

For both chambers, the main objective of the *socio-economic intervention* was broken down into a number of specific *intermediate objectives*, which are summarized in Table 7.2. In both cases, the approach aimed to improve internal cohesion and external coherence to:

- create *a shared identity* on all levels (by employees, by elected officials, and by employees and elected officials) through the introduction of common tools and the same management approach; and
- define *a strong and federating strategic project* to meet changes in the external environment and pressure on budgets, and to reinforce internal coherence and cohesion.

Table 7.2.   Intervention Objectives of the Two Chambers

	Chamber 1	*Chamber 2*
Main objective	Implement a strategic and organizational plan mobilizing all staff in new projects and optimizing the use of the chamber's resources.	Federate the chamber's actors through shared tools and a common management project.
Intermediate objectives	• Introduce a policy of preventive and global change integrating the initiatives in progress. • Ensure the *visibility* and coherence of the actions already in progress. • Highlight dysfunctions and hidden costs to better control overhead and optimize budgetary and human resources. • Improve the productivity of the organization while maintaining the diversity of the range of skills available. • Transfer the management tools to the staff to allow it more and better control in the implementation of the chamber's strategic project. • Obtain the resources to create and implement a project mobilizing all chamber employees. • Adopt concrete tools and methods to express the vocation of the Chamber to the world outside • Guarantee the deployment and sustained use of management tools and procedures.	• Introduce an overall-progress approach integrating new and existing projects. • Avoid dispersions in the progress plan. • Improve the chamber's productivity by the use of management tools applied by all. • Highlight and improve performances for users. • Gradually improve team cohesion and operational performance. • Optimize the chamber's financial resources to prepare budgets for the future, reflecting the current economic situation. • Train a team of internal interveners to improve control of the effective adoption of the approach and as the chamber's relay for progress implementation.

Source: © ISEOR 2004.
Note: The information in the table is based on a summary of the ***intervention-research agreements*** signed by the ISEOR with both chambers of commerce and industry.

This double target is characteristic of proactive internal strategies. To implement this type of strategy, the common factor linking the projects designed for each of the two chambers of commerce consisted of initiating an approach that was simultaneously strategic and economic in nature to forecast budget needs in the current context of economic pressures, while providing for human and social needs in order to gain participation from all chamber employees toward achieving these goals.

In both cases, the ***Horivert process*** was applied during the first year of intervention. Chamber 1 decided to limit the intervention to the ***initial***

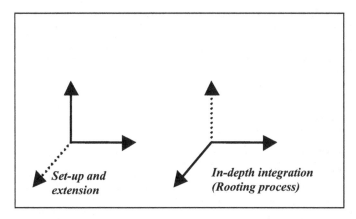

Source: ©ISEOR 1985.

Figure 7.2.   *Set-up and rooting processes.*

set-up phase, whereas Chamber 2 decided to continue to improve its operations and consolidate the tools and approaches introduced during the first year by an in-depth *rooting process* (described in the mains results section). The *initial set-up phase* concentrates on developing appropriate *processes and tools*. Within this context, political decisions have a lower priority (without being neglected), because the main focus is to fully understand every aspect of the system to be implemented.

In the *rooting phase*, in contrast, political decisions are the focal point of the approach, since they accompany in-depth *rooting products* concerning the chamber's strategic decisions. This second stage integrates all the tools in the organizational strategy (see Figure 7.2). Chamber 2 decided to adopt the in-depth *rooting approach* and to apply theoretical socio-economic organizational concepts such as *socio-economic marketing*, *management audit*, *periodically negotiable activity contracts (PNAC)*, and the development of extensions to "basic" tools.

The *Horivert* approach is two-dimensional:

- *Horizontally*, the approach aims to reinforce staff management cohesion and the interface with elected officials. This approach was particularly important in Chamber 2, because of "open" warfare between elected officials and chamber employees. For Chamber 1, the problem of cohesion, less visible at the start, appeared progressively as the implantation progressed. In fact, it both revealed and contributed to the internal cohesion problem.
- *Vertically*, the objective was to involve all layers and all levels of chamber actors through "capillarity" (the *rooting principle*). The ver-

Table 7.3.   *Socio-Economic Intervention* Schedule

Actions	Months											
	1	2	3	4	5	6	7	8	9	10	11	12
Policy-decision axis												
Steering group	1		2				3					4
Management tools axis												
Collaborative training (5 *clusters*)	1	2	3		4		5		6			
Personal assistance		1	2			3				4		
Improvement-process axis												
Horizontal diagnostic	├———————┤											
Horizontal project				1	2		3		4			
Vertical diagnostics					├———————┤						
Vertical projects							1		2	3	4	

Source: © ISEOR 1999.

tical approach resulted in in-depth *vertical diagnostics and projects* in every service (e.g., sales, training, equipment).

The pace of the intervention was based on a schedule including all three *socio-economic intervention axes*, involving tools, the *dysfunction* diagnostic, and resolution and politico-strategic decisions (see Table 7.3). *The horizontal diagnostic* was presented at the second *steering committee* meeting (during the third month of the intervention) and in the third session of the various *collaborative training clusters*. The *vertical diagnostics* of the different sectors were presented at the third meeting of the steering committee and the firth session of the various *collaborative training clusters*.

Each axis analysis is based on the *intervention architecture* developed for both chambers (Figure 7.3) and presents the specific range of each component: the *steering committee*, the *horizontal diagnostic* and *project*, the *vertical diagnostic* and *project*, *collaborative training* meetings, and the constitution of the *internal-intervener* team. The steering committee is a strategic and political group responsible for the orientation of the initial action plan, information and communication flows, process evaluation and the harmonization of the approach in comparison with other interventions, and actions and major projects in progress. Due to the need to ensure the coherence of the intervention with the strategy of the organization, it appeared essential to co-opt elected members responsible for chamber policy. In both cases, the president of the chamber attended *steering committee* meetings and participated in the analysis of the dys-

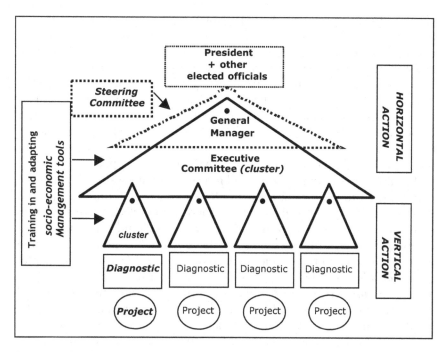

Figure 7.3.   *The intervention architecture.*

functions detected in the **horizontal diagnostic**. In the second case, because of the chamber's budgetary problems, the treasurer was also made a member of the steering committee and participated in the horizontal diagnostic. Additionally, a dozen elected members contributed to the horizontal analysis of chamber dysfunctions. The opposition, even conflict, among elected officials made it necessary to widen the vision of the chamber's dysfunctions and to associate highly contrasted reports from many different actors.

The involvement of key personalities among governance system personnel, in particular members on the **steering committee**, can be a critical factor in (1) guaranteeing the political and strategic coherence of the horizontal approach and (2) unifying actions by the full-time staff with the political project, especially since the latter are defined by the president and the elected officials of the chamber. In this respect, the role of the **intervener-researcher** is to systematically make the full-time staff, and in particular the general manager, aware of the necessary "reporting" of actions concerning the members executive committee (the "core" of the elected officials assembly). In both cases, each major stage that was in progress was fully reported (e.g., at the start of the approach, during the results of the *horizontal diagnostic* and *vertical diagnostics, assessment of achievements*). The success of the approach depends on creating a climate of mutual confidence and respect through the commitment of elected officials to projects implemented by chamber employees. This dynamic is essential because the nature of the relationships between the two groups, by definition, is imbalanced. Any situation where volunteer policy-makers from private enterprise define what is "good" for the chamber with field implementation by a salaried staff of "civil servants" is open to conflict.

The *diagnostic* and the *horizontal project* made it possible to bring the operational project propounded by the chamber's full-time managerial staff closer to the political project defined by the elected officials. This dialogue and partnership approach led to internal and external strategy meetings between the members of the executive committee and the elected officials and the development of a 5-year *internal-external strategic action plan (IESAP)*. The *vertical action* plan (i.e., the *vertical diagnostic* and *project*) aimed to relay the total plan, and the operational orientations defined, to all levels of the chamber, respecting the specific nature of each sector. The intent was also to motivate the participation of each actor in the implementation of the political project defined at the senior levels of governance. The vertical action plan consisted of implementing a *vertical diagnostic* to highlight the principal dysfunctions in each sector selected and initiate dysfunction resolution in line with the policy and axes chosen at the end of the horizontal project.

In the case of Chamber 1, this in-depth analysis addressed three main sectors: administrative and financial management, the business school, and the directorate of commercial and territorial action. The logic behind this choice, which was negotiated between the consultants and the chamber, was that there should be: one central service, both geographically (in terms of a central headquarters) and strategically (in terms of a financial center in the context of budgetary pressure and reduction in government funding); one major chamber service (the business school); and one service representing the traditional services provided by the chamber, the diverse nature of which often complicates overall operations.

Many actors, both those concerned by a vertical action who asked "why us?" and others who asked "why not us?" expressed their doubts and resisted the intervention, preferring to mark their opposition. This was notably the case of the Graduate School of Commerce, the analysis of which is presented below. The choice of services to be analyzed was mainly guided by budgetary and time considerations, but did little to improve problems of cohesion and synergy between the various services inside the chamber.

In Chamber 2, all sectors were included in the vertical approach. Two had already been selected when the intervention was initially negotiated due to their fundamental problems: the chamber's core business (providing company support services) and Training, which was in financial difficulty. When the first *steering committee* meeting was held, the president decided to extend the analysis to the other sectors provided by the chamber, including all *infrastructures* (the port and the airport) and all central and general services. This political decision by the President to go beyond the chamber's traditional services (e.g., company support and training) demonstrated that it was possible to improve *productivity* in other sectors. This decision was unanimously approved by chamber executive committee members. Although it appeared that the decision was coherent, the down-side was that it lengthened the time required for diagnostics. The last two diagnostic interventions, for example, were completed more than six months after the first, which reduced the significance and timeliness of the findings. The benefit was that it increased interest in the *vertical project* and, in fact, the impact of the results that were obtained.

This "process" approach is supplemented by *collaborative training cluster* sessions and individual *personal assistance*, the aims of which are to introduce management tools destined for common usage among management personnel. In both chambers, these meetings involved the entire managerial team—approximately 100 people in Chamber 1 and 40 in Chamber 2. The principal socio-economic *management tools* utilized during these sessions were:

- *time management tools*, to develop awareness by the services and actors of the need to measure their activities and control of their costs;
- the *competency grid*, to answer the challenges of evolution and change in the Chambers' businesses;
- an *internal-external strategic action plan* (*IESAP*) and *priority action plans* (*PAP*), to formalize the future of the organization;
- *strategic piloting (indicators) logbook* and the architecture of the management audit ratios and indicators, with the objectives of developing transparency, simplifying decision-making and strengthening vision of activities and performances; and
- the *periodically negotiable activity contract* (*PNAC*), to meet the problem of modernizing the chambers' human resources policy.

All these tools were implemented in each service, even if the *periodically negotiable activity contracts* were only discussed on an informational level. Indeed, implementation of the *project* to modernize human resource management and the remuneration system that was used by the Chambers was sufficiently difficult that few of them dared to introduce an innovative approach to pay packages designed to stimulate job performance. A typical excuse was that it was virtually impossible to remunerate individual merit in a civil service-type pay system.

To assist with the transfer of these new approaches, both chambers created *internal-intervener* teams. This decision was made because of the need to cover all the sectors of activity in which chambers are involved. For this purpose, at least one internal intervener per service attended the ISEOR *socio-economic management* consultants' training course at the Institute's headquarters. Nevertheless, although both chambers adopted the same basic approach, trainee profiles were different. In the first case, general management preferred to use specialist managers, whereas the second chamber initially preferred to train every member of its executive committee. This difference is due to the fact that Chamber 2 wanted to use its executive committee as an example for the rest of the organization. In both cases the consultant teams participated in the implementation of the diagnoses and vertical projects as well as providing training on the tools.

LESSONS LEARNED

To highlight the results obtained in both intervention-research analyses, the following discussion emphasizes *differences* across the two chambers while considering the convergent points of both cases and also the specificities of each organization.

Governance System Challenges

For both chambers the *horizontal diagnostic* revealed strong convergences, translating the need to address the implementation of strategies, management style, human resource management, *work organization*, team cohesion and cross-sectional activities (elected members, executive committee, paid staff). Table 7.4 summarizes the topics and key ideas of the *expert opinions* on which the *horizontal project* of each chamber was centered.

Both Chambers set up *task group*, run by the members of the Executive Committee with co-opted management representatives. The complexity of the relationships with elected officials, which characterized Chamber 2, required a wider base for the analysis. As such, each working group co-opted an elected official. Without detailing the steps in *project* elabora-

Table 7.4.   A Comparison of *Expert Opinion* Across the Two Chambers

Chamber 1	*Chamber 2*
1. *Strategic implementation* • Failure in relaying ***strategic objectives*** • Atrophy of the strategy management tools and measurement indicators • Absence of clear, shared strategic vision **2. Cross-sectional activity and synergy** • Failures in ***communication-coordination-cooperation*** across actor categories • Failure in cohesion between the satellites and HQ **3. *Work organization*** • Partially unsatisfactory working conditions • Excessively heavy work organization **4. Management and *human potential*** • Unsatisfactory pay policy • Management lacking in managerial skills • Low motivation to change	**1. *Work organization*** • Partially unsatisfactory working conditions • Unclear and excessively heavy work organization **2. Internal customer-supplier interfaces** • Internal customer-supplier quality problems • Lack of cohesion between paid managers and elected officials **3. *Strategic implementation* and management** • Lack of management performance measurement ratios and indicators • Low strategic pressure and low motivation to change • Unsatisfactory management style and human resource management

Source: © ISEOR 2004.

tion, the task group developed a 5-year ***internal-external strategic action plan*** (***IESAP***), constituting the core of their organizational strategy. The IESAP was relayed to each service in a semi-annual ***priority action plan*** (***PAP***) detailing the actions to be carried out to implement the chamber's strategy. In both cases, the strategic axes defined and focused on the main problems of the chamber: (1) the chamber's role with respect to the local economic environment and the development of new services; (2) definition of activity performance management ratios and performance measurement indicators; (3) tasking and productivity of the ***work organization***; and (4) relationships with elected officials. The two chambers also proposed a strategy based on the social and economic coherence, relevance and responsibility of their actions, both in terms of their internal target groups and their ***external environments***.

The case of the second chamber was particularly interesting, because the strategic project approach revealed a true evolution in the behavior patterns of each part of the governance system. Joining forces, the elected officials were more supportive of the employees in the implementation of their actions and grew closer to the positions of the chamber's salaried decision-makers. In both cases, the effect of the project was to reinforce the cohesion of the executive committee and restore the confidence of the managers as they felt involved in and committed to the implementation of the strategy adopted by "their" chamber. The methodology focused on listening to the actors (therapy) and bringing them together (mediation), roles that the *intervener-researcher* was in charge of, leading to two important outcomes—cohesion and confidence. The *horizontal diagnostic* and *project* led to a unified approach to the future—the first step in *rebuilding* the identity of the chamber.

Apart from unification, the two chambers reacted differently to the intervention. The following discussion reviews some of these differences, especially in the context of Chamber 1's business school and the internal reorganization action plans and successful financial outcomes in Chamber 2.

Chamber 1 and Its Business School

The vertical action implemented at the business school in Chamber 1 is particularly interesting for a couple of reasons. First, it illustrates the behavior patterns of the actors in the sector during the intervention and the results of the diagnostic. Second, it demonstrates the way the project was managed and the results of the project.

The behavior patterns of the actors in the business school constituted, initially, an obstacle to the approach to change. From the start of the intervention, the school's teaching body expressed their opposition to the approach and their refusal to cooperate during the diagnostic process. While the *vertical diagnostic* showed a significant amount of *hidden costs* (varying from 12,000 to 37,000€ per person per year in the sectors examined in both chambers), the *hidden costs* calculated for the business school appeared to be quite low (roughly 8,000€ per person overall). These initial hidden cost results, however, are most probably underestimated due to the large number of dysfunctions with financial impacts noted during the diagnostic and the lack of the actors' *cooperation* in the analysis of the problems and possible solutions.

This lack of cooperation appears to be partly explained by the fact that the School failed to obtain Equis accreditation[6] at the moment at which the intervention began. Proposing the socio-economic approach, often referred to as an *integral quality* approach, seemed to "offend" the actors. To complicate matters, competition exists between university intervener-researchers in the field of management consulting and management teachers working in business schools (in France, higher education has been characterized for many years by competition between the public university system and the tuition-paying system practiced in business schools). In addition, the lack of involvement by the actors in the vertical approach in their sector stems from a conflicting relationship between the business school campus and the chamber's headquarters. The increasingly "independent" behavior of school actors created a reciprocal feeling of mistrust between the campus and the headquarters' services represented by the general manager and chamber president.

In an attempt to work around the lack of participation, the ISEOR team of intervener-researchers proposed a project based on key work-related topics, developing a series of key-ideas based on the analysis of the dysfunctions explicitly evoked by the actors and the expression of the *non-dit (unvoiced comments)* evoked by the interveners (see Table 7.5). *Focus group* participants addressed these topics listed below and five project *task groups* were set up to find solutions to the problems observed:

- improving communications about the school's strategy to school employees;
- improving the ways in which the school executive board operates;
- obtaining Equis certification and developing recertification planning;
- measuring performances at the School; and
- setting up a motivating pay system.

The main tangible contributions and outcomes from these project task groups included:

- introduction of a management check-list for school activities;
- officialization of a strategic plan for the school and a semiannual presentation of results to all school staff;
- organization of the teaching staff in centers and the preparation of research and teaching activity forecasts;
- introduction of cross-sectional course interoperability;
- definition of executive committee rules and operating procedures (decision-making authority);
- a proposal for a profit-sharing system adapted to the operation of the school; and
- introduction of modus operandi with the chamber based on a *synchronized decentralization* socio-economic model.

While the school failed to qualify for its initial Equis application (just before launching the socio-economic intervention), after completion of

**Table 7.5.   Chamber 1 Business School Intervention:
Key Ideas and *Improvement-Process Axes***

Key Ideas From the Analysis	Improvement Actions Proposed
• Lack of relevant indicators for the evaluation of activities and curricula • Weakness in the School's strategic choices	• Develop the strategic ambition of the school and provide the conditions for its implementation, by the introduction of the relevant management control and assessment indicators of the activities and teaching programs
• Lack of staff direction and objectives • Lack of interest in fault-finding and dysfunction trouble-shooting • The school executive committee never acts as an arbitrator or decision maker	• Change the role of management in team control, transversal, descending and ascending communication and decision making
• The school does not feel itself to be part of the chamber • No positive or negative sanction system to stimulate performance • Sluggishness of the central human resource management procedures	• Reinforce the links between the chamber and the school by improving the management of the interfaces between the two entities in terms of communication and shared ideas and goals
• Partially unsatisfactory working conditions • Lack of anticipation and concerted time management reflexes • Lack of synergy between school courses	• Improve organization and working conditions, reduce internal and external strife and mitigate time management problems

Source: © ISEOR 2000.

the socio-economic project in June 2000 it was able to attain its Equis certification.

Chamber 2

In Chamber 2, the intervention made it possible to undertake an in-depth reorganization. In the first phase, particularly strong action was taken to restructure the executive committee and harmonize relationships among its members. Many chambers in France had operated without an executive committee for years. To rectify this deficiency, they all recently tried to rebuild this decision-making authority, but were all confronted with the same problems: proliferation of the number of participants, weakness in decision making, and, in general, an ineffective body with collegial overtones. In Chamber 2, the executive committee rediscovered its decisional role, thanks to a "spring cleaning" which restricted membership to people hierarchically responsible for operational and functional services, and excluded everyone who was on the committee simply because he or she had particular affinities with the elected officials.

Change in Internal Organization

This action clarified the organizational structure and the chamber's hierarchical links. In this way, corporate support services, which, when our intervention began, faced a triple set of problems were restructured into two services with two managers—a center for small and medium-sized industrial enterprises and a local development center dedicated to commercial structures. This reorganization simplified the vision for employees who thus, unambiguously, reported to a single, more powerful, superior. Before the intervention no one really controlled corporate support services, and this action made it possible to reinforce *productivity* and create true dynamics for change in these traditional services.

The second example of organizational clarification related to infrastructure. To make their management more coherent, it was decided to group the port and airport. Thus, a director of infrastructures was appointed as head of the two entities and a common mode of operation was set up (e.g., *coordination* meetings, commercial development committee). In addition, finances, human resources, quality, forecasting and communications were grouped in a single "general services" department.

Change and Improvement in Profits

Following the first socio-economic management *initial set-up phase*, some of the chamber's elected officials appropriated the results, to such an extent that the treasurer of the chamber, who was responsible for rebalancing chamber finances, presented the economic and financial results obtained at the end of the first year in public. His remarks suggest the extent of the commitment to change and its outcomes: "Over the last 20 months, we have introduced major organizational, behavioral and financial changes in our organization and strategies."[7]

Improvements in the way the chamber operated made it possible to set new *game rules* for greater homogeneity in working methods, increase rigor in service operations, and ensure that actors assumed responsibility for the quality of their personal inputs. As the treasurer stated, these changes were fundamental success factors: "We all knew that to achieve our goals, everyone involved had to adopt our objectives and our business plans. One of the most significant results was the shift from resistance to change to the commitment of all the actors. This shift contributed to the chamber's financial successes. The actors adopted the *vertical diagnostic and project* problem-solving procedure, and managers began to discover tools that enabled them improve their control over the activities and personnel for which they were responsible.

Instead of a chronic deficit of approximately 1 million Euros at the end of 2001, the chamber showed a profit of over 1 million Euros. External purchases and charges have been reduced by almost 76% and the payroll was stable, despite the addition of 14 employees possessing the skills needed to meet the new challenges facing the chamber. In addition, a working capital deficit of 1.8 million Euros in 2001 was halved by the end of 2002. In 2004, working capital is positive for the first time in 25 years. Explaining these results, the treasurer noted that over the course of one year, "we changed from a totally passive chamber unable to act in any way, to an active chamber with its own action plans, able to allocate resources where they are needed and now able to recruit." These results were facilitated by socio-economic innovation *focus groups* targeted to transforming identified **hidden costs** into visible and tangible **value-added**. The hidden costs identified during the diagnostic phase in each sector showed the improvement potential and focused the actors' actions on resolving operational and strategic dysfunctions.

The chamber's general manager qualified the initial results by arguing:

> What is essential is that employees now have complete confidence in what they are doing, in the sense that they can now concentrate on the jobs they perform for the general public and local business ... because they know that the chamber's financial problems are being resolved. This is the best possible return on investment since it improves conditions both for people working in the chamber and simultaneously assists local businesses.

These positive results have encouraged the chamber to build on the first phase to sustain its socio-economic performance and improve competitiveness through enhanced morale and internal managerial enthusiasm. The chamber has decided to undertake several in-depth analyses of various aspects of socio-economic management, starting with activity and team management tool consolidation meetings addressing: activity planning and **scheduling**; skills management and an **integrated training plan**; and a **priority action plan (PAP)** and **socio-economic management control** system. In addition, **focus groups** were set up to focus on three sets of themes of major importance for the chamber:

- The introduction of **socio-economic management control** (Savall & Zardet, 1992) to strengthen control of core activities, measurement of the cost of the activities and control of the chamber's expenditure and charges. This concern is basic to the current evolution of chambers of commerce and industry in general where cost measurement of activities is essential to fix fees for some services. Chambers must be perceived as competitive solutions in increasingly open markets. The project consists in developing **socio-eco-**

nomic management control system compatible with chamber management style characteristics.

- A project focused on the *socio-economic marketing* of corporate support and training services to improve the sale of traditional chamber services. This project focuses on the central missions of the chambers and addresses the problems of how to sell services which were previously free of charge without a clearly identified sales force. In addition emphasis was placed on how to set up an appropriate "commercial think-tank" to develop the deployment of the vital selling skills of each actor at every level. This project aims to provide solutions that more closely meet the internal evolution (personnel) and external evolution of customers' (previously called "users" or "constituents") requirements.

- *Periodically negotiable activity contracts (PNAC)* made it possible to define better solutions for change in areas that many employees and external specialists considered to be totally obsolete.

CONCLUSION

The intervention-research carried out in these two chambers of commerce and industry highlighted the importance of *internal* **cohesion** and coherency as success factors in strategies adopted to improve impact on the external environment. The intervention-research in this chapter demonstrates that internal fortification is a necessary prerequisite to legitimizing a chamber's sphere of activity. It also illustrates that financial results reflect the organization and behavior patterns of each actor through greater proactivity, commitment and contribution to *discriminating value-added*. In both case-studies, socio-economic intervention was deemed to be an effective and efficient approach to rebuilding a chamber's identity and reinforcing its legitimacy and competitiveness as a driving force in an increasingly complex local economy.

NOTES

1. This chapter draws on several ISEOR publications (ISEOR, 2004a, 2004b) and earlier conceptual work by Savall and Zardet (1995).
2. In France a district is a territorial perimeter. France is divided into districts called "circonscriptions consulaires."
3. CNPF was the National Council of French Employers, now the Mouvement des Entreprises de France (MEDEF); CGPME is the National Federation of Small and Medium-Sized Enterprises.

4. The urban community concept regroups several municipalities, which join forces to prepare and implement a joint urban development project for their territory. France is divided into 100 administrative areas called *départements*, which are grouped together into *régions*.

5. This quote is an excerpt from the letter sent by the general manager to chamber employees in April 1999, informing them about the adoption of the socio-economic approach (Babin, 2000).

6. Equis, which stands for the (European Quality Improvement System, is a European quality improvement certification process.

7. The quote is an excerpt from the paper "Analysis of the financial results at the end of the first year of the *progress* approach in a Chamber of Commerce," given at the ISEOR (2004a) Annual Conference titled "*Strategic Change in Chambers of Commerce and Industry.*"

REFERENCES

Babin, D. (2000). Mise en oeuvre d'un projet stratégique et organisationnel pour mobiliser et optimiser les ressources internes d'une grande organisation: Le cas de la Chambre de Commerce et de l'Industrie de Bordeaux [Implementing a strategic and organizational project to mobilize and optimize the internal resources of a large organization: The case of the Chamber of Commerce and Industry of Bordeaux]. In ISEOR (Ed.), *Le Notariat Nouveau* (pp. 131-140). Paris: Economica.

Charreaux, G. (1997). *Le gouvernement des entreprises: Théories et faits* [Governing enterprises: Theories and facts]. Paris: Economica.

de Trogoff, M. (2003). Démarche de progrès dans une organisation consulaire: Le cas de la Chambre de Commerce et de l'Industrie du Morbihan [Progress procedure in a consultative organization: The case of the Chamber of Commerce and Industry of Morbihan]. In ISEOR (Ed.), *L'université citoyenne: Progrès, modernisation, exemplarité* (pp. 255-276). Paris: Economica.

Eggrickx, A. (1999, September). Un exemple type d'organisation incontrôlable, paradoxalement contrôlée: Le cas des chambres de commerce et d'industrie [A typical example of an uncontrollable organization, paradoxically controlled: The case of Chambers of Commerce and Industry]. *Revue Comptabilité Contrôle Audit, 2*, 151-170.

Gibert, P., & Thoenig, J. C. (1993). La gestion publique: Entre l'apprentissage et l'amnésie [Public management: Between apprenticeship and amnesia]. *Revue Politiques et Management Public, 11*(1), 2-23.

ISEOR. (Ed.). (2004a). *La Mutation Stratégique des Chambres de Commerce et d'Industrie* [The strategic transformation of Chambers of Commerce and Industry]. ISEOR Annual Conference Proceedings. Lyon, France: ISEOR.

ISEOR. (Ed.). (2004b). *Analyse des résultats financiers de la démarche de progrès à l'issue de la première année, au sein d'une CCI* [Analysis of the financial results of the progress procedure at the end of the first year]. Paris : Economica

Krief, N. (1999). Les pratiques stratégiques des organisations sanitaires et sociales de service public. Cas d'expérimentation [Strategic practices of health and social public service organizations: Experimental cases]. Unpublished doctoral dissertation, University of Lyon, France.

Larçon, J. P., & Reitter, R. (1979). *Structures de pouvoir et identité de l'entreprise* [Structures of power and identity in businesses]. Paris: Nathan.

Laufer, R., & Ramanantsoa, B. (1982). Crise d'identité ou crise de légitimité? [Identity crisis or legitimacy crisis?]. *Revue Française de Gestion, 37*, 18-26.

Puaux, P. (1998). *Les chambres de commerce et d'industrie* [Chambers of commerce and industry]. Paris: Presses Universitaires de France.

Savall, H. (1974, 1975). *Enrichir le travail humain dans les entreprises et les organisations* [Work and people: An economic evaluation of job enrichment]. Paris: Dunod.

Savall, H. (2003a). An updated presentation of the socio-economic management model. *Journal of Organizational Change Management, 16*(1), 33-48.

Savall, H. (2003b). International dissemination of the socio-economic model. *Journal of Organizational Change Management, 16*(1), 107-115.

Savall, H., & Zardet, V. (1987). *Maîtriser les coûts et les performances cachés: Le contrat d'activité périodiquement négociable* [Mastering hidden costs and performances: The periodically negotiable activity contract]. Paris: Economica

Savall, H., & Zardet, V. (1992). *Le nouveau contrôle de gestion: Méthode des coûts-performances cachés* [New management control: The hidden cost-performance method]. Paris: Éditions Comptables Malesherbes-Eyrolles.

Savall, H., & Zardet, V. (1995). *Ingénierie stratégique du roseau, souple et enracinée* [Strategic engineering of the reed, flexible and rooted]. Paris: Economica.

Savall, H., & Zardet, V. (2004). *Recherche en sciences de gestion: Approche qualimétrique. Observer l'objet complexe.* Unpublished English translation: *Research in management sciences: The qualimetric approach. Observing the complex object.* Paris: Economica.

Savall, H., & Zardet, V. (2005). *Tétranormalisation: Défis et dynamiques* [Competitive challenges and dynamics of tetra-normalization]. Paris: Economica.

Savall, H., Zardet, V., & Bonnet, M. (2000). *Releasing the untapped potential of enterprises through socio-economic management.* Geneva, Switzerland: International Labor Office-ISEOR.

PART II

CROSS-CULTURAL INTERVENTIONS WITH SEAM

CHAPTER 8

SOCIO-ECONOMIC INTERVENTION IN DEVELOPING COUNTRIES IN AFRICA AND ASIA

Emmanuel Beck

As the *socio-economic intervention* method has spread throughout France and other European nations, international demand has also grown, especially from a variety of Asian and African institutions (see, for example, International Labor Organization, 1998; Savall, 2003a, 2003b). In order to explore the utility of the *socio-economic approach to management* (*SEAM*) (Savall, 1974, 1975, 1979, 1987, 2003a; Savall & Beck, 1980; Savall & Zardet, 1987, 1992, 1995) to respond to the needs of these organizations, a series of *intervention-research projects* have been underway since the 1980s that have resulted in an international dissemination procedure of SEAM methodology in approximately 20 countries in Africa and Asia. This effort has included direct and indirect intervention by ISEOR researcher-consultants in Asian and African organizations and companies (e.g. Mafra, Hidalgo, Eraers, & Beck, 1999) as well as direct intervention by postgraduate researchers and consultants from those

Socio-Economic Intervention in Organizations: The Intervener-Researcher and the SEAM Approach to Organizational Analysis, pp. 199–214
Copyright © 2007 by Information Age Publishing

countries who have been trained by ISEOR (e.g. Bensalem, 1999; Bousso-fara, 1993; Buoy, 2001; Defdouf, 1995).

The ILO (International Labor Organization), under the direction of Pierre Hidalgo, a senior officer at the ILO, officially recognized the merit of the **SEAM** approach by publishing a book dedicated to the method simultaneously in three major languages (English, French and Spanish) (Savall, Zardet, & Bonnet, 2000). This trilingual report[1] was disseminated throughout the member countries of the ILO to help entrepreneurs to implement **SEAM**.

The central hypothesis guiding these efforts is based on the precept that the needs and expectations for guidance and management in human organizations are universal, limitless and continually changing. They are guided by the will to survive in a hostile and competitive environment, where *resources* are expensive and hard to find and people are underprivileged in terms of energy, food, education and health.

GENERAL CHARACTERISTICS OF THE
ASIAN AND AFRICAN COUNTRIES

ISEOR has been working in Africa for more than 20 years and in Southeast Asia for approximately 10 years. The first links were forged through young junior consultants coming to France to pursue their higher education, notably students from North Africa. Most of these international students trained in SEAM methodologies and became consultants, heads of businesses, or teacher-researcher-consultants (Savall, 1989). In the majority of cases, such opportunities provided the occasion to implement SEAM methodologies in their own countries. The main areas and countries referred to in this chapter include Central Africa (Angola, Burundi, Cameroon, Ivory Coast, Ghana, Malawi, Nigeria, Senegal, Togo), North Africa (Algeria, Morocco, Tunisia), and Southeast Asia (Cambodia, China, Thailand, Vietnam).

All of these countries are considered emerging economies with certain key characteristics in common (Clare, Thiébaut, Albert & Baudu, 2004):

- *Low Gross National Product* (GNP) per capita. As an example, the GNP per capita (in U.S. dollars) for the Ivory Coast (630), Malawi (160), Vietnam (410) and Cambodia (270) can be compared to the GNP for the U.S. (34,280) and France (22,730) per capita. This data (provided by the OECD [Organization for Economic Cooperation and Development], IMF [International Monetary Fund] and EBRD [European Bank for Reconstruction and Development] demonstrates the gulf between the industrialized world and the developing world.

- *Population growth, density and location.* Although population density is generally lower in many of these countries than in France, the population growth rate is much higher. In Burundi, for example, there are 270 people per square kilometer with a population growth rate of 1.90%. In Algeria there are 13 people per square kilometer, with a population growth rate of 1.56%. In comparison, France has 108 people per square kilometer with a population growth rate of only .42%. Despite the existence of huge urban centers, such as Bangkok in Thailand and Ho Chi Minh City in Vietnam, the populations of these countries remain essentially rural.

- *Age and life expectancy.* The percentage of young people is often proportionally very high. For example, in Algeria roughly 35% of the population is under 14, compared with just 18.7% in France. Life expectancy is also lower in these countries than in France. In Cambodia, the figure is 53.91 years, compared with 79.16 in France.

The history of these countries is marked by nineteenth century colonialism and twentieth century wars of independence, some extremely violent and leading to genocide (as was the case in Cambodia). Following such troubled times, some of these countries have taken years to regain a balance that remains precarious even today (e.g., Burundi). Relationships between political or military forces and their cohorts over frontier rivalries and territorial ambitions often involve conflict (as in Cambodia and Vietnam), despite the efforts of international organizations like the Association of Southeast Asian Nations (ASEAN). Against this background, the consequences are often dramatic, precipitating a loss of governmental autonomy with a reduced role for parties and less democracy, the increasing power of financial markets, the activities of multi-national corporations, the proliferation of tax havens, and rising third-world debt and environmental alteration.

Although some of these countries possess significant natural and mineral resources, they are under-utilized and often exploited and the target for unfair trade tactics with industrialized nations (e.g. the outrageous exploitation of Cambodia's timber *resources* that is damaging the country's natural climate and ecology). Very often, the health balance of these countries is upset and hard-to-treat illnesses create significant demographic shifts in the way populations evolve (e.g. AIDS in Central Africa and Southeast Asia) in difficult climatic conditions that make working difficult.

The developing countries of Africa and Asia where SEAM interventions have been conducted all share a low level of schooling. Teachers have very few teaching resources and are often poorly paid in comparison with those doing other jobs. Many, like those in Cambodia, need to take a second job to feed their families properly.

Organizational *structures* are extremely hierarchical, with highly accentuated leadership traits and a predominance of ethologically masculine behavior whose result is to devalue women and their place in the economy. Men and women play very different and unequal roles. We have also observed high levels of unemployment that are not declared in official statistics, when these statistics exist at all. In this context, the populations concerned have very little purchasing power. In many cases, the basic wage allows *survival*, but very little else. Despite all of this, labor is abundant, but poorly trained, inexpensive and at the mercy of the company.

While the emergence of nongovernmental organizations (NGOs) has responded to dispersed needs for aid and supports *development*, such efforts are often undertaken without a focused, long-term strategy. Where NGOs do exist, they are confronted by the steamroller of today's unrestrained globalization, a characteristic of contemporary society, which unwittingly reinforces a social-economic dichotomy. Similarly, governments do not always channel tax revenue towards *infrastructure* construction and maintenance (e.g., roadways, waterways, power generation) and aid provided by industrialized countries remains at too low a level to enable real change to get underway. Tax collection is still too often beset by corruption (e.g., Cambodia, Vietnam, Africa). Mafia-like networks over-burden economic circuits that require serenity to run smoothly, to say nothing of the effects of civil wars (like those in Burundi and Ivory Coast). Lastly, inflation and erratic currency exchange rates create imbalances that are sometimes irreversible and require the cyclical writing-off of debts which these countries, with their high inflation rates, simply cannot manage. These themes have been regularly addressed during the Annual François Perroux Colloquium, co-organized by ISEOR, the Association of the Friends of François Perroux and the François Perroux Foundation.[2]

ISEOR's decade-long (2003-2013) research program is intended to push for new ways to overcome this societal *dysfunction* (see Savall & Zardet, 2004, 2005). Even in its early stages and against this background, socio-economic intervention has been linked with notable improvements in the economic and social situations of the organizations and companies where they have been carried out.

SOCIO-ECONOMIC INTERVENTION IN ASIA AND AFRICA

There have been several different types of *socio-economic intervention* conducted in Asian and African countries since the beginning of the 1980s, including *intervention-research* conducted and guided by teacher-researchers, consultants and postgraduate students trained at ISEOR headquarters. Socio-economic intervention in these countries has been guided by SEAM policies and procedures and the *implementation* of the six key *SEAM tools* to varying degrees. All interventions have involved *integrated training* sessions for company actors and a *HORIVERT*-type process guided by an *intervener-researcher*.

Given the distances involved, the training of local relay consultants is vital for the effective operation of these interventions. This approach was validated through direct intervention in 10 Vietnamese companies, with distant support, and relayed by consultants (teacher-researchers) trained in the socio-economic methodology (Mafra et al., 1999). Figure 8.1 indicates the periods during which the actions were conducted by professor-consultants and actually accomplished in the field.

Another approach involves training external consultants in socio-economic methodology. For example, there are organizations such as the Malawi Institute of Management (MIM) in Malawi that want to acquire the tools of consultancy and adapt them for use in their own activities. Where such organizations did not already exist, it has been possible, with the support of international organizations such as the International Labor Organization (ILO), to set up Training And Consultancy Centers in existing universities (e.g., the case of Viet Nam and Ho Chi Minh City University). Within this context, ISEOR's role is to train university teacher-researchers in a new role so that they can act as consultants for local companies as well as train their own students.

In a slightly different approach, ISEOR has been involved in training foreign postgraduate junior consultants and PhD students with contractual commitments to write their dissertations and theses on interventions in their own countries, with or without the support of their own governments or France (scholarships). This tactic has facilitated the implementation of the SEAM procedure over the long term. In these cases, missions are guided and managed at a distance using permanent synchronized contacts to answer questions and deliver appropriate responses. The academic assessment of this work incorporates reports from the managers.

Over the years, this approach has resulted in the creation of an extensive global network of former students, which in turn has led to the creation of supportive institutions (e.g., a Tunisian Association of University teacher-researchers and consultants performing SEAM). Such projects have provided a basis for discussion at many symposia, underscoring the universal appeal and applicability of the SEAM approach.

After SEAM has been carried out, managers become increasingly aware of other, more inclusive, management methods by applying the socio-economic approach more systematically (Savall & Beck, 1980). They no longer focus their attention solely on costly production factors such as particular raw materials, but understand that by training and valorizing

	1998		1999		2000		2001	
	1st Semester	2nd Semester	1st Semester	2nd Semester	1est Semester	2nd Semester	1st Semester	2nd Semester
1. Internal SEAM training of 5 Vietnamese professor-consultant-researchers and of 1 ILO consultant-coordinator.	↕ ———————————		———— ↕					
2. Integrated training of 5 professor-consultant-researchers in Viet Nam by the ISEOR.			↕ ————	———— ↕				
3. Socio-economic intervention (SEAM) carried out by the 5 professor-consultant-researchers in 10 Vietnamese companies.						↕		
4. Creation and implementation of a Vietnamese Training and Consulting Center by the ILO and autonomous continuation of SEAM interventions by the Vietnamese relays.						↕ ————————————		———— ↕

Source: ©ISEOR 2002.

Figure 8.1.   Examples of a *socio-economic intervention plan*: Vietnam (1998-2002).

the role of their staff, their people will pay attention to the factors over which they have control and deliver better quality products, goods and services with improved *competitiveness*.

Technology Transfer and the Multiplier Effect

An important dimension of the consultant training mission that ISEOR has conducted with international organizations is the transfer of management knowledge and processes. As an example, in the 1990s teacher-researchers traveled abroad to train consultants on short (2-week) missions in Malawi and at the ILO in Turin, where teacher-researchers from Ho Chi Minh City University in Vietnam were trained in the SEAM approach to create an ILO-supported Training and Consultancy Center. The resulting technology transfer made it possible for a relatively small number of ISEOR teacher-consultant-researchers to promote the SEAM approach in their courses and draw on case studies in Vietnamese companies.

Figure 8.2 illustrates the architecture for the general, international dissemination of the Horivert transorganizational process (see Savall, 2003b). The intervention schedule presented in the figure illustrates the major

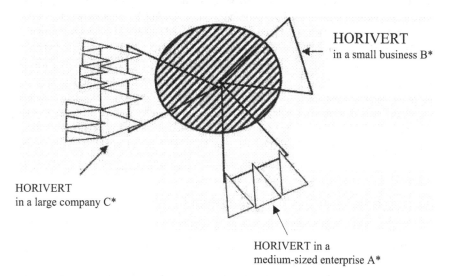

Source: Savall (2003b). ©ISEOR 1997.
Note: *HORIVERT process* implemented in each organization by *external* (professional consultants) and *internal-interveners* with the support of their respective managing directors.

Figure 8.2.   *Horivert process.*

phases of the intervention process, from their beginning in 1998 with the training of professor-consultants until the creation of the Center for Vietnamese Training and Consulting by the ILO, to enable autonomous continuation of the actions carried out by Vietnamese correspondents.

Today, international consultants are trained directly at ISEOR headquarters in short 2-week sessions with a goal of returning home to apply the method in their own countries, with periodic audits of their practices and results. When a country has a sufficient number of trained consultants, a related objective is to create an association of socio-economic consultants. The Association of Tunisian Socio-Economic Managers (AGSET), for example, achieved this stage of formation in 2005. The aim is to conduct research, gather socio-economic data, discuss and publish the results of the research, and organize symposia with the ISEOR's support and sponsorship.

Training international teacher-researchers (Tunisian, Vietnamese, Cambodian, etc.) also enables these individuals to introduce socio-economic methodology in their own University courses, which requires close control of teaching support resources and a contractual commitment to respect copyright. The problem of translation requires the greatest *vigilance* on the part of ISEOR and it is important to create a glossary of key concepts in the local language. For example, an agreement was reached between ISEOR and a Tunisian university Arabic translation of its expert system software, the SEAM Expert System (*SEAMES® software*), for processing semi-directive interviews as part of the MED-CAMPUS Operation.

The *socio-economic management* expert system software *SEAMES® software* includes more than 3,500 types of qualitative dysfunctions, identified through inventories of research projects carried out since the beginning of the ISEOR (1978). These key ideas on *dysfunctions* have been enriched and updated regularly by intervener-researchers following every intervention and added to the SEAMES expert system database after being validated by the director of the ISEOR (see chapter 1 Appendix "SEAMES®: A Professional Knowledge Management Software Program").

Transferring Expertise and Language Challenges

When conducting qualitative *socio-economic diagnostics* and *integrated training sessions* in companies, *intervener-researchers* often encounter difficulties in translating general scientific and specific socio-economic concepts into the local language. In many instances, there is an ancestral vernacular language transcribed into written systems that are sometimes complex and based on a linguistic structure that is very different from that of the second languages more recently learned, practiced and acquired (English and French for the most part). Yet, despite monotheistic and polytheistic religions, philosophies with deeply-rooted traditions and very ancient beliefs all influence behavior within families and the wider society (e.g., Buddhist philosophy, African Shamanism), in most of the cases encountered thus far these local contexts do not present any significant barriers to SEAM introduction. After several months into an initiative, there is a gradual, *heuristic* formalization of socio-economic methodology, without any difficulties in identifying and quantifying *hidden costs*. This requires no more than reprocessing the available data and, in some cases, interpreting obscure and opaque organizational practices.

There are still, however, some lingering challenges. Following *integrated training sessions*, the actual *implementation* of socio-economic tools typically advances only to a limited degree, even though the basic understanding and results are not essentially different than those seen in Europe. There is, however, greater difficulty in gaining acceptance of some tools, such as the *periodically negotiable activity contract* (*PNAC*), which is a keystone of the entire system, while implementing others pose few problems (e.g. the *competency grid*; see Beck, 1979). To some extent, such difficulties can be attributed to the complexity of the tool in question rather than the technology transfer process per se (e.g., even in France, certain interveners experience difficulties in setting up the *PNAC* tool).

These difficulties are caused by both the formal appropriation of the PNAC concept, which requires in-depth explanation of the theory of hidden costs-performances developed by SEAM, and of the reticence of directors to establish a charter of values. Indeed, the distribution of economic surplus gained through *socio-economic projects* constitutes a delicate problem, both in developing countries as well as in advanced economies. Major involvement by outside consultants made it possible to set up the PNAC in other enterprises.

Heterogeneity, Ethnicity and Conflict: Intervention Challenges

In the African and Asian countries where ISEOR has carried out SEAM interventions, we frequently observe strategies that are rather limited in ambition, guided in many instances by instincts of simple *survival* (Aman, 2001; Dinh, 2000; Frika, 2000; Hor, 2000; Kammoun, 1998; Lahlali, 2001; M'Rabet, 1997; N'Kanagu, 1992, 1996; Tamo, 1998; Yamuremye, 2001). Yet, while a great deal of effort has still been invested in their development, the broader political context contributes to a lack of *confidence* in one's capacity to survive, especially where major conflicts have dominated the environment (e.g., Cambodia, Burundi, Ivory Coast).

The origins of these conflicts are often grounded in ethnic and religious differences and their consequences can readily shape and influence how people handle information flows (e.g., withholding information, sharing partial bits of information), even long after a conflict has ended.

We have also observed this phenomenon in countries that were governed by communist regimes for many years (e.g., Vietnam). Indeed, the right to express oneself has often been stifled in such countries, sometimes leading to practices of denunciation when rules were not respected.

A related problem concerns corruption. Accounting transparency and accuracy, in many instances, are not as systematic as they should be. Yet, since a shared language, priority objectives and formalized socio-economic strategies tend to emerge several months after an intervention has been implemented, it is possible to collect approximate data that allow for the calculation of *hidden costs*. Generally, it is possible to carry out *extra-accounting* calculations resorting to classic audit techniques, which enable collecting the necessary data to determine the *hourly contribution to margin on variable costs* (*HCMVC*).

Awareness Building: The Impact of Labor on Productivity and Hidden Costs

Compared with labor costs, raw materials account for a significant component of production cost prices, especially in manufacturing and food processing companies. Awareness of this fact is heightened through the calculation of hidden costs. For example, in a factory producing industrial beverages, the cost of raw materials was considerably higher than the cost of labor. This induced high *over-consumption* hidden costs, caused by stocks of edible products being damaged by insects and rats due to poor storage conditions, while labor represented only a small portion of the production costs of beverages produced.

The *HCMVC* and the *hourly contribution to value-added on variable costs* (*HCVAVC*) are also useful strategic indicators that are easily calculated and can be used to measure *productivity* trends within organizations. On every occasion where it has been possible to calculate this indicator, it has contributed to management awareness as we have been able to observe less wastage in terms of raw materials and working hours. An example, encountered in Central Africa, illustrates how deficient electrical supply forces the cutting of wooden boards destined to manufacturing furniture to be done with hand saws. Thus, the absence of electricity does not incite artisan enterprises to invest in the purchase of electric saws, more productive and less tiring for employees.

In some developing countries—especially those in Africa—basic energy supplies (electricity, gas and water) are often unreliable, which further contribute to low levels of productivity. Key organizational assets such as vehicles, industrial equipment and office equipment are often obsolete or have been salvaged by the industrialized donor countries. Computerized networks and data processing systems are few in number and poorly structured, compounded by a lack of trained information technology staff and instructors. Industries therefore generate relatively little *value-added*, require few skills and are deprived of long-term development leverage.

Using an appropriate, participative approach, *SEAM* can help to improve this situation by better economizing (e.g., *efficiency* and *effectiveness*, sources of competitiveness) the resources that are available. It also improves the use of those resources through greater profitability and *productivity* while stimulating an awareness of relevant investment-making decisions through the use of *economic balances*. These balances are determined by taking into account the basic-equipment investments and the subsequent hidden costs reductions. In the cited examples, better stock management, together with pest control using appropriate insecticides and raticides, proved that these low-cost actions immediately contributed to a substantial reduction of hidden costs.

The Attraction of Skills as a Vector for Growth

The need for training remains very high in the SEAM interventions conducted to date (Beck, 1980; Savall & Beck, 1980). Illiteracy and innumeracy are widespread and *personnel turnover* rates are very high, due in part to low labor costs and the lack of formal qualifications. Management personnel, often trained in the former colonial countries responsible for their political and economic domination, have little inclination to return to their native countries, resulting in a brain drain from the very countries that have the greatest need (Bonnet, Agnese & Pegourie, 1991; Moulette, 2002).

The SEAM approach enables the implementation of a range of socio-economic tools. For example, in implementing socio-economic tools in Vietnamese enterprises, we can stress that the *resolution chart* has, from the start of the intervention, paved the way for all projects conducted by the enterprise. It facilitated, in very concrete manner, setting up (in order of importance and ease of implementation) the competency grid, *behavior grid, priority action plan (PAP), internal-external strategic action plan (IESAP), strategic piloting logbook, time management, resolution chart,* and *periodically negotiable activity contract (PNAC)* in enterprises in collaboration with Vietnamese professor-consultants. The latter benefited from prior *personal assistance* during their training session conducted by ISEOR teacher-researchers at the International Training Center of the ILO in Torino, Italy.

Managers, better trained and more efficient, were able to make better use of resources as a result of structured training plans included in the *priority action plan* (*PAP*) and based on the *integrated training manual* (**ITM**). For example, the analysis and audit of *competency grids* enabled easy identification of operations in the production process where actors suffer from an insufficient level of competency. This permitted developing for those operations a well-targeted training plan, grouping together by levels the employees to be trained in the framework of the biannual *priority action plan*. The resulting reduction in hidden costs leads to an improved ability to compete and new trading opportunities in both national and international markets.

CONCLUSION

All the experience gained over a 20-year period shows that SEAM can be successfully implemented in the developing countries of Africa and Southeast Asia. The major problems resulting from the success achieved arise essentially as a result of distance and the fact that guidance is not always sufficiently provided, because there are simply not enough skilled intervener-researchers available. The training of local relay consultants and the formation and *negotiation* of technology transfer contracts should go a long way towards reducing these obstacles.

Many of the political and economic barriers are easily lifted once a climate of trust has been built between the various parties involved. This comes about through better mutual knowledge, which in turn requires a high level of commitment from the intervener-researcher and a will to achieve socio-economic targets requiring a great deal of cooperation with local actors. To facilitate this level of *confidence* between intervener-researchers and local actors, the former should not hesitate to involve themselves in the field, providing work methods through inductive education and through exemplarity. It is by "doing with" the actors and at the same time explaining to them what the intervener-researcher is doing that actors comprehend, step by step, the usefulness of *socio-economic management tools* and methods.

The results gathered have contributed to verifying the central SEAM hypothesis and have been incorporated into the ISEOR Research Center's databases to generate new knowledge and new generic intervention methodologies to help businesses and organizations develop *sustainable global performance*. They thus illustrate the *generic contingency principle* on the community of certain business problems and principles of action, despite very real national and cultural differences.

NOTES

1. See Savall, Zardet, and Bonnet (2000).
2. See, for example, the Association of the Friends of François Perroux & François Perroux Foundation's (2001), "Taming Globalization."

REFERENCES

Aman, A. (2001). *La formation intégrée, méthode de gestion des compétences: Expérimentation dans une administration publique marocaine* [Integrated training and competency management method: Experimentation in a Moroccan public administration]. Unpublished doctoral dissertation, University of Lyon, ISEOR.

Association of the Friends of François Perroux & François Perroux Foundation. (2001). *Taming globalization*. Lyon, France: ISEOR Editions.

Beck, E. (1979). *L'amélioration de l'équilibre formation-emploi par action sur la variable emplois qualifiés* [Improving training-job equilibrium through action on the qualified-jobs variable]. ISEOR Report.

Beck, E. (1980). *Équilibration formation-emploi et changement de structure des qualifications en milieu industriel.* [Training-job equilibrium and changing qualification structures in the industrial milieu]. Unpublished doctoral dissertation, University of Lyon, ISEOR.

Bensalem, R. (1999). *Des incidences des coûts cachés sur le système d'information de comptabilité et leurs répercussions sur les décisions stratégiques et opérationnelles des entreprises et des organisations: Cas d'expérimentations.* [The incidence of hidden costs on the accounting information system and its repercussion on strategic and operational decisions in businesses and organizations: Cases of experimentation]. Unpublished doctoral dissertation. University of Lyon, ISEOR.

Bonnet, M., Agnese, V., & Pegourie, S. (1991). *Management socio-économique des actions de lutte contre l'illettrisme au niveau des collectivités territoriales. Expérimentation dans une agglomération et élaboration de grilles de compétences.* [Socio-economic management actions in the fight against illiteracy at local government levels and the development of competency grid]. Research report for the Permanent Working Group on Illiteracy, with financial support from the French Ministry of Labor, Employment and Professional Training.

Boussofara, S. (1993). *Acquisition et maîtrise technologique dans les entreprises tunisienne* [Technological acquisition and mastery in Tunisian enterprises]. Unpublished doctoral dissertation, University of Lyon, ISEOR.

Buoy, V. (2001). *L'amélioration de la compétitivité par action sur les variables socio-économiques: Cas d'une entreprise agro-alimentaire khmère* [Improving **competitiveness** through action on socio-economic variables: The case of a Khmer agri-food enterprise]. Unpublished master's thesis, University of Lyon, ISEOR.

Clare, B., Thiébaut, D., Albert, E. & Baudu, L. (2004). *Atlas économique et politique mondial* [World economic and political atlas]. Paris: Le Nouvel Observateur.

Defdouf, A. (1995). *Contribution du Management socio-économique à l'amélioration du pilotage des entreprises publiques algériennes* [Contribution of socio-economic management to improving the management of Algerian government enterprises]. Unpublished master's thesis, University of Lyon, ISEOR.

Dinh, T.M. (2000). *Amélioration de la qualité du fonctionnement interne-externe des entreprises étatiques vietnamiennes: Cas d'expérimentation* [Improving internal/external operational quality in Vietnamese state enterprises: cases of experimentation]. Unpublished master's thesis, University of Lyon, ISEOR.

Frika, M. (2000). *Management socio-économique et développement de l'ingénierie du marketing—essai de modélisation: Cas d'une compagnie aérienne tunisienne* [Socio-economic management and market engineering development—modelization experiment: The case of a Tunisian airlines company]. Master's thesis, Université de Lyon, ISEOR.

Hor, S. (2000). *Amélioration des modes de gestion du marketing d'une entreprise agro-alimentaire cambodgienne* [Improving the modes of marketing management in a Cambodian agri-food enterprise]. Unpublished master's thesis, University of Lyon, ISEOR.

International Labor Organization. (1998). *Le conseil en management: Guide pour la profession* [Management consulting: A guide for professionals] (3rd ed.). Geneva, Switzerland: ILO Editions.

Kammoun, H. (1998). *Équilibration du rôle des cadres : un essai de modélisation pour une conduite du changement et du développement de l'efficience de fonctionnement des PME (Tunisia)* [Balancing the role of managers: Modelization experiment for change management and the development of SME operational efficiency (Tunisia)]. Unpublished doctoral dissertation, University of Lyon, ISEOR.

Lahlali, M. (2001). *Le management des organismes de formation professionnelle au service du pilotage des systèmes de formation. Cas d'expérimentation au Maroc* [The management of professional training organizations dedicated to piloting training systems]. Unpublished doctoral dissertation, University of Lyon, ISEOR.

Mafra, A., Hidalgo, P., Eraers, R., & Beck, E. (1999). *La promotion du conseil à l'international: Transfert de savoir-faire en management socio-économique au Vietnam.* [Promoting international consulting: Socio-economic management knowledge transfer in Viet Nam]. Paper presented at the symposium *Les conseils aux entreprises*, organized by ISEOR, Editions Economica.

Moulette, P. (2002). *Contribution à la gestion de la remédiation aux situations d'illettrisme en entreprise: Cas d'expérimentation* [Contribution to the management and remedy of illiteracy situations in the enterprise: Cases of experimentation]. Unpublished doctoral dissertation, University of Lyon, ISEOR.

M'Rabet, M. (1997). *L'implication modulaire de l'individu dans les processus stratégiques –expérimentation dans les entreprises et les organisations tunisiennes* [Modular involvement of the individual in strategic processes: Experimentation in Tunisian enterprises and organizations]. Unpublished doctoral dissertation, University of Lyon, ISEOR.

N'Kanagu, G. (1996). *Contribution de la méthode de formation intégrée à l'amélioration des performances des entreprises de l'Afrique Centrale: cas d'intervention socio-économique dans les entreprises du Burundi* [Contribution of an integrated training method for improving the performance of Central African enterprises]. Unpublished doctoral dissertation, University of Lyon, ISEOR.

N'Kanagu, G. (1992). *Diagnostic socio-économique dans une brasserie du Burundi. La grille de compétence et la formation intégrée: Application et limites* [Socio-economic diagnostic in a Burundian restaurant. Competency grid and integrated training: Applications and limitations]. Unpublished master's thesis, University of Lyon, ISEOR.

Savall, H. (1974, 1975). *Enrichir le travail humain dans les entreprises et les organisations* [Work and people: An economic evaluation of job enrichment]. Paris: Dunod.

Savall, H. (1979). *Reconstruire l'entreprise: Analyse socio-économique des conditions de travail* [Reconstructing the enterprise: Socio-economic analysis of working conditions]. Paris: Dunod.

Savall, H. (1987). Les coûts cachés et l'analyse socio-économique des organisations [Hidden costs and the socio-economic analysis of organizations]. *Encyclopédie du management* (pp. 599-628). Paris: Economica.

Savall, H. (1989). Point de vue: Professeur-consultant. Le bilan d'une expérience [Point of view: Professor-consultant, summing up experience]. *Revue française de gestion, 76*, 93-105.

Savall, H. (2003a). An updated presentation of the socio-economic management model. *Journal of Organizational Change Management, 16*(1), 33-48.

Savall, H. (2003b). International dissemination of the socio-economic model. *Journal of Organizational Change Management, 16*(1), 107-115.

Savall, H., & Beck, E. (1980). *Méthode d'expérimentation d'actions de restructuration des emplois avec formation intégrée* [Experimentation method for job redefinition actions using integrated training]. Research report, ISEOR.

Savall, H., & Zardet, V. (1987). *Maîtriser les coûts et les performances cachés: Le contrat d'activité périodiquement négociable* [Mastering hidden costs and performances: The periodically negotiable activity contract]. Paris: Economica

Savall, H., & Zardet, V. (1992). *Le nouveau contrôle de gestion: Méthode des coûts-performances cachés* [New management control: The hidden cost-performance method]. Paris: Éditions Comptables Malesherbes-Eyrolles.

Savall, H., & Zardet, V. (1995). *Ingénierie stratégique du roseau, souple et enracinée* [Strategic engineering of the reed, flexible and rooted]. Paris: Economica.

Savall, H., & Zardet, V. (2004). *Recherche en sciences de gestion: Approche qualimétrique. Observer l'objet complexe.* Unpublished English translation: *Research in management sciences: The qualimetric approach. Observing the complex object.* Paris: Economica.

Savall, H., & Zardet, V. (2005). *Tétranormalisation: Défis et dynamiques* [Competitive challenges and dynamics of tetra-normalization]. Paris: Economica.

Savall, H., Zardet, V., & Bonnet, M. (2000). *Libérer les performances cachées des entreprises par un management socio-économique* [Releasing the untapped potential of enterprises through socio-economic management]. Geneva, Switzerland: International Labor Office-ISEOR.

Savall, H., Bonnet, M., Beck, E., & Yamuremye, H. (2001). Formation des employeurs au management socio-économique dans les pays en voie de développement [Training employers in socio-economic management in developing countries]. *Proceedings: Recherche-intervention et création d'entreprises—Accompagnement et évaluation*(pp. 207-214). Paris: Economica.

Tamo, K. (1998). *Proposition d'un modèle interprétatif du développement humain susceptible d'améliorer les performances des organisations africaines: Application à l'administration publique angolaise* [Proposal of an interpretive model of human development capable of improving performances in African organizations: Application to Angolan public administration]. Unpublished doctoral dissertation, University of Lyon, ISEOR.

Yamuremye, H. (2001). *Contribution de pratiques de délégation concertée à l'amélioration du fonctionnement des équipes de direction: Cas d'expérimentations innovantes dans les PME burundaises* [Contribution to delegated cooperative practices to improve management team operations: Cases of innovative experimentation in Burundian SME]. Unpublished doctoral dissertation, University of Lyon, ISEOR.

CHAPTER 9

SMALL BUSINESS CONSULTING IN NEW MEXICO

The Theatre of
Socio-Economic Intervention-Research

David M. Boje, Mark E. Hillon, and Yue Cai

New Mexico State University has a Memorandum of Understanding with Jean Moulin/Lyon 3 University and ISEOR to provide *socio-economic management (SEAM)* training to students in our small business consulting program. Over the past 6 years, our undergraduate business and MBA students have completed more than 85 consulting projects focused on improving the outlook for small business survival in our state. To offer some perspective, there were more than 41,500 small businesses with fewer than 500 employees in New Mexico in 2005. Adding self-employed entrepreneurs into this category raises the number of small businesses in the state to approximately 153,800 (U.S. Small Business Administration, 2006). Nationally, just over 50% of the U.S. workforce is employed by small businesses with fewer than 500 workers and 99.7 % of U.S. businesses with a payroll have fewer than 500 employees. These businesses account for slightly more than 45% of the total U.S. payroll (U.S. Census

Socio-Economic Intervention in Organizations: The Intervener-Researcher and the SEAM Approach to Organizational Analysis, pp. 215–227
Copyright © 2007 by Information Age Publishing
215

**Table 9.1.   Examples of
New Mexico Businesses Involved in *SEAM Interventions***

Retail Sales and Service	Skilled and Professional Services	Custom/Light Manufacturing	Not-for-Profit
• Auto sales & parts	• Landscaping	• Motorcycles	• Educational pro-
• Purified water	• Healthcare	• Clothing & home	grams
• Day care	• Upholstery	Décor	• Arts foundation
• Car Rental	• Welding/machin-	• Furniture	• Academic publica-
• Photography	ing	• Computer hard-	tions
• Flowers	• HVAC	ware	
• Wine	• Engineering		
• Restaurants	• Accounting		
• Comic books	• Financial planning		
	• Insurance		
	• Real estate		

Bureau, 2002). The theatrical SEAM methodology thrives in a small business context because it is often possible to interview all actors, penetrating to the psychological core, rather than just a superficially representative sample of informants. As Table 9.1 indicates, our experience over the past 6 years shows that **SEAM intervention-research** (Savall & Zardet, 2004) can be applied effectively over a wide range of industries.

SOCIO-ECONOMIC ORGANIZATIONAL THEATRE
IN SMALL BUSINESS CONSULTING

"Organization *is **theatre**,*" says Henri Savall, the founder and director of ISEOR's *socio-economic approach to management* (Boje & Rosile, 2003b, p. 21). We have emphasized the theatrical side of SEAM in our New Mexico consulting and have found that small businesses as theatres are filled with actors, scripted **dysfunctions**, and hidden potential. Drawing on the **SEAM** process, this theatrical framework and its accompanying qualitative data reflect a concerted attempt to observe and understand the **metascript** of the organization, which is defined as the "multiplicity of contending and fragmented [organizational] scripts" (Boje & Rosile, 2003b, p. 23). The metascript evolves and emerges in a dynamic and continuous process of scripting and attempts at re-scripting. Over time, organizational scriptwriters and editors can be quite prolific, thereby producing a fragmented, overlayering effect as scenes are performed on multiple stages (Boje & Rosile, 2003a).

One of the major differences between **SEAM** and other management consulting approaches is that it requires a thorough contextual understanding of relationships among all actors and the detailed staging and scripting of organizational performances (Boje, 2000). Consultants are not the solutions to problems; rather, they are guides in channeling the clients to discover the roots of their own problems and reflecting on possible answers (Buono, 2003; Savall, Zardet, Bonnet & Moore, 2001). If the process is carried out correctly, the *root causes* of the symptoms will emerge as each actor reveals his or her connection to the metascript. This realization comes through a *mirror-effect* performance by the consultants, a replay of collected scenes delivered in the organizational actors' own words (Bonnet & Cristallini, 2003).

A typical semester may involve as many as 10 consulting teams working with 10 different types of small businesses. The *socio-economic dysfunctions* which constitute barriers to efficient use of material and human resources (Savall & Zardet, 2004, 2005) that we find most often in New Mexico small businesses involve two main areas. First, entrepreneurs rarely stop to acquire the basic business training that they will need to manage their day-to-day operations. Most lack business plans, accounting systems, reliable procedures, inventory controls, and management communication skills. The second area is quite different from the first, but it intensifies the dysfunctions due to these deficiencies in business skills. Many of the small businesses that have asked for our consulting services are family owned, a management *theatre* with its own unique set of *dysfunctions*. Distrust of nonfamily employees restricts growth, market expansion, and the release of *human potential* (Hillon, 2006).. Sibling rivalries and a paternal style of management reduce communication flows, productivity, and commitment to firm success. Overall, taboo topics such as these find their way into many of the organizational scripts that structure the relations and performances of actors in small family-owned businesses. For example, conservative estimates over the past 6 years of *hidden costs* and unrealized potential due to these two major areas of common dysfunctions have ranged from $3,000 for small restaurants to more than $500,000 per year for an engineering firm.

Metascript Mirror-Effect Theatre

Figure 9.1 illustrates our theatric view of the SEAM methodology. Instead of directing formal mirror-effect presentations, our consultants-in-training prepare theatric role plays of the *dysfunction* stories that they have collected from interviews and observations. This performance makes explicit the reflexive nature of the learning process in two ways. First, the tacit assumption of rationality in business processes and decision making runs counter to the practical reality that "Organization *is theatre*." Savall

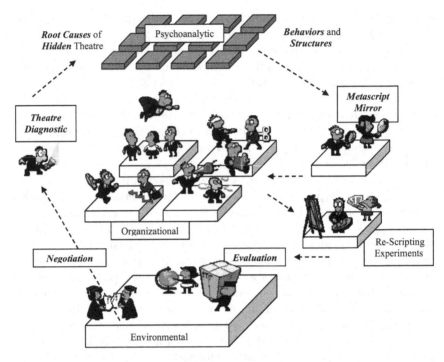

Figure 9.1.   The *socio-economic approach* to organizational *theatre*.

was not speaking metaphorically when he offered this illustration. Thus, a dramatic reenactment of some of the dysfunctions observed in the *diagnosis* helps to firmly connect the underlying scripts to the performances. Second, although our consultants work as a team, most of their interviews are conducted one-on-one with the managers and employees of the business. Therefore, the voices were heard at different times, by different people, with perhaps different themes. As a result, it is often difficult to sort out the polyphonic voices and dysfunction themes to discover how they are connected to the *root causes* of problems. Reflection through the *mirror-effect theatre* helps the consultants to immerse themselves in the context of the client's metascript, thereby feeling the interaction of the dysfunctions compiled and described in direct quotes from the original actors. Gaps in meaning or interpretation of events become apparent when this collection of quotes is reanimated and restaged. Through this public reflection and interaction with spectators, the consultants often gain a more profound understanding of their client's business, and also discover areas in need of more study. Three examples are offered to help illustrate these learning points.

Health Support Service Agency

Imagine the following brief scenes from the *mirror-effect* theatre of a small business consulting intervention. These scenes were collected from the actors in a not-for-profit health support service agency:

- there was a lady who had worked here since the beginning ... no one knew what she did, but when she went on maternity leave, everything was covered without any problems. When she came back, there was nothing for her to do ... she resigned in December.
- [the glass wall] "creates a gap between the employees and clients."
- "I feel pressured, but I am not sure if it is me or the agency."
- A client noted, "I feel that they should let them [the clients] know that the money will run out and help us learn how to live on our own."

The student-consultants dramatized a scene with the glass wall as a representation of the many barriers between managers, employees, and clients of the agency. A former director had put up a glass wall out of fear to protect her employees from the mistaken threat of contagions from clients. This wall was the visual tip of the iceberg for many employees, but quite a few other barriers kept the agency from serving its clients. The three employee *fieldnote quotes* above point to inadequately assigned responsibilities, pressures in *time management*, and a palpable feeling of distance between them and their clients. High turnover, poor communication, disorganized office space, and a chronic lack of resources were also mentioned in interviews with the employees (Macías & Macías, 2001). The combined effect was a helpless sense of dissociation, frustration, and doubt about the agency's survival. Without such a deep psychological understanding of these actors, any intervention to improve performance would likely only address the symptoms or surface manifestations of the organization's hidden *dysfunctions*.

Restaurant

Another set of *mirror-effect* scenes reflected the underlying *dysfunctions* in a family-owned and operated (mother and son) restaurant identified specific fragments of the *metascript*:

- Director with no formal business training: "I am attending the branch college and studying restaurant hospitality."
- Mission: "No, I don't have anything."
- Duties and responsibilities: "We have no specific training for employees."

- Job performance: "I don't tell them [employees] about job performance. I used to have meetings but not anymore."
- Inventory: "I just go back and look and if we are low we order some more."
- Employee: "He [the manager] doesn't know how to talk to people, he corrects you on the spot and makes it hard to work.... Getting yelled at when I don't do something right."
- Customers: "I have been coming here for years, I love it" or "I will never come back, the service was horrible."

The student-consultants dramatized a collection of *dysfunctions* that resulted in dissatisfied customers. Menu items were often either not available or were of inconsistent quality due to experimental substitutions of ingredients, both due to the lack of an inventory control system. Inadequate training and poor job performance feedback for servers meant that customers were polarized into a love/hate dichotomy. The difficulty in attracting repeat business was a clear indication of serious dysfunctions. Although humorous, the dramatized scene of customers returning regularly just to be neglected clearly reflected clear connections between meta-script, dysfunctions, and organizational performance in the *mirror-effect theatre*.

Hidden Cost Theatre in a Small Furniture Factory

As a more detailed example, this section will look into the theatre of a small furniture business, owned and managed by one of the sons of the founder. Located near the U.S.-Mexico border, this family-owned small business had been manufacturing wood furniture for more than 15 years. It had approximately 50 production workers and five managers (general, production, marketing, purchasing, and accounting). At the first meeting with the consultants, the managing owner of the company seemed to be happy that his business would receive the objective scrutiny of an *external consultant*'s evaluation. However, he emphasized in the discussion that there were "no problems at all" in the business.

Two student consultants conducted 12 visits to the company to interview managers and employees. They found *dysfunctions* throughout the company's production, administrative and employee service areas (break room and canteen). The production employees—who were mostly young, illiterate, and poor—were working in a hazardous environment. The production shop was dirty, the air was dusty, and temperatures were either extremely hot in the summer or extremely cold in the winter. The ventilation system was also poor, a highly significant factor in the work environment because the production of furniture produces unhealthy dust and fumes from wood cutting, sanding, and painting. Employees reported that they were unhappy and uncomfortable with the *working conditions*. Some stated that they often felt sleepy, the painters reported that they had to take breaks from time to time in order to breathe fresh air, and others said that throughout the day they needed to eat something in order to keep from fainting. The equipment used in production was old, in bad condition, and was not properly maintained. As a result, most machines broke down from time to time causing down time, delays, and scrap in production. Additionally, the low quality of materials often resulted in rework and unpaid overtime.

In the administrative area, the owner/general manager had little trust in his managers and production employees and he was constantly checking on the workers personally to determine whether they were "working hard." He also monitored the work of the production manager and helped the workers fix equipment whenever it broke down. All of these unnecessary activities took time from his daily schedule, time that could have been better spent on concern for the future of the company and on developing strategies to reduce its financial insecurity.

In the next instance, the purchasing manager carried out functions that could easily have been done by an assistant, such as driving to suppliers to make sure that orders were correct in quantity, type and quality. These trips required several hours in transit as well as additional time spent waiting to receive the merchandise. The purchasing manager reported that he was unhappy with this situation and that the excessive time wasted in picking up supplies left insufficient time for him to develop a purchasing strategy that would reduce the constant problems in quality and material specification. Relations with suppliers were also in need of attention, as the inconsistent but often low quality of production materials resulted in an 80% postsale rework level, customer complaints, and follow-up service calls.

The production and accounting managers did not seem to have any complaints. However, they had only been employed by the company 2 months prior to the consultants' arrival. Finally, the marketing manager (the owner/general manager's brother) was almost never present because he worked in a small showroom located far from the company's production workshop. His expense account seemed to serve as a proxy for direct surveillance and was added to the long list of distractions that constantly drew the owner/general manager's attention away from effective long-term management of the business. Overall, there was no strategic thinking among any of the managers and the company lacked a vision of a desirable future direction. In the words of the owner, "we are concentrated on surviving."

As the consultants began to analyze their interview transcripts, more questions and insufficient answers about the *metatheatre* directed their inquiry toward the psychological level—the source of hidden theatre. The collected story fragments can be organized into a *root cause* chart, a visual interpretation of the *SEAM method*'s grouping of related *dysfunctions* into *baskets* (Savall & Zardet, 1987). The resulting causal story chart seemed to indicate that many of the company's problems originated from the failure to create and implement a strategic plan to enable the employees to produce furniture of a consistently high quality to meet customer needs and expectations. Material specifications and processes were not recorded or standardized and employee training was nonexistent, therefore success was haphazard and difficult to duplicate.

However, if the *root cause* was so obvious to everyone, then why hadn't anything been done about it already? The company was not new or inexperienced, for they had maintained the same size workforce and production levels for more than 15 years. Their sales were also healthy enough to financially support an extensive strategic quality initiative, and the lack of competition in the local custom-designed furniture market kept the factory working at full capacity.

At this juncture, the consultants decided to approach the company from a psychoanalytic perspective to analyze the backstage of the *theatre*. They decided to examine the *metascript*, to determine the individual roles and motivations of the actors, and to calculate the effects of these factors on the behaviors and *structures* at work in the company. The first thing they noticed was that the owner/general manager was not as open to suggestions as he appeared to be. He avoided meeting with the team, but whenever he did talk with them, he controlled the opening of the conversation with phrases like "any problem you find, I already know," or "any solution you propose, I already know or have tried it." The hidden theatre investigation revealed taboo or unspoken situations that financially affected the company, such as the simmering feuds between the male members of the family. Not only were the owner/general manager and his brother the marketing manager constantly fighting over money (they had no fixed salaries) and the pride of being the bigger workaholic, they also had three other brothers in addition to their father, the founder of the business. These three brothers and their father had been forced out of the company some years ago, causing a great deal of tension within the family.

Feuding aside, the most significant barrier to change was that the owner did not want to manage or even to be part of the company. He revealed to the consultants that his true aspiration was to earn a graduate degree in psychology, a dream that had been thwarted by obligations to the family business. Likewise, his brother the marketing manager also had the dream of returning to school. Neither of them liked the business, but they kept running it because they had no other income, no other choice to provide for their families. Their wives did not work, both men were highly interested in the education and development of their teenage children, and their status as upwardly-mobile middle class families was costly to maintain. Lastly, the other three brothers and their father all continued to collect monthly paychecks for "not being there," so, the two brothers were burdened with providing a living for the entire extended family.

The psychological hidden *theatre diagnostic* showed that both brothers suffered from addictions, the most visible of these was a constant paranoia of being robbed or hustled by everyone around them. This feeling was evidenced in the owner/general manager's behavior of constantly checking up on the employees, the other managers, and even his brother's expense account. The paranoia also extended to external advisors or consultants, due to bad experiences in the past.

Frustrated hidden desires for different lives, counterproductive workaholism to support the extended family, and fear and paranoia in top management conspired to socially construct the *metascript*. First, the owner/general manager's lack of interest in the company prevented him from developing viable strategies for a better future. Second, his lack of trust in everyone hindered the production and purchasing managers' creative thinking and the desire to change. Third, his actions contradicted his words that "the workers are the most important thing in the company." He made them work in a hazardous environment, he did not give them health or retirement benefits, he failed to pay them overtime, he did not give them a decent place to have their breaks and meals (workers had to sit and eat on the floor), and he did not listen to their petitions to improve conditions.

The employees, in contrast, had a very different story to tell. Their role in life was surviving and fulfilling the basic needs of their families, but the organizational *theatre* in which they were trapped was not poised for the long run and the scripting was beyond their control. From the workers' perspective, the vision was to find a better job, one with a safe environment, better salary, and better hours. Although both the owner's and workers' psychological needs authored the same alternative script of escape to compete with the dominant scripts of *survival*, there was no possibility for an agreement between the two long-term visions. Thus, the consultants had discovered the true nature of the psychological barriers to change that had doomed all previous interventions to failure. Rescripting to correct the company's *socio-economic dysfunctions* posed a monumental task, but the task was finally moved into the realm of possibility. Psychological reflection from the consultants' research validated one of the fundamental tenets of *socio-economic management*, that work and life outside of work are inseparable in the minds of both worker and

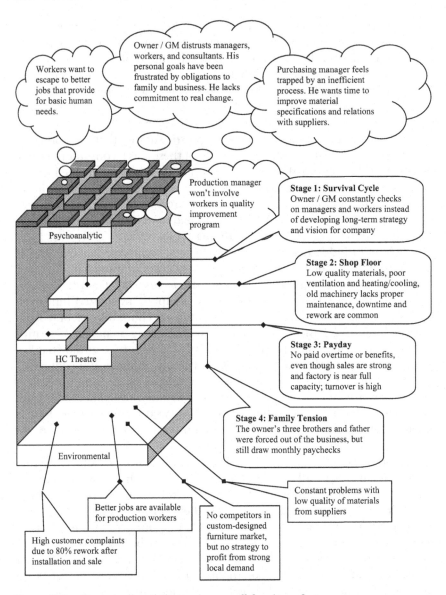

Workers want to escape to better jobs that provide for basic human needs.

Owner / GM distrusts managers, workers, and consultants. His personal goals have been frustrated by obligations to family and business. He lacks commitment to real change.

Purchasing manager feels trapped by an inefficient process. He wants time to improve material specifications and relations with suppliers.

Psychoanalytic

Production manager won't involve workers in quality improvement program

Stage 1: Survival Cycle
Owner / GM constantly checks on managers and workers instead of developing long-term strategy and vision for company

Stage 2: Shop Floor
Low quality materials, poor ventilation and heating/cooling, old machinery lacks proper maintenance, downtime and rework are common

HC Theatre

Stage 3: Payday
No paid overtime or benefits, even though sales are strong and factory is near full capacity; turnover is high

Stage 4: Family Tension
The owner's three brothers and father were forced out of the business, but still draw monthly paychecks

Environmental

Better jobs are available for production workers

No competitors in custom-designed furniture market, but no strategy to profit from strong local demand

Constant problems with low quality of materials from suppliers

High customer complaints due to 80% rework after installation and sale

Figure 9.2.   Organizational *theatre* in a small furniture factory.

manager (Savall, 1974, 1975, 2003a, 2003b; Savall & Zardet, 2005). Rescripting would require a realignment of both workers' and owner's visions, thereby providing the dreamed of better life out of the organization that actively denied its realization. Figure 9.2 summarily illustrates this example of *SEAM* organizational *theatre*.

The difficulties of mirroring and rescripting are, of course, readily transparent, especially when the psychological barriers are firmly entrenched against obviating the context of *root causes* to clients so that change may proceed. As an example, in the fall of 2002, we were approached by a small geotechnical engineering firm for assistance. The owner/manager, the firm's only professional engineer (PE), had 12 employees and the business had been successful over its 20-year history largely due to the lack of any real local competition. In essence, the local business environment had helped them to remain profitable in spite of dysfunctional *socio-economic performance*.

When we arrived for our first meeting to discuss the consulting program and to negotiate assess for our team of student consultants, the owner/manager PE was engaged in office/secretarial work. During this first conversation, he noted that he did not have an office manager and consequently, too much of his time was spent running the office instead of on revenue generating engineering work. Our team confirmed from time studies that he spent approximately 2.5 hours per day on nonengineering tasks. This *hidden cost* alone amounted to $71,500 per year—more than twice the local average salary for a competent office manager. However, a number of more significant hidden costs accompanied this instance of mismatched compensation. The owner/manager's 2.5 hours of office work per day was insufficient to process contract paperwork, schedule lab workers, and to respond to all potential customer phone calls. Additionally, the firm had been in business long enough to have a reliable estimate of the revenue generated by each client. Thus, each missed phone call, each delayed contract, and each unscheduled work order could easily be converted into a quantitative dollar figure representing unrealized potential. By conservative estimates, the total cost of hidden theatre to the firm was shown to be greater than $503,000 per year. However, when our calculations were greeted with irrational disbelief, we realized that the ultimate barrier to change in the *metascript* of the organization had a familiar theme, one encountered in many of our *small business* consulting *interventions*. The PE had not hired an office manager because of his own inability to trust another to run the day-to-day operations of his business. Reflection of this *root cause* was essential before he could move forward toward a solution.

CONCLUSION

Improving organizational performance is a difficult proposition even when the *root causes of dysfunctions* can be clearly traced back through a multitude of scripts, actors, and scenes. Consultants often have preoccupations with quick action based upon "objective" performance measures such as financial statements and technology utilization. Yet, genuine change depends on the ***intervener-researcher*** delving into the psychological context that not only determines how a business functions, but more importantly, that maintains the dysfunctional patterns of the organizational culture (Péron & Péron, 2003).

A ***socio-economic diagnostic*** follows a relatively straightforward process and produces generally indisputable findings supported by a mix of ***qualitative and quantitative*** data. Yet, it is the pivotal reflection of the ***mirror-effect*** that opens the door to ***conversion of hidden costs into value-added*** and realization of untapped potential. This performance of collective reflection—as an actual series of reenacted dramas—can help both consultant and client to access the deeper recesses of organizational memory. Through six years of experience, we have found the ***theatre of SEAM*** to be quite effective in exploring the psychological and ***metatheatrical*** contexts of ***intervention-research*** that must be understood in order for organizational change to proceed.

REFERENCES

Boje, D. (2000). Socio-economic analysis of management: An export from France. In *Transorganizational Development Gameboard*. Available: http://web.nmsu.edu/~dboje/TDseam.html

Boje, D., & Rosile, G. A. (2003a). Comparison of socio-economic and other transorganizational development methods. *Journal of Organizational Change Management, 16*(1), 10-20.

Boje, D., & Rosile, G. (2003b). Theatrics of SEAM. *Journal of Organizational Change Management, 16*, 21-32.

Bonnet, M., & Cristallini, V. (2003). Enhancing the efficiency of networks in an urban area through socio-economic interventions. *Journal of Organizational Change Management, 16*, 72-82.

Buono, A.F. (2003). SEAM-less post-merger integration strategies: A cause for concern. *Journal of Organizational Change Management, 16*(1), 90-98.

Hillon, M. (2006). *A comparative analysis of socio-ecological and socio-economic strategic change methodologies*. Doctoral dissertation in management sciences, University Jean Moulin Lyon 3, France & New Mexico State University, USA.

Macías, E. & Macías, E. (2001). *Cocinas Integrales de la Sierra* [Sierra Custombuilt Kitchens]. New Mexico State University Small Business Consulting Service Report. Las Cruces, NM.

Péron, M., & Péron, M. (2003). Postmodernism and the socio-economic approach to organizations. *Organizational Change Management, 16*(1), 49-55.

Savall, H. (1974, 1975). *Enrichir le travail humain dans les entreprises et les organisations* [Work and people: An economic evaluation of job enrichment]. Paris: Dunod.

Savall, H. (2003a). An updated presentation of the socio-economic management model. *Journal of Organizational Change Management, 16*(1), 33-48.

Savall, H. (2003b). International dissemination of the socio-economic model. *Journal of Organizational Change Management, 16*(1), 107-115.

Savall, H., & Zardet, V. (1987). *Maîtriser les coûts et les performances cachés: Le contrat d'activité périodiquement négociable* [Mastering hidden costs and performances: The periodically negotiable activity contract]. Paris: Economica.

Savall, H., & Zardet, V. (2004). *Recherche en sciences de gestion: Approche qualimétrique. Observer l'objet complexe.* Unpublished English translation: Research in management sciences: The qualimetric approach. Observing the complex object. Paris: Economica.

Savall, H., & Zardet, V. (2005). *Tétranormalisation: Défis et dynamiques* [Competitive challenges and dynamics of tetra-normalization]. Paris: Economica.

Savall, H., Zardet, V., & Bonnet, M. (2000). *Releasing the untapped potential of enterprises through socio-economic management.* Geneva, Switzerland: International Labor Office-ISEOR.

Savall, H., Zardet, V., Bonnet, M., & Moore, R. (2001). A system-wide, integrated methodology for intervening in organizations: The ISEOR approach. In A. F. Buono (Ed.), *Current trends in management consulting* (pp. 105-125). Greenwich, CT: Information Age.

U.S. Census Bureau. (2002). Employment size of employer and nonemployer firms, Washington DC: Author. Retrieved November 10, 2006, from http://www.census.gov/epcd/www/smallbus.html

U.S. Small Business Administration. (2006). *Small business profiles for the states and territories.* Springfield, VA: Author. Retrieved November 10, 2006, from http://www.sba.gov/advo/research/profiles/

CHAPTER 10

BRINGING SOCIO-ECONOMIC INTERVENTION TO THE UNITED STATES

The Able Plastics Case

Randall Hayes, Lawrence Lepisto, and Debra McGilsky

The chapter is based on the first consulting engagement performed by the collective faculty of the Institute for Management Consulting (IMC) at Central Michigan University (CMU). The IMC was developed to administer the management consulting concentration in the master of business administration (MBA) program that employees the *SEAM* methodology and philosophy. Prior to launching the new program, the CMU faculty team had received training in the SEAM methodology (Savall, 1974, 1975, 2003a, 2003b) from the Institute for Socio-Economic Research in Organizations (ISEOR) at the University of Lyon.

A CMU consultant had earlier worked with the Able Plastics Company, which subsequently indicated its interest in working with the CMU consulting faculty using the *SEAM* methodology. Able Plastics wanted to expand its operations and realized that the company would have to

Socio-Economic Intervention in Organizations: The Intervener-Researcher and the SEAM Approach to Organizational Analysis, pp. 229–250
Copyright © 2007 by Information Age Publishing

change the way it was managed if it was to be successful in its expansion plans. The management team at Able Plastics felt that a **SEAM intervention** would be an ideal mechanism to provide them with a thorough assessment of the company by outside consultants. In turn, the intervention would also provide the IMC faculty with an opportunity to systematically apply the SEAM methodology.

The CMU consulting faculty embraced the SEAM approach for several reasons: it has a sound theoretical and empirical foundation; it approaches the client with no preconceived expectations; and it requires the client to participate in the engagement, thereby increasing "buy-in" by the client and a greater likelihood of implementing long-term change. However, there are likely several reasons why consultants in the United States may utilize only parts of the SEAM approach: it does not promise quick findings, which is not appealing to U.S. firms; it does not promise easy formulae for the client that will have an immediate effect; and clients often do not even thinking about long-term change (Lepisto, Hayes, McGilsky, Love, & Bahaee, 2005).

ABLE PLASTICS: BACKGROUND

Able Plastics is a manufacturer of shipping containers for the auto industry located in the small town of Roscommon, Michigan. The town is in central Michigan, and Saginaw, the nearest metropolitan area, is about 30 miles south along Interstate 75. The area is characterized by small manufacturing companies and tourist firms serving local hunting and fishing activities. Other than a state prison, no large employers exist in the area, and the Michigan Economic Development Corporation (a state organization that assists businesses) considers Roscommon County a "special-needs" area. This designation makes companies located in the county eligible for special economic development assistance and training programs provided by the State of Michigan.

Rob Stewart, his brother Larry, and Carl Browning, a recently retired engineer from a large chemical manufacturer, started the company. The three individuals comprise the board of directors of the company. Rob and Carl work exclusively for Able Plastics, while Larry has a full-time job in the insurance industry. Rob previously worked at Resolution Plastics, another shipping container firm, as an engineer working in production and sales. He left Resolution along with four other employees to form Able Plastics in 1996. Over the initial 3 years of existence, the company's sales grew to about $4 million but revenue growth leveled out in 1999 and has remained relatively flat for the last 2 years.

Products and Strategy

The company's strategy was to provide high-quality, custom-made shipping containers to the automotive industry. Its major customers, General Motors, Chrysler, Ford, and Delphi, accounted for about three quarters of total sales. Other than occasional shipments to Canada, Able Plastics does not have any international sales.

The containers that the company sells are used to ship large parts and components (e.g., transmissions, engine blocks, fuel lines) between "Big-three" (GM, Daimler/Chrysler, Ford) automotive plants and their suppliers. The products are custom-engineered to fit particular components. The auto companies draw up the specifications for the containers they need and most contracts are sent out for bid. There are three competitors in the Midwest that bid for the contracts, and Able Plastics has a "fulfillment rate" (percentage of successful bids) of about 20%. The average size of contracts varies widely, with the largest about $250,000 although most contracts are for significantly less money.

The company's sales strategy is to be the most responsive supplier to the customer. Able Plastics seeks to work with the logistics specialists within the auto companies to help design containers that provide superior utility. The product characteristics that create utility for a container go beyond just providing shock and moisture protection for the contents. A well-designed container will also exhibit ease of use, that is, it is easy for the worker to open or close, to insert or extract the contents, and to handle the container in shipment. In addition, a good container will have convenient and legible labeling locations, a shape that is conducive to efficient stacking and storing, and easy recycling at the end of its useful life. Able Plastics designers seek to work with the customer to design these features into the container. As a result, the company often secures contracts that are not sent out for bids. Initially the company was successful in this effort, but expense-containment efforts on the part of the auto companies have meant that fewer of these sole-sourced contracts have been available in recent years.

Organization Structure

Figure 10.1 contains the organization chart of the company as of March 2001. The company has about 30 production employees and one production supervisor. Three designers produce many of the designs, and these individuals have some sales and customer-relationship responsibilities. Carl Browning serves as chief engineer and head of quality control. The administrative staff consists of an administrative assistant to Rob

Stewart, and three other individuals who handle the bookkeeping and track production orders and shipments. A local CPA firm, Regent Accounting LLC, provides most accounting functions, and the staff accountant assigned to Able Plastics effectively serves as the chief financial officer (CFO) of the company.

Peterson Manufacturing produces the molds used for the containers. The company is located in Colbert, Michigan, a town about 40 miles from Roscommon. The molds are custom-made out of aluminum to Able Plastic's specifications, and they require a sophisticated fabrication effort to produce. Sales to Able Plastics constitute about 90% of the manufacturer's total sales. The owner of Peterson Manufacturing is often a partner in the design work, and occasionally visits clients with Rob Stewart to work on more complex container projects.

The Production Process

Five thermoforming machines produce the containers. Each machine requires one worker. The worker begins production by securing a square sheet of plastic (which are large and heavy) in support brackets inside the machine. The machine then rotates the plastic over gas heaters that bring the temperature of the plastic close to its melting point. When the appropriate temperature is reached, the machine extracts the sheet from the heaters and lowers it over a mold. A vacuum is formed between the mold and the sheet and this causes the plastic to conform to the shape of the mold. After the plastic is shaped, cold water is pumped through pipes inside the mold to harden the plastic. Once the plastic is cool enough to handle, the machine extracts the plastic from the mold and the worker removes the finished container from the machine. The worker then repeats the process with a new sheet of plastic.

The heating, molding, and cooling process for one sheet takes about five to ten minutes. During this time, the worker uses a rotary cutting tool to trim excess plastic away from the container that was just removed from the machine. Using various scrapers and sanders, the worker also finishes the edges. Depending on the particular design, the worker may also insert metal supports or handles and apply labeling. After a certain number of containers are produced, the worker stacks the containers on a dolly and moves them to the shipping area.

While the production supervisor spot checks the finished containers, most quality assurance is the responsibility of the production workers. The worker is also responsible for filling out the paperwork at end of production run. Because Able Plastics is ISO 9000 certified, this paperwork does take some time (usually about 10 minutes per production run).

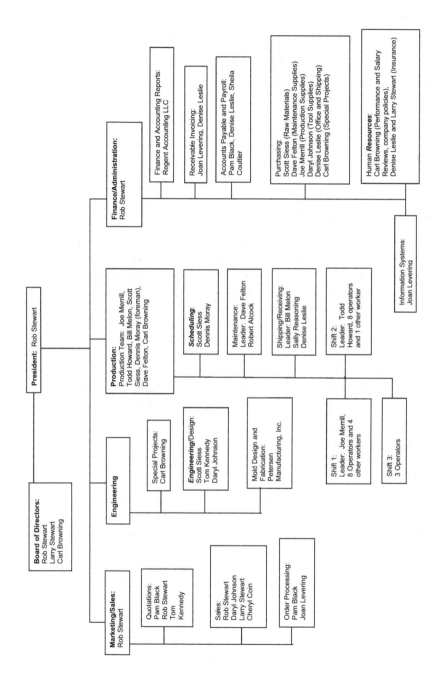

Figure 10.1. Able Plastics organization chart.

As of March 2001, production ran over two shifts, 5 days a week. A three-man third shift did maintenance, janitorial, and set-up work. *Overtime* work was unusual, but the production supervisor was often called in to handle scheduling or machine problems on the second and third shifts.

Since the production jobs are largely unskilled, Able Plastics conducted minimal training. New workers spend a week observing or working with an experienced worker, and then are assigned to a machine to work on simpler jobs by themselves. During this period, the production supervisor keeps close tabs on the new worker and is there to offer assistance and coaching. Turnover among production workers has been averaging about 25% per year.

The Sales Organization

Most sales are due to the efforts of Rob Stewart and, to a lesser extent, the in-house designers. The company had a relationship with two manufacturing representatives in the Detroit area, but sales acquired through these representatives constituted less than 25% of total revenue. Virtually all sales were acquired through personal selling and responding to requests for bids. While two individuals assisted in the quotation process, Rob Stewart was responsible for the pricing of all bids.

The company did not engage in advertising and only occasionally visited trade-shows as an exhibitor. The company provided a brochure and promotional items (e.g., mugs, pens, and pads of paper with Able Plastic's name and address) as part of sales visits to potential customers.

Financial Results

Figure 10.2 contains the financial results for the fiscal years 2001 and 2002. Regent Accounting LLC prepared the statements and, other than abbreviated profit and loss statements prepared each quarter, the financial statements constituted all the financial reporting and cost accounting within the company. Able Plastics had a moderate annual increase in sales and maintained a modest profit margin. The balance sheet showed a substantial investment in inventory. The firm did not have a sophisticated cost accounting system, which prohibited it from allocating costs and profits on the basis of customers, product category, and so forth.

THE CONSULTING INTERVENTION

Rob Stewart and the board were concerned that Able Plastics had reached a plateau in its growth. The company had achieved almost $5 million in sales in 2002, and Rob expected about the same level over the next few

ABLE PLASTICS, INC. FINANCIAL STATEMENTS

Balance Sheets
September 30, 2002
and 2001

Assets:	2002	2001	
Current Assets:			
Cash	85 333	23 209	62 124
Accounts Receivable	845 409	798 376	47 033
Inventory	403 509	287 873	115 636
Prepaid Expense	23 004	24 421	-1 417
Total Current Assets	1 271 922	1 133 879	138 043
			0
Fixed Assets:			0
Machinery and Equipment	988 943	902 329	86 614
Furniture and Fixtures	34 449	31 234	3 215
Leasehold Improvements	73 845	73 845	0
Less: Accumulated Depreciation	788 587	701 374	87 213
Net Fixed Assets	308 650	306 034	2 616
			0
Other Assets:			0
Organization Costs (net of amortization)	45 367	48 088	-2 721
			0
Total Assets	1 625 939	1 488 001	137 938
			0
Liabilities and Stockholders Equity:			0
			0
Current Liabilities:			0
			0
Accounts Payable	569 923	428 385	141 538
Notes Payable	59 569	68 344	-8 775
Accrued Payroll Taxes	9 588	10 342	-754
Current Portion of Long term debt	87 009	100 485	-13 476
Total Current Liabilities	726 089	607 556	118 533
			0
Long Term Debt	57 984	122 349	-64 365
			0
Stockholders' Equity:			0
			0
Common Stock	138 900	138 900	0
Additional Paid-in Capital	62 837	62 837	0
Retained Earnings	640 129	556 359	83 770
Total Stockholders' Equity	841 866	758 096	83 770
			0
Total Liabilities and Stockholders' Equity	1 625 939	1 488 001	137 938

Figure 10.2.   Able Plastics financial statements (page 1).

years. He realized that achieving higher sales levels would mean require
systemic changes within the company. The barriers to growth seemed to
exist on virtually all levels: sales, production, logistics, design, and information systems. Rob and the board believed they needed help in identifying the critical issues and prioritizing their importance. They contacted a
consulting group from a local university, called CBA Consulting, and

Statements of
Operations
For the Years Ending
September 30, 2002
and 2001

	2002	2001	1 0
Sales	4 124 405	4 539 884	-415 479
			0
Cost of Sales:			0
Material	2 309 214	3 241 228	-932 014
Regrind credit	#########	#########	202 125
Regrind Costs	6 488	5 664	824
Production Components	603 339	200 887	402 452
Labor	623 383	628 448	-5 065
Payroll Taxes	56 944	63 448	-6 504
Commissions	42 568	82 224	-39 656
Tooling Costs	322 338	320 446	1 892
Prototype Materials	122 323	154 668	-32 345
Freight	12 442	44 555	-32 113
Manufacturing Overhead	644 848	698 553	-53 705
Total Cost of Sales	3 521 564	4 015 673	-494 109
			0
Gross Profit	602 841	524 211	78 630
			0
General and Administrative Expenses	456 456	445 334	11 122
			0
Operating Income	146 385	78 877	67 508
			0
Other Income (expense)			0
Bad Debts	-4 181	0	-4 181
Interest Expense	-24 136	-40 586	16 450
Gain on Sale of Asset	4 288	0	4 288
Total other Income (Expense)	-24 029	-40 586	16 557
			0
Income Before Taxes	122 356	38 291	84 065
			0
Provision for Taxes on Income	38 586	13 224	25 362
			0
Net Income	83 770	25 067	58 703

Figure 10.2.   Able Plastics financial statements (page 2).

asked the group to "help us to sort out the problems." One of the main reasons Rob approached this group was their focus on process improvement within the context of a larger organizational change environment. Rob had hired consultants before who had tried to improve certain processes or systems within company, but he had always found the results of these efforts disappointing. He had found that while the recommended processes or systems might be technologically improved, the company had difficulty incorporating and using the new capabilities. He was not exactly sure why this problem was occurring, but he suspected that the transformation efforts were too narrow in scope.

Statement of
Cash Flows
For the Years Ending
September 30, 2002
and 2001

	2002	2001
Net Income	83 770	25 067
Adjustments to Reconcile Income to		
Net Cash Provided by Operations:		
Amortization and Depreciation	89 934	90 345
Gain on Sale of Asset	-4 288	
Change in:		
Accounts		
Receivable	-47 033	20 002
Inventory	-115 636	110 386
Prepaid Expenses	1 417	290
Accounts Payable	141 538	-240 234
Note Payable	-8 775	-9 000
Payroll Taxes	-754	-231
Current Portion of LT Debt	-13 476	-2 000
Net Cash from Operations	126 489	145 854
Financing:		
Decrease in Long Term Debt	-64 365	-56 455
Investing Activities:		
Increase in Machinery and Equipment	-86 614	-56 365
Increase in Furniture and Fixtures	-3 215	-2 344
Change in Cash	62 124	89 399

Figure 10.2.   Able Plastics financial statements (page 3).

The Initial Diagnostic

After touring the plant and surveying the industry, the CMU consulting faculty conducted an in-depth interview with the Rob in a session that lasted almost three hours. The focus of the discussion centered on the general difficulties faced by Able Plastics, in any order that made sense to him. All six SMU Consultants attended the session, and based on this initial contact, a number of problem areas were identified.

Marketing

Able Plastics lacked an organized marketing effort. Two sales representatives located in Detroit, and Rob himself, represented the sales effort. Rob has tried to include his designer in customer development, but thus far that effort has not yielded any significant results. The company does not advertise, and it has few brochures and other sales materials. Rob maintained the customer list on a spreadsheet, which provided details on

**Balance Sheets
For the Years Ending
September 30, 2002
and 2001**

	2002	2001
Assets:		
Current Assets:		
Cash	85 333	23 209
Accounts Receivable	845 409	798 376
Inventory	403 509	287 873
Prepaid Expense	23 004	24 421
Total Current Assets	1 271 922	1 133 879
Fixed Assets:		
Machinery and Equipment	988 943	902 329
Furniture and Fixtures	34 449	31 234
Leasehold Improvements	73 845	73 845
Less: Accumulated Depreciation	788 587	701 374
Net Fixed Assets	308 650	306 034
Other Assets:		
Organization Costs (net of amortization)	45 367	48 088
Total Assets	1 625 939	1 488 001
Liabilities and Stockholders Equity:		
Current Liabilities:		
Accounts Payable	569 923	428 385
Notes Payable	59 569	68 344
Accrued Payroll Taxes	9 588	10 342
Current Portion of Long term debt	87 009	100 485
Total Current Liabilities	726 089	607 556
Long Term Debt	57 984	122 349
Stockholders' Equity:		
Common Stock	138 900	138 900
Additional Paid-in Capital	62 837	62 837
Retained Earnings	640 129	556 359
Total Stockholders' Equity	841 866	758 096
Total Liabilities and Stockholders' Equity	1 625 939	1 488 001

Figure 10.2.   Able Plastics financial statements (page 4).

the sales histories of the respective customers and how the price of the shipping containers was calculated. The spreadsheet also listed quantities purchased, the dates manufacturing took place, and where the goods were shipped. When bids on new jobs were undertaken, the spreadsheet was used to determine how similar jobs were bid in the past. The costs of the custom details of the new job were then computed, with a 35% gross

Statements of Operations
For the Years Ending
September 30, 2002
and 2001

	2002	2001
Sales	4 124 405	4 539 884
Cost of Sales:		
Material	2 309 214	3 241 228
Regrind credit	########	########
Regrind Costs	6 488	5 664
Production Components	603 339	200 887
Labor	623 383	628 448
Payroll Taxes	56 944	63 448
Commissions	42 568	82 224
Tooling Costs	322 338	320 446
Prototype Materials	122 323	154 668
Freight	12 442	44 555
Manufacturing Overhead	644 848	698 553
Total Cost of Sales	3 521 564	4 015 673
Gross Profit	602 841	524 211
General and Administrative Expenses	456 456	445 334
Operating Income	146 385	78 877
Other Income (expense)		
Bad Debts	-4 181	0
Interest Expense	-24 136	-40 586
Gain on Sale of Asset	4 288	0
Total other Income (Expense)	-24 029	-40 586
Income Before Taxes	122 356	38 291
Provision for Taxes on Income	38 586	13 224
Net Income	83 770	25 067

Figure 10.2.   Able Plastics financial statements (page 5).

margin to arrive at a price to bid. The fulfillment rate (the number of jobs acquired as a percent of the number of jobs bid) was roughly 20%.

The owner was dissatisfied with the fulfillment rate and the length of time required to bid on job. The time between a request for a bid and the time the bid is delivered to the customer was oftentimes two to three weeks. The delay was blamed on a lack of time on Rob's part.

Production

Production was another troublesome area. The rate of rejected parts ["parts" is the jargon for the individual shipping containers in a batch] varied between 5 and 25%. In addition, there were instances of customers returning parts because they were faulty. Although there was no written documentation of the nature of the defects, the problems were believed to largely related to poor trimming of the parts. The high reject rates led the company to purchase 125% of the final number of sheets needed to complete the job, and any unused sheet were placed in inventory for possible use in the future.

Scheduling of production was also problematic. Some rush orders did not get into the queue quickly enough, and oftentimes orders were not matched correctly with the appropriate thermoforming machines.

Finally, the company was working toward becoming ISO 9000-certified, something that the auto companies demanded of their suppliers. The effort generated a lot of paperwork and, unfortunately, did not translate into significant process or quality improvement. The company did, however, successfully pass a recent audit of its ISO procedures.

The Work Force and HR Support

The work force did not seem to care about the quality problems, despite what amounted to a paternalistic management approach in the company. *Absenteeism* and *personnel turnover* were also problems. Relationships between production supervisors and the workers manning the machines were occasionally strained and unproductive. Rob, however, believed his relationships with the workers and supervisors were "good."

The back office staff was described as "thin." As noted earlier, Regent Accounting did all the company's accounting on a contract basis. Although Able Plastics had a bookkeeper, this individual was not trained as an accountant. Two women handled the payroll and benefits package for the staff. The company provided health insurance for its employees, but it has no retirement program. Job descriptions existed for the office staff, but were none for the production supervisors or workers. Rates of pay were low in comparison to other industries in Michigan, but are competitive in the area in which Able Plastics is located.

Design

The company's key competitive strength was its design work. Peterson Manufacturing was also a highly skilled organization that can produce superior molds. Some molds are intricate and have to accommodate multiple accessory attachments, latches, and hinges. The cooling coils in the molds worked well, and they accurately met the water injection ports of the thermoforming machines. The owner of Peterson Manufacturing does not like sales work, and he does little work for anyone other than Able Plastics. The company works well with Daryl, Able Plastic's most experienced designer, an individual who Rob believes is "very capable."

The Board

The board of directors (Rob Stewart, Larry Stewart, and Carl Browning) met every other Saturday at Rob's house. The financing for the company came from these three individuals. While they did not have any long-range plan, their goal was to double the company's size in the next 5 years, and then double it again in the next 5-year period. The board was concerned, however, with the auto companies' demand for price concessions from their suppliers and their efforts to reduce the number of suppliers that worked with.

The Engagement

As mentioned earlier, one of the reasons Rob called CBA Consulting was the consulting-firm's focus on improving processes within the context of an organizational change effort. The firm had adopted ISEOR's *socio-economic intervention* methodology in an attempt to bring balance to its organizational change efforts. Most organization change techniques either put weight on a single set of factors, either (1) economic and technological rationality or (2) the individual's needs and feelings. *SEAM* represents an effort to incorporate both approaches by combining the procedures of each approach in a series of ordered diagnostic and guided change techniques (see chapter 1; also Savall, Zardet, Bonnet, & Moore, 2001).

As captured in this volume, there are four primary phases within the SEAM approach: *socio-economic diagnostic*, the use of *expert opinion*, *priority action plans*, and the *assessment* of results.

Socio-Economic Diagnostic

This initial phase utilizes a *mirror-effect* through which summary interview data, collected by the consultants from all levels of the organization, are fed back to top management. The mirror effect describes the organization through the eyes of employees at all levels, without any evaluative commentary by the consultants. This direct feedback process is meant to make management aware of the problems, or (more often) to provide objective information that prevents management from denying their existence. As part of this diagnostic phase, the consultants estimate the direct and *hidden costs* associated with the identified *organizational dysfunctions*, and these *cost estimates* are presented to management as part of the *mirror-effect* process.

After management reviews the data, management team shares the interview data and *cost estimates* with employees. The extensive sharing of information with management and employees is a crucial step designed to overcome resistance to change, and it differentiates SEAM from other consulting approaches. The aim is to validate the existence of the dysfunctions within the organization and to develop awareness of their costs, which will lead to greater acceptance of the need to change the organization's culture, strategies, or operational processes.

In the Able Plastics intervention, the consultants spent a significant amount of time with the senior manager, explaining the SEAM procedure and working out a time schedule. Rob understood that during the *mirror-effect* phase the firm would lose roughly two hours of productive time from each employee, and that more time would be lost when project teams began operation. He also understood that he would have to devote a significant amount of his own time to analyzing the data, working with the consultants, and actively coordinating the *project teams*. This was cause of some concern to Rob because he was the main source of sales for Able Plastics, and he was worried about servicing his customers. He was also concerned about his ability to supervise ongoing production and company administration. The requirement for Rob's time, therefore, represented the main source of contention in the *negotiations* over the engagement. The final agreement specified that the consultants would make every effort to minimize their demands on Rob, while at the same looking for ways to make Rob more efficient.

CBA Consulting planned to interview all of the employees of Able Plastics. The consultants sought to conduct semi-directed interviews in which employees were asked both open-ended and specific questions about the direction of Able Plastics and what it was like to work there. The Appendix contains the interview form the consultants used to interview the production workers. Similar questions were asked of supervisors and the office staff. The consultants planned to conduct these interviews over a three-week period and spend the next three weeks organizing the interview data for the *mirror-effect*. The engagement plan assigned one consultant to examine the production and staffing costs of the company and to collect external data that could be used to help estimate the costs of the identified *dysfunctions*.

The consultants encouraged Rob to share the data from the mirror effect with all employees. He agreed to share the data with the board, the engineering staff, and the lead individuals in the administration area. Because he was concerned about personal criticisms of individual supervisors in the production area, he reserved judgment in distributing unedited interview data to the production staff. He did agree, however, to distribute summary interview data and the cost analyses of the identified problem areas to all organizational members.

Expert Opinion

Based on the *mirror-effect diagnostic*, calculation of *hidden costs*, and top management's reaction to the data, a document is created by the consultants that describes areas that may be targeted for improvement. Referred to as *expert opinion*, this step is part of the overall attempt to drive needed organizational change during the intervention. The document objectively states the consultants' conclusions regarding the *organizational dysfunctions* evidenced in the mirror effect. In an effort to reduce tendencies to minimize or deny the importance of these dysfunctions, the expert opinion explicitly focuses on problems and dysfunctions. As part of the process, possible countervailing positive attributes within the organization are not discussed. Expert opinion, therefore, is not a balanced portrayal of the status of the organization, mentioning both good and bad aspects. Rather, the consultants prepare a document that inventories the apparent dysfunctions within the organization's culture, strategies, and/or internal processes.

CMU Consulting planned to prepare this document within two weeks of the mirror effect discussion. In accordance with the SEAM protocol regarding the content of the experts' opinion, the document would only tabulate the dysfunctions of Able Plastics. It would make no mention of any positive aspects to the company. A summary of the document is provided in Table 10.1.

Priority Action Plans (PAP)

After reviewing the *mirror-effect* and *expert opinion*, the next step is for management to form *focus groups* to address the dysfunctions and create an oversight committee to organize this effort. Normally, the oversight

Table 10.1. Summary of *Dysfunctions*

Dysfunctions	*Manifestation of Dysfunction*
High *absenteeism*	Overstaffing, low morale
High scrap rate	High material and labor costs, extended production time
Overordering (to accommodate high scrap rate)	High inventory costs
High inventory of unused sheets	High inventory costs, space inefficiencies
Poor *working conditions* (heat, dirty)	Low morale, *personnel turnover*, higher scrap rate
Slow set-up for new production runs	Downtime, higher initial scrap rate
Inefficient use of top *management's time*	Expenses, opportunity costs
Lack of coherent strategic plan	Little synergy, frustration
Some problematic supervisors	Low morale, employee turnover
Lack of available cost data	Pricing problems, inability to allocate expenses, profits
Inability to link computer systems	Duplication of work, other inefficiencies

committee assigns a member of top management to each team to ensure *coordination*, and the committee may also assign a consultant to facilitate the individual team efforts. With the direction and coordination provided by top management, the respective teams each develop a *priority action plan (PAP)*, which lists the *dysfunction* and the strategies and specific techniques that will address the problem. The *PAP* also specifies the performance targets for the new strategies and operational procedures. The process for designing the PAP is best described as semi-participative, i.e., the teams prepare the PAP but top management plays an active role in an effort to keep the strategies and plans of the organization *synchronized*.

The engagement plan at Able Plastics estimated that this phase would last at least eight months. The consulting team and Rob believed it would take a significant amount of time just to form the teams, educate them about the tasks expected of them, and establish their procedures. In addition, the employees of Able Plastics were not used to working in teams, and they were not used to having input in the design new operational processes. It was decided that the CMU consultants would facilitate the efforts of the individual teams, particularly in the production area, and a representative of top management would be present at the meeting of every team to provide direction and *coordination*.

Results Assessment

In the final phase, the management oversight committee assesses the outcomes of the performance measures set by the teams in their respective *PAPs*. The aim is to measure the *effectiveness* of the change initiatives. The consultant normally assists in this process in an effort to ensure an objective appraisal of the new processes. From an organizational development perspective, the assessment seeks to legitimize the changes that have occurred and provide a rationale for making them permanent.

The Able Plastics' assessment was scheduled for 12 months following the start of the engagement. The plan called for the consultants to be part of the assessment, paying particular attention to measurement reliability. Each team was scheduled to receive its own assessment and have the opportunity to comment on the results. The engagement plan specified that follow-on work after the assessment phase would be required but the plan specified no details.

Tables 10.2 and 10.3 contain the company's cost data collected by the consultant assigned to estimate the costs of the dysfunctions at the company. The goal of computing the *hidden costs* of *dysfunctions* was to help the client appreciate the significant negative impact of the dysfunctions on the organization. By understanding the extent of damage that a dysfunction can cause, the client becomes more committed to making the necessary, and often difficult, long-term changes to address these prob-

**Table 10.2.   Cost Data Relating to *Absenteeism,
Personnel Turnover,* and Reject Rates**

1	Wage rate for workers	$8/hour + $2/hour in benefits
2	Number of production workers	30
3	Average hours per year worked	2,000
4	Absenteeism: average daily rates	Between 2-3 people on the first shift; 2-3 people on the second shift; usually none on the third shift
5	Average daily absenteeism rate in Michigan	Published sources say 5% for manufacturing firms
6	Number of extra workers assigned per shift to make up for expected no-shows	2 on the first shift and 2 on the second shift; none on the third shift
7	Number of missed deadlines due to absenteeism and turnover	None but frequently jobs have had to be rushed which upsets schedules on the shop floor
8	Turnover per year	About 7-8 workers per year
9	Average annual turnover rate in Michigan	Published sources ay about 10% per year for manufacturing firms
10	Training time for a new worker	About 16 hours
11	Recruitment costs of a new worker	About $500 to $1,000 per worker
12	Other training costs	About $200 to $500 per worker
13	Piece reject rate for new hires	About twice that of experienced workers (20% for new hires versus 10% for experienced workers)
14	Time required to become an "experienced worker"	About 1 month
15	Average cost of a sheet of raw plastic	$20 per sheet

lems. Thus, this procedure assists the client in moving from the ***diagnostic*** phase of the intervention (what is wrong) to the ***implementation*** phase (implementing the changes to fix the problems), which is often difficult for many clients.

The process of calculating ***hidden costs*** begins with the consultant identifying the potential ways that a dysfunction can impact the client's operation. This is done to prepare for an extensive meeting with the client to work through these calculations together. The client is involved in this process because (1) the client understands the subtleties of his firm and (2) the practice forces the client through the often painful process of understanding the extent of the costs of the ***dysfunctions***. This process does not require precise cost calculation, but realistic, conservative estimates that the client cannot dispute. During the meeting on hidden costs, the hidden costs of several dysfunctions were computed.

Table 10.3.   Data for Reject Rate Cost Estimation

1	Reject rate on average in the plant	About 10%, sometimes lower, sometimes higher
2	Average cost of a sheet	About $20 per sheet
3	Number of thermoforming machines	Five
4	Average number of machines running per shift	Four
5	Number of shifts	Two 8-hour production shifts; one maintenance shift
6	Average parts per hour per machine	About 10
7	Average order	About 200 parts per order
8	Number of orders per year	500 to 600 on average
9	Number of extra sheets purchased per order to make up for rejects	Buy about 125% of the ordered number of parts
10	Number of sheets in sheet inventory	About 5,000
11	Percent of inventory area occupied by leftover sheet inventory	50%
12	Square feet of inventory space	2,500 square feet
13	Cost of capital	Never computed
14	Interest rate on debt	About 10%
15	Rent on plant per year	About $12,000
16	Percent of plant devoted to inventory space	About one third of plant area is devoted to inventory space
17	Out of spec parts in the shipped and returned by customers	Happens occasionally but company just remanufactures the orders and ships
18	Percent of orders that can use the excess sheet	About 5% of orders can use the leftover sheet from earlier jobs

Tables 10.2 and 10.3 were used to assist in the calculation of hidden costs. For example, the hidden costs of *absenteeism* include a variety of expenses, including the costs of keeping replacement workers available to "fill in," even when they are not needed; the costs of increased scrap rate and reduced quality; the costs of morale when workers must cover for those who do not show up for work; the *nonquality* costs associated with moving workers out of their normal jobs when forced to juggle work assignments; the costs associated with the additional management time spent daily to find ways to adjust to the absent employees; and the costs of recruiting new employees because of *personnel turnover* resulting from poor morale. This process incorporates the direct and indirect expenses and/or opportunity costs that result from a dysfunction and usually generates a cost much higher than the client realized. In the Able Plastics case, the client later reported feeling alternately stunned that these costs were

so substantial and hopeful that, when properly addressed, significant savings could be realized.

REFLECTIONS ON THE INTERVENTION

The *SEAM intervention* was the first step in a long relationship between Able Plastics and the CMU consultants. Realistically, consultants cannot hope to "fix" a client but rather help the client begin the *process* of improving their ability to recognize problems and issues and increasing their capacity for change and improvement. This initial relationship has continued for nearly five years and has included numerous meetings, seminars, and other informal *communications*.

Able Plastics is a much improved organization. As an illustration, the following is a partial list of changes that have occurred since the initial SEAM intervention:

- Sales have doubled in a difficult economic climate.
- An ERP system has been installed (with CMU consultants' guidance).
- Numerous managerial changes have occurred, including a new production supervisor, a new price quoter, and other changes of responsibilities and assignments. These changes have freed up the company's founder and president, who was able to devote much more time to sales.
- An augmented ISO 9001:2000 certification has been received.
- A plastics regrinding service has been expanded and is now a significant profit center.
- The sales force has been restructured and expanded.
- The physical plant has doubled in size and new thermoforming machines have been added.
- A comprehensive HR policies and procedures have been developed and disseminated.

The *SEAM intervention* proved to be appropriate for Able Plastics because the company's management understood they needed help, they wanted to improve, and they recognized that there were no quick fixes. They were looking for a long-term relationship with the CMU consultants, which they recognized was necessary to make the changes for continual growth and improvement. In fact, at the time of this writing the CMU consultants have agreed on another project to assist Able Plastics as they seek to *improve efficiencies* in their production processes. In follow-up meetings, they have underscored the value of the SEAM approach because it helped change their way of thinking about their business and it gave them the tools to help them to continually improve their operations and their strategy.

The **SEAM methodology** worked well for Able Plastics but would not likely be appropriate for many U.S. companies that expect brief interventions and quick improvements, or for companies that are not sufficiently "introspective" to more thoroughly understand their organization. The Able Plastics intervention was a "pure" **SEAM intervention** that followed the SEAM in toto. It is also possible to utilize various tools in the SEAM methodology to match the different needs and requirement of different clients. By understanding the SEAM approach, consultants can use their philosophy of involving the client in the diagnostic phase of the intervention to help facilitate the **implementation phase**. The end result is a consultant with more capability to meet the needs of his or her clients.

APPENDIX

Able Plastics Interview Guide
CBA Consulting
June, 2001

Good afternoon. My name is _____ and this is my colleague _____. As you may know we are part of a management consulting team from Central Michigan University that is assisting Able Plastics by evaluating some of its operations to help the company improve and grow. An important part of our work is to gather information from as many employees as possible.

Today we're going to be asking for your thoughts and opinions about some of the things you do at Able Plastics and how things operate in general. We'll be taking some notes so that we remember what you say, but we're not recording names, only general comments. We will put together all the information we gather from the people we interview in order to identify some common issues. Everyone at Able Plastics will see a summary of what we find at some point down the road.

For today, please help us by being as candid and honest as possible. We are interested in any and all information you wish to share if you think it's important. Feel free to respond to a question directly, build on a comment made by someone else in the group, or provide a specific example.

Do you have any questions about what we're doing before we get started? (*Address any concerns here.*)

1. **In your opinion, what things seem to work particularly well at Able Plastics?** (Prompts: What's good about Able Plastics? *Working Conditions, Work Organization & Coordination, Communication, Training*)

   - Can you give me an example of something (e.g., a project, a job order, etc.) that *worked really well?*

2. **Now consider the opposite. What things don't work particularly well at Able Plastics?** (Prompts: What are some problems at Able Plastics? Working Conditions, Work Organization and Coordination, Communication, Training)

   - Can you give me an example of something (e.g., a project, a job order, etc.) that *didn't work very well?*

3. **Where do you think Able Plastics is going? In particular, tell me how you see Able Plastics as a company 5 years from now?** (Prompts: What does Able Plastics have to do to "make it" in the future?)

REFERENCES

Lepisto, L., Hayes, R., McGilsky, D., Love, K., & Bahaee, M. (2005, August). *An assessment of three management consulting diagnostic strategies in a U.S. corporate environment.* Paper presented at the Academy of Management annual meeting, Honolulu, Hawaii.

Savall, H. (1974, 1975). *Enrichir le travail humain dans les entreprises et les organisations* [Work and people: An economic evaluation of job enrichment]. Paris: Dunod.

Savall, H. (2003a). An updated presentation of the socio-economic management model. *Journal of Organizational Change Management, 16*(1), 33-48.

Savall, H. (2003b). International dissemination of the socio-economic model. *Journal of Organizational Change Management, 16*(1), 107-115.

Savall, H., & Zardet, V. (1987). *Maîtriser les coûts et les performances cachés: Le contrat d'activité périodiquement négociable* [Mastering hidden costs and performances: The periodically negotiable activity contract]. Paris: Economica

Savall, H., & Zardet, V. (1992). *Le nouveau contrôle de gestion: Méthode des coûts-performances cachés* [New management control: The hidden cost-performance method]. Paris: Éditions Comptables Malesherbes-Eyrolles.

Savall, H., & Zardet, V. (1995). *Ingénierie stratégique du roseau, souple et enracinée* [Strategic engineering of the reed, flexible and rooted]. Paris: Economica.

Savall, H., Zardet, V., & Bonnet, M. (2000). *Releasing the untapped potential of enterprises through socio-economic management.* Geneva, Switzerland: International Labor Office-ISEOR.

Savall, H., Zardet, V., Bonnet, M., & Moore, R. (2001). A system-wide, integrated methodology for intervening in organizations: The ISEOR approach. In A. F. Buono (Ed.), *Current trends in management consulting* (pp. 105-125). Greenwich, CT: Information Age.

CHAPTER 11

SOCIO-ECONOMIC APPROACH TO MANAGEMENT IN MEXICO

Margarita Fernandez Ruvalcaba

Since 1997 a group of six professors at Universidad Autónoma de México (UAM), one of Mexico City's universities, has taken part in *experimental research* initiated by ISEOR, under the direction of Henri Savall, Véronique Zardet and Marc Bonnet. A basic goal of this work was to create generic knowledge in management by constructing common principles through interaction among a wide diversity of problematics, contexts, roles, visions, cultures and interests of actors committed to overall performance-improvement processes in organizations.

This chapter presents the perspectives and challenges that *socio-economic approach to management (SEAM)* applications in Mexico have brought to light. To this end, we describe and analyze the context of organization consultancy in Mexico, the positioning of SEAM in the larger consultancy offering, and major *SEAM* applications and outcomes. The chapter also examines significant behaviors observed in an array of some 15 organizations, featuring both industrial (Enriquez Galvan, 1982) and commercial enterprises, both small- to medium-sized (SME) and large public organizations (Fernandez Ruvalcaba, 1984; Fernandez Ruvalcaba & Peñalva Rosales, 2004). SEAM application in the *Institución de Educación*

Socio-Economic Intervention in Organizations: The Intervener-Researcher and the SEAM Approach to Organizational Analysis, pp. 251–278

Superior Pública, a public institution of higher education **internal-external strategic action plan (IESAP),** is presented in greater detail (Fernandez Ruvalcaba, 1998), with illustrations of several SEAM tools. Throughout the chapter, special attention is devoted to the context of this analysis, since it reveals some critical points for successful application of the unique contributions offered by this **intervention-research** method (Savall & Zardet, 2004), especially **cognitive interactivity** and **contradictory intersubjectivity** between internal and **external interveners,** executive management, middle management and operational personnel, customers and suppliers, and university research workers and practitioners.

THE MEXICAN CONTEXT

From 1950 to 1972 Mexico implemented a strong economic policy of domestic substitution for importation, including closing borders to consumer products and services, both consumables and durables. The idea was to avoid competition that was deemed to be unfair from more experienced producers. With this protection, the time necessary to consolidate national enterprises would be gained thanks to a captive internal market.

In the 1960s, once the need for manufacturing equipment had been satisfied in a large number of enterprises, performance improvement efforts were focused on staff training and **development.** Legislation also provided encouragement by requiring that enterprises devote a percentage of their revenues to these issues. Pioneer management consulting firms offered services in recruiting, selecting and training personnel, as well as in evaluating jobs, remuneration systems, organization and methods.

By the end of the decade, organizational development (OD) had been introduced in Mexico and could be considered a transitional phase in consultancy training. The OD approach advocated planned change, taking into account the needs and expectations of both individuals and organizations, thus endeavoring to offer something more than just meetings and training sessions. It was during this period that "sensitivity training" labs and "T Groups" began to appear. Inspired by industrial psychology, these approaches were experimented in large enterprises, as were other subsequent formulas including Total Quality Management (TQM), Just In Time methods (JIT), Management By Objectives (MBO) and, in the past decade, downsizing, reengineering and outsourcing.

Today, with the opening of the market and globalization, quality certification and guarantee through international ISO norms have become common practice in the on-going search for improved organizational performance. Over this almost 50 year period, some models can be characterized by predominantly technical factors (especially machines, equipment, processes, inputs), while others can be characterized by predominantly "people" factors. Some models focus on both factors: the less material human aspect and the more material technological aspect. Like a pendulum, it seems that efforts to improve business and organization performance sway from one extreme to the other. Today, efforts tend to be more moderate, synthesizing elements that traditionally have been opposed to one another in the production process: the social and the technical, the strategic and the operational, the task and the person, the short-term and the long-term, the internal and the external, and so forth.

Currently, among consulting firms that focus on developing organizational members' potential as a lever for improving organization performance, one can distinguish:

- The services offered by large *transnational consultancies*, most of which have become worldwide firms with their main offices in the United States or England;
- *National consulting firms*, some of which have enough prestige to seriously compete with international firms; and
- Numerous *freelance consultants* who have begun organizing into flexible networks as a means of more effectively responding to client demands. These consultants have certain competitive advantages when they enjoy local prestige, even if it cannot be compared to the worldwide prestige of international firms, and when client decisions are based solely on the criterion of price.

These consulting offers are enriched by the diversity of techniques accumulated in programs that bill themselves as "custom-tailored." Given this broad panorama, the discussion turns to why the socio-economic approach to management (SEAM) has been of interest in Mexico.

SEAM AND PERFORMANCE
IMPROVEMENT PROCESSES IN MEXICO

Mexico's import substitution policies were not accompanied with plans to actively stimulate the **development** of managerial capacities or to strengthen Mexican enterprises, raising their competency levels closer to those of their foreign competitors. In fact, having eliminated the obligation to contend with competitors beyond their borders, national producers grew accustomed to a patient, undemanding market, forced to consume whatever domestic enterprises chose to offer. The result was a weak industry **structure**, sustained by import taxes that penalized foreign products. When the North American Free Trade Agreement (NAFTA) with North America, Canada and Mexico came into effect at the beginning of the 1990s, there was no transition to allow Mexican enterprises to obtain the conditions necessary for improving their **competitiveness** (Savall & Zardet, 2005).

Mexican executives are also accustomed to believing that by paying a consultant with a sterling reputation, they can disown problems and simply await notification that "everything has been solved!" Hence, there is an urgent need for reliable methodologies that can stimulate competency. One of the pillars of **SEAM** (Savall, 1974, 1975, 1979, 2003a; Savall & Zardet, 1987, 1995) is developing fundamental management competencies, since these bring to life and support strategic decisions and actions, setting them into action, such as **time management, communication-coordination-cooperation** and identification of the competencies that personnel implement to create value. The intervention method proposed by SEAM is devised to foster client responsibility and autonomy and to discourage simulation by demonstrating that feigning is less profitable and less gratifying than actually doing the job. Postulating that top management is a key lever for change, working closely with them and involving them in the **intervention architecture** with line personnel in **focus groups**, the SEAM method succeeds in constructing a platform for authentic change.

The SEAM intervener is an expert in accompanying the change process, during which members of the client organization *coproduce* alternative solutions for improving performance by means of actions they conceive and coordinate, with **implementation** support and involvement by the SEAM intervener. This expertise is based on the three axes of the change methodology proposed by Henri Savall (see chapter 1): (1) the **policy-decision axis** that establishes the direction and orientation of change efforts; (2) the **improvement-process axis** whose four stages permit a common definition of the problem, generate **projects** for solving it, implement them and evaluate the results; and (3) the **management-tool axis**, which gives greater **visibility** to the current state of **resources** invested in the change action, and the ramifications for the time and competency levels of organizational members.

The SEAM approach promotes the development of *systemic thinking*. During the diagnostic, it places the **dysfunctions** in relation to the causes evoked by the organization's members, suggesting root causes from the intervener's viewpoint. This provokes thought about prevailing mental structures in the enterprise. An in-depth approach of this type not only inspires more effective **regulation** of dysfunction effects, it also develops systemic thought reflexes. Within the SEAM methodology, a decision is evaluated both in terms of its short-term, deliberate effects in the specific organizational space of the intervention and in terms of its other possible counterproductive effects in different organizational spaces and longer time periods.

SEAM capitalizes on resources that are already present within the organization, but are wasted due to mismanagement. By identifying non-planned modes of *dysfunction* regulation and by calculating their qualitative, quantitative and financial impact, SEAM demonstrates that the problem is not lack of resources, but lack of good judgement in their use. This constitutes one of the most powerful action-levers developed by SEAM for confronting a frequent obstacle and lament faced by the advocates of performance improvement: "we don't have enough resources."

Training and SEAM Interveners

SEAM explicitly positions its foundation in reference to frameworks that are well known to organizational interveners, such as the stages of a change process: the *diagnostic*, *project*, *implementation*, and *evaluation*. These stages are based on a set of logical and empirical concepts that are linked to a theory on overall or *integral performance* in organizations. They are also linked to a set of tools that afford greater *visibility* to the intangible resources of the enterprise, for example focusing on (1) personnel competency (*competency grid*), (2) quality of time management (*self-analysis of time grid*), (3) clarification of a 3 to 5-years medium-term plan (*IESAP* tool), which stimulates commitment at divisional and departmental levels and among individuals responsible for these actions, and (4) periodic cooperative *negotiation* of contributions and subsequent rewards (e.g., *priority action plan* and *periodically negotiable activity contract (PNAC)*).

The professional expertise of the intervener-researcher/consultant is attained through well-adapted training. This is contrary to the prevailing idea that consultants can only be forged through long years of experience (Fernandez Ruvalcaba & Andrade Romo, 2000; Fernandez Ruvalcaba & Estrada, 2001; Fernandez Ruvalcaba & Paramo Ricoy, 2004; Fernandez Ruvalcaba, Andrade Romo & Chávez Cortez, 2001; Fernandez Ruvalcaba, Martínez Vázquez & Zardet, 2003). The SEAM approach is taught through a formal educational program of consultant training, with specific objectives, schedules and practical application exercises. It is a systematic program based on inclusive criteria, such as participant age (both young and old), background (teachers and practitioners), sequence and intensity of training sessions, multimedia educational technology, and instructors. Yet, the program remains flexible thanks to the manner in which the training sessions are conducted, encouraging every participant to introduce his or her own context as pedagogical material. *Personal assistance* is provided throughout the training process, extending over a 6-month period. Practical application of materials are studied at intensive 2-day sessions every 2 months focused on the problems encountered by each participant in his or her mission as an intervener in the enterprise—which has been noted to be a key factor for success. This pedagogical practice, together with SEAM theory and methodology utilized for processing case studies, is one of the fundamental ingredients of the *integrated training* of *socio-economic management* consultants.

The SEAM methodology is based on *confidence* and respect for organizational members as a means of achieving comprehensive, systematic, qualifiable and continuous change. As Enriquez (2001) notes, "It is a methodology that does *not* compete with other techniques; it can be applied with reengineering, Just-In-Time and a wide variety of other techniques that target short-term results, strengthening and sustaining them in the long-term." The approach is also accessible to a wide range of professionals, from engineers to educators, made possible by the way in which the systemic approach to organizations and their environment is presented. For example, an ostrich is used to represent evasion behavior regarding *hidden costs*, and a nutcracker represents the methodology which, when applied with expertise, does not harm the organization and permits uncovering what lies beneath its outward appearance. The quantification of an *hourly contribution to margin on variable costs (HCVAVC)* (see Table 11.4), as the basis for calculating hidden costs, requires no more than basic mathematical skills (addition, subtraction, multiplication, division).

Diffusion Strategy: SEAM Technological Transfer-Assimilation in Mexico

What strategy was pursued in the case of SEAM diffusion in Mexico? There were two principal dimensions that constituted the criteria for diffusing SEAM in Mexican organizations: (1) the diversification of actors involved and (2) alternation of geographical locations (Savall, 2003b). This strategy consisted of working with local agents who had a personal vision and a professional situation that favored their involvement in SEAM diffusion and *implementation* actions. ISEOR undertook a number of forums that were organized by institutions from different sectors of activity. For example:

- In Mexico, the first presentation of the SEAM method, which as introduced by its founders Henri Savall and Véronique Zardet, took place in 1995 during a 25-hour seminar titled "Socio-Economic Management *Engineering*." Approximately 100 people attended the seminar, among which were graduate students from an economics and technological change management program, senior managers and staff from the administrative sector, researcher-teachers from the host establishment and other universities, and independent consultants. Meticulous note-taking during the session, focused on questions raised by the audience, permitted a preliminary identification of the heterogeneity of the Mexican public's centers of interests and reactions to the SEAM approach.

- In 1996 the SEAM method was presented at the 11th Conference of the Mexican Association of Directors of *Applied Research* and Technological Development (ADIAT). This association brings together nearly 1,000 directors of applied research and technological development centers (CIAyDT), both public and private, as well as representatives from the financial sector and university professors involved in Mexico's technological development. Further presentations of SEAM in the ADIAT network took place in the cities of Guadalajara, Monterrey and Mexico City.

- From 1998 to 2002, the EDUFRANCE event in Mexico, in the presence of the president of France, brought together several hundred French and Mexican universities and graduate schools showcasing their educational offers. ISEOR introduced its management science masters and doctoral degree programs, and presented three conferences on *socio-economic management* during the event.

- SEAM was also present in 2000 at the International Conference on Organizational Analysis in the city of Zacatecas, and in 2001 in Queretaro, on the occasion of the homage organized in honor of Michel Crozier by the University of Queretaro.

- From 1998 through 2002, thanks to support from the ECOS-NORD program established between the governments of France and Mexico, six Mexican professors continued in-depth SEAM training as intervener-researchers by participating in the consultant training program in Lyon (Brindis Almazán, 2007; Gonzalez Herrera, 2004; Gonzalez Perez, 2006; Martínez Alvarez, 2000; Martínez Alvarez & Monroy, 2001; Martínez Vázquez, 1999, 2001, 2005; Peñalva Rosales, 2001, 2006; Pomar Fernandez, 2007; Ramírez Alcantara, 1999, 2001, 2006).

- During the same time period, 19 France-Mexico missions and 16 Mexico-France missions were carried out, representing a total 1,248 days of direct interaction between Mexican intervener-researchers and the ISEOR's experienced ("senior") consultants.[1]

- In 2003, ISEOR and the Metropolitan Autonomous Universities (UAM) obtained ECOS-NORD program support for the second time and on exceptional terms for a 4-year period in order to consolidate the results attained with the first project. This second project was entitled "Socio-economic development of Mexican enterprises and organizations and proactive change strategies," one of whose objectives was establishing a doctoral program in Management *Sciences* focused on developing the profession of teacher-intervener-researcher and socio-economic management.

• *Cooperation* agreements were signed by ISEOR and the University Jean Moulin Lyon 3 with three UAM units of Xochimilco and Iztapalapa for the application of the methodology in one faculty, two research departments, one administrative sector and their libraries, and participation in the doctoral program in organizational studies as research seminar directors.

A network was also created with five regional universities (Sinaloa, Hidalgo, Aguascalientes, Guerrero and Mexico) in an attempt to promote the Management Science doctoral programs with the orientation proposed by SEAM—sticking close to the real problems of organizations with intervention utilized as research technology. As this brief listing of events underscores, there was a continuous effort to diffuse the SEAM approach. Relationships were also consolidated in the academic sector, where ISEOR's concept of the teacher-intervener-researcher could be truly developed.

The State of SEAM Applications in Mexico: An Illustrative Case

Applications carried out in Mexico since 1997 are presented in Table 11.1. The content of these applications is illustrated by the case of a state university.[3] The university was founded in December 1973 in response to the growing demand for higher education in the metropolitan area of Mexico City. The university was organized in three units, each one endowed with a great deal of autonomy and with decentralized modes of administration and operation. Each campus (unit) was placed in a different geographical zone chosen on the urgency of social, economic and cultural development needs, which is one of the reasons each unit adopted its own organizational *structures* that do not necessarily correspond to those of the other two units. The university has carried on with the application of socio-economic management in the units where it was set up and has created a research center dedicated to SEAM (LIMSE). The latest results of *socio-economic intervention* obtained by the UAM University were presented at the Merida Colloquium, co-organized with the ISEOR in November 2004.

Approximately a year and a half after the 1995 training session (conducted by H. Savall and V. Zardet), the rector and campus secretary signed a cooperation agreement to set up SEAM methodology in the University. This decision began a 4-year experiment during which SEAM was applied to the education and research departments of the Faculty of Bio-

Table 11.1.   *SEAM* Applications in Mexico

States Where Interventions Took Place	Organizational Units of the Intervention	Intervention Scope	Intervention Period
Veracruz	7 enterprises	5 ***diagnostics*** 2 complete interventions	1996-2001
Yucatan	Oficialia mayor (government ministry of the state of Yucatan)	Complete interventions	Since 1999
	4 enterprises in the textile and agro-food sectors.		
Hidalgo	3 enterprises in the domains of tourism and the food industry.	Diagnostic	Since 2004
Distrito Federal (Mexico City)	12 small and medium enterprises in clothing and service sectors	In the enterprises: 11 diagnostics and 1 project	1997-2000
	1 state university	At the university: complete process in 3 of the 6 intervention sectors; 2 diagnostics	In 2004 university authorities requested resuming intervention in one sector.

logical Sciences and Health (BSH) and four administrative sectors, among which were the three university libraries of the three different units (campuses).

Among the contextual forces that contributed to a better understanding of the intervention's objectives (see Table 11.2), there were four key factors:

- The energy, owing to the design of the organizational structure, which emanated from a key person (the full-time teacher-researcher) who was called upon to assume two of the three functions of the university: research and education, as well as knowledge diffusion.

- The University-Enterprise relationship which, at this particular university, had fallen behind in the social science sector. The Educational Planning and Coordination Committee, officially responsible for interfacing that relationship, was separate from the teacher-researchers who, being most apt to discuss their research project outputs with potential users, should have constituted one of the poles of the interface. In essence, the professors lacked the experience to successfully connect with the organizations in their environment (i.e., potential users of their research findings). Yet, the institutional structures that would authorize full-time professors to establish contact with real-world organizational problems had not been created, the only exception being a professor's private initiative.

- Student social services, which could have constituted another point of contact between the organizations, were restricted by *regulation* to enterprises in the public sector or the university. Many students enrolled in the social service programs only to fulfill the legal requirements for obtaining their diplomas.

- Voluntary or mandatory training internships in enterprises for students did not exist.

Intervention Objectives

Following SEAM methodology, intervention proposals gave detailed descriptions of the objectives, *methods and services to be provided (OMSP)*, the intervention architecture and the *provisional intervention schedule*. Four agreements on these proposals were signed by the consultants, the teacher-researchers and the university:

- The first agreement was signed by the rector and the campus secretary, covering interventions in a library, an academic coordination department and a faculty.

- The second agreement was concluded with the director of the Biological Sciences and Health Faculty (BSH), concerning coordination of eight educational programs. Over the course of the procedure, the heads of the research teams in all four departments that comprised the BSH faculty joined the intervention. The third agreement was signed with the head of an academic department for a vertical intervention.

- The fourth agreement was signed by the secretary general and covered the university libraries on all three campuses.

The learning acquired by the group of intervener-researchers being trained by carrying out the first agreement provided the base on which a number of related interventions were constructed. For example, core knowledge ("know-how") was transferred on ways of assisting organizational heads in expressing and clarifying their expectations and needs and transforming them into intervention objectives. Gradually these services were almost entirely taken over by internal interveners as the external consultants transferred their know-how. Thus, for the fourth agreement, whose intervention space was the BSH Faculty, 80% of the

**Table 11.2.   Overall Objectives
of the *Intervention-Research* in the State University**

	Intervention		
Cycles	First Cycle 2 years	Second Cycle 2 years	Evolution and State of Intervention at the end of 2004
Unit X Department Library Academic Systems Faculty	*Set-up* a process of *socio-economic* change and *innovation* in 2 administrative services and 1 faculty	Make the intervention profitable regarding first cycle performance *improvement* of *dysfunctions* thanks to concrete strategies among hierarchical heads of the university libraries	The hierarchical heads of the 3 libraries pursued their cooperation on strategic decisions. A new request for intervention was made at the end of 2004 to set up a quality control system for evaluating administrative support services to academic education.
Library system of the three Units		Cooperate in building a federated library system that constitutes a national reference in the disciplines of the university.	The same database management system was adopted in all three libraries
		Become involved in the *communication-coordination-cooperation (3C)* frameworks of the libraries of the 3 units. Form a permanent collegial body having a composition analogous to that of the *core group* of the *socio-economic project*. This framework permits submitting to the union proposals for "academizing" the libraries and responds, at the same time, to the expectations of the administrative personnel.	Cooperative personnel management strategies shared by the hierarchical heads of all three libraries.

(Table continues on next page)

intervention services were carried out by internal interveners. Table 11.2 presents a synthesis of the overall *objective-products* that were broken down into *method-products* and *service-products*.

**Table 11.2. Overall Objectives
of the Intervention-Research in the State University**

| Cycles | Intervention | | |
	First Cycle 2 years	Second Cycle 2 years	Evolution and State of Intervention at the end of 2004
Biological Sciences and Health Faculty		In the framework of the 4-year development plan, identify potential performance **improvement** for the faculty. Special attention is given to graduate schools and to their relationship with research and support services.	Request by the heads of research teams to resume intervention. Doctoral research focused on better understanding and encouraging the university-enterprise relationship is currently in progress.
Biological Sciences and Health Faculty		In the framework of the 4-year development plan, identify potential performance **improvement** for the Faculty. Special attention is given to graduate schools and to their relationship with research and support services.	Request by the heads of research teams to resume intervention. Doctoral research focused on better understanding and encouraging the university-enterprise relationship is currently in progress.
Economics and Management Faculty	Form a group of **internal researcher-interveners**, experts in the concrete application of the socio-economic method. Adapt the **SEAM engineering** method to the "culture" of state universities in Mexico.	Integration of the **SEAM engineering** laboratory. Recomposition of the intervention group as new members arrive and certain initiators leave	5 professors have completed the doctoral training program in **socio-economic management**.

Note: An intervention cycle includes four process stages: **diagnostic, project, implementation** and **evaluation**.

Intervention Architecture

As the second element of the intervention proposal, the architecture shows how focus groups are set up within the organization's structure, taking charge of transformative actions resulting from the interaction between structures and behaviors. The architecture illustrated in Figure 11.1 depicts an intervention carried out in the libraries, the goal of which was to "academize" the library personnel, that is, help them to function in ways that served faculty and research needs, and not in the "ways and customs" that the administrative personnel had managed to impose on teachers and students with the support of unions and the permissiveness of university authorities. These ways and customs, inspired by 1970s discourse and adopted by the administrative personnel, compared the university to a profit-making enterprise that exploited its workers and considered university directors as "bosses." In the socio-economic *intervention architecture*, these interest groups were incorporated at the *project phase* of the process.

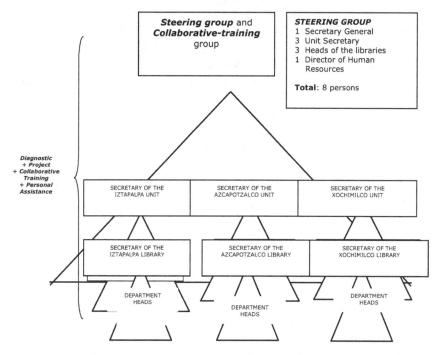

Source: ©ISEOR 2000.

Figure 11.1.   *Intervention architecture* in the libraries of the state university.

The heads of all three libraries enjoyed a great deal of autonomy and took decisions concerning personnel management in autonomous fashion, even though it directly affected operations in the other two libraries. In contrast, information from all three libraries converged toward the union concerning these different policies and management practices. This situation strengthened the union, which adopted the strategy of fighting for the extension of the most favorable *working conditions* negotiated in one library to all three. Thus, the context was that of authorities in a state of noncommunication without common strategies. Their only shared strategy to avoid conflict with the union seemed to be toleration and even compliance with practices that corrupted the quality of administrative services intended, in principle, to support the activities of the academic community, in particular those of undergraduate students (95% of the student population). This situation grew even worse during the transition to a system of automatic information processing.

The intervention architecture brought the key parties together for regular meetings over an 18-month period: (1) the secretary general of the university, the three campus secretaries, and the directors of the three libraries forming the project *core group*, 2) and the directors of the three libraries and the department heads from each library in the *plenary group*. One of the most revealing outcomes produced by the intervention was insight into the problem that the differences that existed between the libraries resulted from lack of *communication-coordination-cooperation* among authorities rather than from demands inherent to the different disciplines served by the three units.

Four years after the intervention, the three libraries continued to evolve toward a federated system with common databases and common user-service policies. The person who at the time was secretary of one of the Units declared,

> Today, as coordinator for the university-enterprise relationship for the Rector of our institution, I truly appreciate the intervention architecture that I learned through SEAM for managing projects involving all sorts of organizations, each one having a considerable degree of autonomy. Such architecture constitutes a structured framework for *cooperation* and implies reflection, communication, coordination, commitment.

Intervention Scheduling: Application to a Faculty

The sequence and duration of the activities that facilitate attaining the objectives of the socio-economic change process, which were agreed to by the intervener and the director of the faculty, were set down in the *provisional intervention schedule* (the third element of the proposal). The architecture shows which individual actors will participate in the different training and focus groups (*core group, plenary group, task group*). The

Month/ Year Activity	Nov. 96	12/ 96	1/ 97	2/ 97	3/ 97	4/ 97	5/ 97	6/ 97	7/ 97	8/ 97	9/ 97	10/ 97	11/ 97	12/ 97	1/ 98	2/ 98	3/ 98	4/ 98	5/ 98	6/ 98
ISEOR trips to Mexico	1						2						3					4		
Steering Group	1		2		3			4					5			6		7	8	
Training team of *Internal Interveners*	1						2						3					4		
Collaborative Training of Management	1		2	3		4	5	6	7	8			9							
Personal Assistance to Management					1					2	3	4	5						←	→
Socio-Economic Diagnostic				←	→															
Socio-Economic Innovation Project								1	2	3	4		5							
Project *Implementation*							←													
Evaluation of socio-economic innovation actions *results*																		←		→

Content of Cooperative Training Sessions

1) Principles of Socio-Economic Analysis
2) Tables of Skills
3) Integrated Training Plan
4) Hidden Cost Calculation and Interpretation Principles
5) Socio-Economic Innovation Project Method
6) Time Management
7) Piloting Indicators Logbook
8) Priority Action Plan
9) Periodically Negotiable Activity Contract

Content of *Personal Assistance*

1. Competency Grid
2. Time Management
3. Piloting Indicators Logbook
4. Priority Action Plan
5. Periodically Negotiable Activity Contract

The underlined numbers indicate services performed by ISEOR *intervener-researchers*.

Source: ©ISEOR 1996.

Figure 11.2.   ***Provisional intervention schedule*** for the first cycle of intervention in two administrative departments and one faculty.

project schedule specifies when and for how long these actors will participate, in order to attain the precise goals of the ***OMSP (objective-method-service-products)***. The schedule presented in Figure 11.2 illustrates the first intervention in the university and Table 11.3 shows that of the Biological Sciences and Health Faculty. By comparing these two schedules, one can measure the increased number of services assumed by internal University interveners.

Table 11.3. *Hidden Cost* Synthesis
by Indicator and by Entity (US Dollars)

	Administrative Services Department					
	Excess Salaries	*Overtime*	*Over-consumption*	*Non production*	*Noncreation of Potential*	*Total*
Absenteeism	NE	NE	NE	62,909	NE	62,909
Work Accidents	NE	NE	NE	NE	NE	NE
Personnel turnover	NE	NE	NE	NE	NE	NE
Nonquality	3,909	4,727	NE	28,364	NE	37,000
Direct productivity gaps	NE	1,364	NE	NE	NE	1 364
Total	3,909	6 ,091	NE	91,273	NE	101,273

Number of persons: 24
(USD) : $4 220 per person/per year

	Faculty					
	Excess Salaries	*Overtime*	*Over-consumption*	*Non production*	*Noncreation of Potential*	*Total*
Absenteeism	NE	NE	NE	NE	NE	NE
Work Accidents	NE	NE	NE	NE	NE	NE
Personnel turnover	NE	NE	NE	NE	NE	NE
Nonquality	NE	1,091	18,118	NE	14,364	33,573
Direct productivity gaps	NE	32,727	NE	9,182	NE	41,909
Total	NE	33,818	18,118	9,182	14,364	75,482

Number of persons: 80
(USD) : $944 per person/per year

Source: ©ISEOR 1998.
Notes: Part of **hidden costs** cannot be reduced; these correspond to the university's flexible operation. Part can be reduced and transformed into **immediate results** and creation of potential. NE: Not evaluated due to the amount of time allotted to the study. Remark: underestimated results.

Hidden Costs

Hidden cost calculation (Savall & Zardet, 1987, 1992) was carried out during diagnostics of the library, the academic systems department and the faculties. A synthesis of this calculation is presented in Table 11.3. The calculated total is underestimated, for only those elements not susceptible

Table 11.4.   Calculation of the *Hourly Contribution to Margin (or Value-Added) on Variable Costs (HCMVC or HCVAVC)*

Margin on variable costs 1996	
Budget (thousand dollars)	30,728
variable costs	–4,612
Sales price to bookstores	–44
Stock	–69
Buildings and installations	–438
Work in progress	–2,811
Available budget	–62
Margin on variable costs	**22,692**
Work hours	
Number of employees (persons)	2,136
Number of workdays (days/year)	223
Number of work hours	1,784
Total hours expected	**3,810,624**
Hypothesis adopted for the calculation:	
workdays at the university per year	**223**

366 total days:
- 104 Saturday and Sunday
- 13 holidays
- 26 vacation days

Hourly contribution to margin (or value-added) on variable costs (HCMVC or HCVAVC) :
22,692 / 3,810,624 = $5.95

Source: ©ISEOR 1998.
Note: Calculated without consideration of the 42-day strike period. Hidden costs are calculated on a yearly basis

to being questioned were taken into account, given the availability of data and the degree of cooperation from actors interviewed.

The effect obtained with these tables and the detailed explanation of the procedure that permitted calculating the *hourly contribution to margin on variable costs (HCMVC)*, presented in Table 11.4, was the conviction that, with the resources already available, the University could produce services of significantly higher quality than was currently the case.

From Diagnostic to Project: Dysfunction Baskets and Pivotal Ideas (Idées-Forces) in the Horizontal Intervention

Across all the interventions, one of the most spectacular moments, for its power to transform behaviors, is the *mirror-effect* during which the different images that different actors had construed of the workplace were displayed, recognized and discussed. Emphasis was placed on identifying

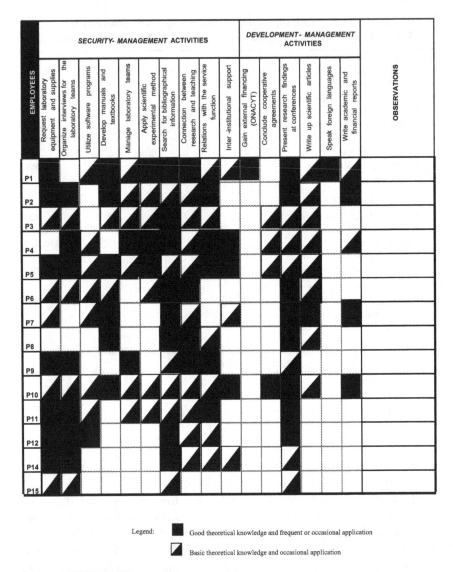

Legend: ■ Good theoretical knowledge and frequent or occasional application

◪ Basic theoretical knowledge and occasional application

Source: ©1997 ISEOR.

Figure 11.3.   *Competency grid* of a faculty.

discrepancies in their expectations and perceptions, and their underlying causes. The mirror effect, which reveals individual and group mental structures, underscores the difficulty in creating a unifying vision of the project that will harness sufficient determination and energy to enable the change.

Competency Grid

Among the applications of the SEAM method, the **competency grid** is a product that was particularly appreciated by the participants. This tool revealed the root-causes of multiple dysfunctions in different domains of action, for example, **work organization** (who does what task), **integrated training** (what are the priorities in personnel training), and **time management** (concentration of work on certain employees who find themselves overloaded which causes **nonquality**). Tracing the relationships between the objectives of the University's development plan, the ability of a given service to carry out its activities and the multiskill level of its members provides important reference points for decisions on the use of training resources. As an illustration, Figure 11.3 presents the Competency Grid for professors in a research unit of the faculty.

The overall intervention objective at the BSH Faculty (see Table 11.2) was broken down into the following specific objectives:

- Develop the competency required for teaching at the undergraduate level, beginning with the core curriculum common to all programs in accord with the principles adopted by the university concerning teaching in modules;
- Update training plans and programs;
- Promote and maintain high quality standards in all support services related to teaching;
- Identify, formalize and circulate the policy and procedures related to services offered by the different departments and laboratories; and
- Know, understand and implement the organizational structure of the university as a whole.

These objectives served as reference points for constructing **dysfunction baskets**; they marked the beginning of the SEAM project development stage. SEAM begins development of the **project phase** based on the definition of these baskets. In the case of the BSH Faculty, the four baskets (see Table

sum up the dysfunctions expressed by members of the Faculty as well as their **non-dit (unvoiced comments)**, expressed by the intervener in his or her role as expert and which were validated by members of the unit where the diagnostic took place.

Major Outcomes

A significant outcome of the present intervention was the creation of a management research laboratory (Laboratorio de Ingeniería en Manage-

**Table 11.5.   Excerpts from the
Dysfunction Baskets of the BSH Faculty**

Basket 1:   Update curriculum for the 8 educational programs offered by the BSH faculty

Work Organization

ND7: Daily activity absorbs almost all of the department heads' energy in conducting and coordinating teaching; no energy left for improvement efforts and for the creation of potential.

Time Management

JEE19: Conflicting priorities between research and teaching, leading to work overload evident in poorly done tasks: professors not supervising student laboratory work, result: equipment damage.

Integrated Training

JEE22: Full-time professors lack competency concerning professional practices in industry, which results in bookish education.

Strategic Implementation

N10: Functions such as social services with very high potential for interfacing the institution and the society are being reduced to bureaucratic functions.

Basket 2:   Develop management competency, particularly those concerning *time management, communication-coordination-cooperation (3C), scheduling,* follow-up and evaluation

Work Organization

JEE6: The workload of the master's program coordinators is judged excessive because of untimely demands from students, professors and authorities.

Time Management

N14: Very strong restrictions in an endeavor to master time use. The idea prevails that this problem is linked to a Mexican idiosyncrasy at the national level, and not only at the level of this university.

Integrated Training

ND4: The institution lacks training policy for its directors. Consequence: a long learning curve of trial and error.

N9: Teachers are very often unfamiliar with regulations and feel unconcerned about them, claiming that is not their responsibility—a situation which induces, for example, procedural demands carried out at the last minute and a bad workplace atmosphere.

Strategic Implementation

N5: Rupture between terms of office between university and faculty authorities. The fear is that accompanying the former administration in order to recover its experience could be interpreted as a lack of competency on that part of the new head.

(*Table continues on next page*)

ment Socio-Económica: LIMSE), where teacher-researchers and doctoral students are able to conduct fieldwork as well as theoretical and methodological reflection, particularly in the *qualimetrics* methodology (Savall & Zardet, 2004) developed by the ISEOR. This group implements socio-economic management principles, theories and methodology in Mexican organizations to improve their performance and to identify convergence and specificity of applications in different contexts. The purpose of this

Table 11.5.   Continued

Basket 3:   Improve knowledge, understanding and management of the *infrastructure* of the Faculty

Work Organization

JEE7: The basic infrastructure is challenged. There exists no shared appreciation on what is implied by the definition of objectives, priorities and time management.

Communication-Coordination-Cooperation

JEE13: Doctoral programs have become strongholds where no one really knows what goes on due to a lack of communication, coordination and cooperation among program directors.

Time Management

JEE15: Heads of departments dedicate much of their time to meetings without always being able to contribute significantly. Meetings where their presence is mandatory should be distinguished from meetings where another person could replace them.

Integrated Training

JEE23: Due to a lack of university scheduling, professors attended training sessions for doctoral students not related to institutional research and/or teaching programs.

Strategic Implementation

JEE25: Even though there is general recognition of the provisional nature of the criteria for attributing budget to research units, those who have received financing refuse to discuss different *game rules*.

JEE27: Certain services provided by administration departments are perceived as threats because of their budgetary repercussion, considered inflated with respect to their actual contribution to research and teaching objectives (travel expenses, airplane tickets, purchase of computers).

Basket 4:   Identity, formalize and circulate procedures and physical locations of functional and operational heads, of laboratories, supply room and laboratory equipment inventory.

Work Organization

JEE9: There is no clear updating of manuals and procedures; consequence: time lost, work overload, same job done twice, job done which is not assigned to the position, excessively centralized decision making.

Communication-Coordination-Cooperation

JEE12: There are no updated publications due to CCC failure among the different departments and offices, and especially with the library.

Strategic Implementation

JEE28: The budget allotted to teaching is not known sufficiently in advance, nor is there sufficient information about the current state of the available laboratory team, thus upon what basis can decisions be made?

Source: ©ISEOR 1999.

---

work is to enhance management theory and methodology and to contribute to improving the overall performance of organizations. The accomplishments thus far include:

- Training a *core group* of professors in social science research methodologies, thus enabling them to acquire the competency necessary for supervising doctoral research.

**Table 11.6.   Comparison of *Dysfunction Baskets*,
Strategic Programs of the Development Plan and *Key Ideas*
for the *Improvement Project* in a University Faculty**

Strategic Programs of the Development Plan		Baskets	1	2	3	4	5
Program for curriculum enrichment	1.	Up-date curriculum and course syllabi.	ND7 ND8		JEE17 JEE19	JEE21 JEE22	JEE32 JEE35 JEE36 ND10
Program for teacher training	2.	***Development*** of management competency: ***time management, communication-coordination-cooperation,*** planning, scheduling, follow-up, evaluation.	JEE6		ND14	ND4 ND6 ND9	JEE31 ND5
Program for organizational ***structure*** management	3.	Knowledge, understanding, manage and ***implementation*** of the matrix organization structure	JEE7 JEE8	JEE13	JEE15 JEE18	JEE23	JEE24 JEE25 JEE26 JEE27 ND16
Program for adequacy between administrative structures and faculty	4.	Knowledge, formalization and deployment of procedures and locations of: • functional and operational hierarchical superiors • laboratories • laboratory equipment inventory.	JEE9	JEE10 JEE11 JEE12			JEE28

Note: Baskets: 1. ***Work organization.*** 2. ***Communication-coordination-cooperation.*** 3. ***Time management.*** 4. ***Integrated training.*** 5. ***Strategic implementation.***

- Establishing a "university-enterprise" relationship necessary to carrying out research in the framework of the ***socio-economic theory of organizations,*** an integral part of the doctoral program "Organizational Studies" organized by the University UAM-Iztapalapa (which began in October 2002).
- The First International Conference (Mexico City, 2000) on the application of ***socio-economic management*** in Mexican organizations (Fernandez Ruvalcaba & Savall, 2004). Participants included: the directors of the various departments where the model had been experimented, the Mexican professor-consultants who participated in the experiment, university faculty and staff members interested

in intervention-research methodology, independent consultants, representatives of employment organizations, representatives of the French Embassy in Mexico, and the International Labor Organization (Savall, Zardet & Bonnet, 2000).

- In 2004, the government of the State of Yucatan, in collaboration with the ISEOR and the Universidad Autonoma Metropolitana de Mexico, organized the Second International Conference (in Merida) on the application of *socio-economic management* in Mexico. Directors of the enterprises and organizations where the approach had been experimented presented their testimonies: cases included the Ministry "Oficialía Mayor" (Yucatan Sate government) and SME's in agri-food and textile sectors. From Europe: the case of Brioches Pasquier, the leading enterprise of the ISEOR which has successfully applied socio-economic management for over 20 years, was presented by its CEO, Serge Pasquier, as well as the case of a Belgian government organization in charge of employment and training (FOREM), presented by its general manager Jean-Pierre Méon (see chapter 5).

- As summarized in Table 11.1, there have been significant reserves for improving performance, largely due to dysfunction diagnostics. Problems related to symptoms, inclinations and behavior patterns (structures) have been identified in a broad array of Mexican organizations, and schemes for overcoming chronic operational management problems have been worked out and applied.

Thanks to the application of *socio-economic management tools*, senior-level managers throughout Mexico have developed their management competency, including: better time management and improved meeting conduct, as resolutions were substantiated with documents and successfully applied and business strategies have become increasingly meaningful as they have been broken down into work units for every workstation, with agreed on priorities by employees and their hierarchical superiors.

BEHAVIOR IN MEXICO

Overall, the introduction of the socio-economic approach to Mexican businesses and organizations has been highly successful, although the approach is still in its introductory stages in approximately 30 enterprises and organizations. An ongoing challenge in Mexico is the difficulty for individuals, once trained as consultants, to perform in rigorous fashion the services provided by socio-economic management methodology. In fact, the number of consultants involved today in socio-economic management in Mexico is still relatively small (approximately 20), composed largely of teachers without university-enterprise experience, and consultants having attained a certain degree of maturity applying their own method based on their own experience.

Hidden cost calculation has been one of the greatest difficulties for professor-consultants. They often fail in negotiating access to information from actors, or in setting up frameworks capable of generating it when information is not available. As far as professional consultants are concerned, some of them demonstrate competitive rather than cooperative behavior. For example, a case of appropriation of the socio-economic method was observed, without proper reference to its original authors. There was also a case of distortion of the method: *socio-economic management* was cited, but another method was implemented, disguised with socio-economic management terminology.

Principal Conditions for Success

An authentic commitment on the part of senior-level managers to attain real performance improvement through *socio-economic intervention* is continuing to grow. Yet, there are a number of lingering concerns. In Mexican public institutions, for example, there is a *risk* of utilizing the *SEAM* approach to develop a pro-active image of *efficiency*, without accepting the need to question one's management style. The fear also exists that socio-economic intervention might reveal new leaders, demonstrating initiative, capabilities and credibility, which could be threatening to the existing hierarchical order. These concerns, however, can be overcome, thanks to the unequivocal commitment of managers to the application of socio-economic methodology.

Today, in 2006, 10 university professors (UAM of Mexico City and states of Sinaloa and Hidalgo) have obtained a PhD in management sciences, with a specialization in socio-economic management prepared at the ISEOR. They now ensure training and socio-economic *intervention-research* in Mexican businesses and organizations. In conjunction, the ISEOR has carried out socio-economic interventions in various Mexican State governments (Districto Federal, Yucatan, Durango, Michoacan, Baja California, Hidalgo et Oaxaca) and has trained approximately 20 internal interveners, in addition to training professional consultants. The partnership between the University Autonoma de Mexico and the University Jean Moulin of Lyon 3 has significantly grown since its beginning in 1995. Two colloquiums coorganized in Mexico City in 2000 and in Merida in 2004 made it possible to introduce SEAM, based on case studies presented by enterprises and public organizations. Appropriately, the Oficialia Mayor of the State of Yucatan recently became the first recipient of the Socio-Economic Management Certificate discerned by the ISEOR.

The continued *improvement* of Mexican organizations through *SEAM intervention* is indeed promising.

ACKNOWLEDGMENT

This project is supported by the ECOS NORD Program and the Universidad Autónoma Metropolitana de México (UAM).

NOTES

1. Funding for 36% of the travel expenses for these projects was provided by ECOS-NORD and 64% by the UAM. One half of the training program costs was financed by ISEOR's own funding.

2. The student population for the bachelor and master degree programs at the state university was 46,845, with an additional 1,714 students enrolled in doctoral programs. Personnel at the university included 2,623 full-time faculty members, among which there were 840 PhDs, 775 graduate assistants and 1,008 undergraduate assistants or specialists. There were also 800 half-time or part-time professors and 5,200 administrative staff. The university offers 62 standard bachelor's degrees and 67 master's and PhD degrees.

REFERENCES

Brindis Almazán, L. (2007). *Structures organisationnelles pour un développement socio-économique local durable: La communauté artisanale* [Organizational structures for sustainable local economic development: The community of artisans]. Unpublished doctoral dissertation, University of Lyon, France/Universidad Autonoma Metropolitana de Mexico.

Enriquez Galván, O. (1982). *Analyse socio-économique des conséquences sur les conditions de vie au travail des décisions stratégiques de transfert de technologie. Cas d'entreprises mexicaines* [Socio-economic analysis of the consequences on work-life conditions of decisions to transfer technology: The case of Mexican enterprises]. Unpublished doctoral dissertation, University of Lyon, France/Universidad Autonoma Metropolitana de Mexico.

Enriquez Galván, O. (2001, November). *Socio-economic analysis of the consequences for working conditions of strategic decisions on technological transfer: Case of Mexican enterprises*. First Colloquium on SEAM Application in Mexican Organizations Conference, UAM, Mexico.

Fernandez Ruvalcaba, M. M. (1984). *Interactions entre conditions de vie au travail et conditions de vie hors du travail. Applications à des cas d'innovation socio-économique* [Interactions between work-life conditions and living conditions outside of work: Application to cases of socio-economic innovation]. Unpublished doctoral dissertation, University of Lyon, France.

Fernandez Ruvalcaba, M. M. (1998). Cas d'intervention socio-économique dans une grande université mexicaine et création d'une équipe de recherche-intervention [Cases of socio-economic intervention in a large Mexican university and creation of an intervention-research team]. In ISEOR (Ed.), *PME-PMI : Le métier de dirigeant et son rôle d'agent de changement* (pp. 193-201). Paris: Economica.

Fernandez Ruvalcaba, M. M., & Andrade Romo, S. (2000). La consultoría en administración como una tecnología de investigación: Alternativa del ISEOR y su aplicación en la UAM-X [Administrative consultancy as research technology: The ISEOR alternative and its application to the UAM-X]. *Revista Contaduría Y Administración, 198*, 29-45.

Fernandez Ruvalcaba, M. M,. & Estrada, A. R. (2001). Transferencia de modelos de consultoría organizacional: El enfoque socio-económico [Transferring models of organizational consultancy: The socio-economic focus]. *Administración y organizaciones, 4*(7), 87-96.

Fernandez Ruvalcaba, M. M., & Páramo Ricoy, M. T. (2004). Heterogeneidad social y sus retos: Procesos de consultoría organizacional [Social heterogeneousness and its challenges: Organizational consultancy processes]. *Revue Sciences de gestion management sciences, 41*, 203-248.

Fernandez Ruvalcaba, M. M., & Peñalva Rosales, L. P. (2004). Uso significativo de la cuantificación en el método de análisis de gestión socio-económica [Significant use of quantification in the analysis method of socio-economic management]. In H. Savall, M. Bonnet & M. Péron (Eds.), *Crossing frontiers in quantitative and qualitative research methods* (pp. 439-454). Lyon, France: ISEOR.

Fernandez Ruvalcaba, M. M., & Savall, H. (2004). *El modelo de gestión socio-económica en organizaciones mexicanas* [The model of socio-economic management in Mexican organizations], Mexico City, México: Editorial UAM.

Fernandez Ruvalcaba, M. M., Andrade Romo, S., & Chávez Cortez, J. M. (2001). Focus on relationships between internal and external management consultants: Experiment in the context of a nonprofit organization. In A. Buono, H. Savall, & G. Trepo (Eds.), *Knowledge and value development in management consulting* (pp. 593-605). Lyon, France: ISEOR-University of Lyon-HEC.

Fernandez Ruvalcaba, M. M., Martínez Vázquez, G., & Zardet, V. (2003). Transfert international technologie de l'immatériel au Mexique [International transfer technology of the immaterial to Mexico]. In ISEOR (Ed.), *Université citoyenne: Progrès, modernisation, exemplarité* (pp. 319-336). Paris: Economica.

Gonzalez Herrera, G. (2004). *Los desafíos para el cambio en organizaciones públicas en Mexico. Dos casos de innovación technológica en el Sistema de Transporte Colectiveo de la ciudad de Mexico* [The challenges of change in public organizations in Mexico: Two cases of technological innovation in the Public Transport System of Mexico City]. Unpublished doctoral dissertation, University of Lyon, France/Universidad Autónoma Metropolitana de Mexico.

Gonzalez Perez, C. R. (2006). *L'organisation créatrice de connaissances. L'organisation synaptique* [The knowledge creating organization: The synaptic organization]. Unpublished doctoral dissertation, University of Lyon, France/Universidad Autonoma Metropolitana de Mexico.

Martínez Alvarez, F. J. (2000). *La Organización Informal y la Empresa Familiar* [The informal organization and the family enterprise]. Unpublished master's dissertation in management sciences, University of Lyon, France.

Martínez Alvarez, F. J., & Monroy, G. S. (2001). Problemas y problemáticas permanentes y emergentes en organizaciones: Su tratamiento sistémico y el aporte del análisis socio-económico [Permanent and emerging problems and problematics in organizations: Their systematic processing and the contribution of socio-economic analysis]. *Administración y organizaciones, 4*(7), 141-151.

Martínez Vazquez, G. (1999). *El Papel de los Cuadros Administrativos en el proceso de Intervención Socioeconómica en una Universidad Pública Mexicana* [The role of administrative managers in the socio-economic intervention process in a public university of Mexico]. Unpublished master's dissertation in management sciences, Universidad Autonoma Metropolitana de Mexico.

Martínez Vazquez, G. (2001). El diagnóstico socio-económico en pequeñas empresas mexicanas [Socio-economic diagnostic in small Mexican enterprises]. *Revista Administración y organizaciones, 4*(7), 97-120.

Martínez Vazquez, G. (2005). *La contribution du management socio-économique à l'aménagement des performances éconimiques et sociales: Cas d'expérimentation dans trois petites entreprises mexicaines* [The contribution of socio-economic management to harmonizing economic and social performances: Cases of experimentation in three small Mexican enterprises]. Unpublished doctoral dissertation, University of Lyon, France.

Peñalva Rosales, L. P. (2001). Un acuerdo de colaboración periódicamente renovable como base para realizar servicios sociales fructíferos y con continuidad [A periodically negotiable activity contract as the basis for providing productive social services and their continuation]. *Administración y organizaciones, 4*(7), 153-168.

Peñalva Rosales, L.P (2006). *Inducción al Aprendizaje organizacional en la Universidad Pública para el desarrollo de estrategias de vinculación con el sector productivo* [Introducing organizational internships in a public university to develop strategies of connection with the productive sector]. Unpublished doctoral dissertation, University of Lyon, France, Universidad Autonoma Metropolitana de Mexico.

Pomar Fernández, S. (2007). *La nature hybride des organisations et le processus de transfert de modèles. Cas des crèches subventionnées au Mexique* [The hybrid nature of organizations and the transfer process of models: The case of subsidized daycare centers in Mexico]. Unpublished doctoral dissertation, University of Lyon, France / Universidad Autonoma Metropolitana de Mexico.

Ramírez Alcantara, H. T. (1999). *Pratiques de Communication Dans un Etablissement d`enseignement Supérieur Public au Mexique. Cas d`expérimentation* [Communication practices in a public establishment of higher education in Mexico: Cases of experimentation]. Unpublished master's dissertation in management sciences, University of Lyon, France.

Ramírez Alcantara, H. T. (2001). La confianza en el management socio-económico [Confidence in socio-economic management]. *Administración y organizaciones, 4* (7), 121-140.

Ramírez Alcantara, H. T. (2006). *Elementos estructurantes de la confianza y su relacion con el desempeño en una universidad publica mexicana* [Structuring elements of confidence and its relationship to performance in a public university in Mexico]. Unpublished doctoral dissertation, University of Lyon, France/Universidad Autonoma Metropolitana de Mexico.

Savall, H. (1974, 1975). *Enrichir le travail humain dans les entreprises et les organisations* [Work and people: An economic evaluation of job enrichment]. Paris: Dunod.

Savall, H. (1979). *Reconstruire l'entreprise: Analyse socio-économique des conditions de travail* [Reconstructing the enterprise: Socio-economic analysis of working conditions]. Paris: Dunod.

Savall, H. (1987). Les coûts cachés et l'analyse socio-économique des organisations [Hidden costs and the socio-economic analysis of organizations]. *Encyclopédie du management* (pp. 599-628). Paris: Economica.

Savall, H. (2003a). An updated presentation of the socio-economic management model. *Journal of Organizational Change Management, 16*(1): 33-48.

Savall, H. (2003b). International dissemination of the socio-economic model. *Journal of Organizational Change Management, 16* (1) : 107-115.

Savall, H., & Zardet, V. (1987). *Maîtriser les coûts et les performances cachés: Le contrat d'activité périodiquement négociable* [Mastering hidden costs and performances: The periodically negotiable activity contract]. Paris: Economica

Savall, H., & Zardet, V. (1992). *Le nouveau contrôle de gestion: Méthode des coûts-performances cachés* [New management control: The hidden cost-performance method]. Paris: Éditions Comptables Malesherbes-Eyrolles.

Savall, H., & Zardet, V. (1995). *Ingénierie stratégique du roseau, souple et enracinée* [Strategic engineering of the reed, flexible and rooted]. Paris: Economica.

Savall, H., & Zardet, V. (2004). *Recherche en sciences de gestion: Approche qualimétrique. Observer l'objet complexe.* Unpublished English translation: *Research in management sciences: The qualimetric approach. Observing the complex object.* Paris: Economica.

Savall, H., & Zardet, V. (2005). *Tétranormalisation: Défis et dynamiques* [Competitive challenges and dynamics of tetra-normalization]. Paris: Economica.

Savall, H., Zardet, V., & Bonnet, M. (2000). *Releasing the untapped potential of enterprises through socio-economic management.* Geneva, Switzerland: International Labor Office-ISEOR.

CHAPTER 12

INTERVENING IN A MULTINATIONAL COMPANY

Marc Bonnet and Henri M. Talaszka

The purpose of this chapter is to present a case of introducing the *SEAM* methodology (Savall, 1974, 1975, 1987, 2003a; Savall & Bonnet, 1988; Savall & Zardet, 1987, 1995, 2005; Savall, Zardet, & Bonnet, 2000) into a large subsidiary of a U.S. multinational company (MNC). The subsidiary, which was part of the MNC since the early 1960s, operated in the French biscuit and pastry market, where it served retail channels under national brands. It employed 2,600 employees in five bakeries. Over the years, the company experienced a series of mergers and acquisitions (M&A) and was faced with a broad array of *dysfunctions*, an issue that has been widely discussed in the M&A literature (e.g., Birkinshaw, Bresman & Håkansson, 2000; Buono, 1997, 2005; Buono & Bowditch, 1989; Marks, 1997; Mirvis & Marks, 1992).

The chapter is written by the company's general manager who oversaw the different approaches to change in the organization, and one of the people in charge of the ISEOR *intervention research* (Savall & Péron, 2003; Savall & Zardet, 2004) team. Our work with this company began in the late 1980s and the discussion focuses on the company's underlying decision processes, the reasons for selecting the SEAM approach, and an assessment of its implementation and outcomes.

Socio-Economic Intervention in Organizations: The Intervener-Researcher and the
SEAM Approach to Organizational Analysis, pp. 279–304
Copyright © 2007 by Information Age Publishing
279

INTERVENTION CONTEXT

During the late 1950s, the parent MNC invested heavily in continental Europe. Unfortunately, a basic lack of understanding of entrenched cultures, diversified consumer habits, and inappropriate human *resources* policies led to failures in a number of countries. In France, for example, an initial acquisition turned sour. In a second acquisition (the focus of this chapter), the MNC gradually learned from its early mistakes, switched its policy, and kept a local team in charge. The result was a turnaround of the ailing target company, a contribution that earned the respect of and full delegation from the MNC. Rather than being rebuffed by the complexity of these new environments, a guiding theme was that a "local team run by men of the art" would guide the strategy (see, for example, Buono & Nurick, 1992).

Over a 10-year time frame, the total staff of the acquired firm jumped from 800 to 1,800 people, including the additional acquisition of five exclusive regional distributors. Over a 4-year period, capital expenditures focused on the construction of a $24 million modern bakery and a $4 million dollar research and development center, roughly 50% of the company's annual revenues. However, poorly planned projects, with little effort to integrate the various operating sites, created a series of "baronies," entities with insufficient technical professionalism and absence of managerial skills. A result was a series of heavy losses, especially when the firm was faced with a very real adverse sociopolitical and economic challenge—the 1973 oil crisis.

In response, the MNC appointed a new president with a strong finance background. The new guiding theme was "entrepreneurial men of the art, take a step backward!" The company then struggled for over a decade, as tension mounted between the accountants and "men of the art," between economics, innovation and quality.

COST REDUCTION VERSUS TOTAL QUALITY: FRUSTRATION AND THE NEED FOR AN INTEGRATIVE APPROACH

The company's strategy consisted of creating innovative high quality products while simultaneously cutting costs in order to increase the profit margin of its existing products. The first axis of the strategy was implemented by the firm's research and development (R&D) department. Implementation of the second axis consisted of a two-sided action plan focused on (1) cost reduction (cost accounting, budgetary control, expenditure requests, payback computations) and cost cutting (restructuring, functional centralization, downsizing, inventory management, capital expenditure freeze) and (2) product portfolio management (including an emphasis on product margins and pricing strategy). After 2 years on this "financial diet," the company showed positive results with increasing profits and the MNC was satisfied with the firm's performance.

Short-Term Profitability Versus Long-Term Innovation

The MNC's reporting system was exclusively financially oriented, and much of the subsidiary's success was based on a "milking strategy." Despite what appeared to be a favorable profit picture, the continued cuts in potential-creation spending were undermining the company's future earnings. While the directed change allowed outstanding profits, it created a technocratic and centralized yoke which froze individual initiatives. It was also becoming apparent that the firm's marketing and research and development (R&D) functions were underdeveloped, and creativity, the company's main asset, and shareholder goodwill were rapidly declining. The subsidiary, which had relied on product innovation and high quality, was becoming stagnant and the most modern bakery in Europe was seen to be at risk.

The underlying problem emerged to the surface through competitive pressures from what had been a modest competitor in the milked-food category. Excessive price increases were placing the subsidiary's products out of the market, opening an opportunity for a price-based entry, a move that was backed by distributors that were fed up with the subsidiary's high prices. The smaller firm rapidly captured a 3% market share. The subsidiary was destabilized by this unanticipated move and it began to reveal the opposition between finance and operations that had largely been masked by the firm's apparent success. It was becoming apparent that the subsidiary's cost-control management orientation was a shortsighted policy, one that was leading the company to failure and under-utilizing plant capacity.

At the same time, the subsidiary's competitors announced a merger of seven companies, which gave the new entity a dominant market share. The subsidiary was now in a distant second position. The competitive environment was also impacted by the concentration of distributors, a move that increased their bargaining power.

The recently nominated European subsidiary's new head and several inside managers became increasingly conscious of the risk. The emerging plan was to turn the situation around through increased product differentiation, without getting trapped in the subsidiary's past cost-cutting strategy. The challenge was to reduce the gap and tensions between finance and operations, achieving a balance between innovation, quality and economic constraints.

The Potential and Limits of Total Quality Management

The subsidiary's initial "solution" was found in an American business school. A management team member, the director of finance and general manager of one of the business units, attended a senior management development program. This program, which was split into two sessions, focused on the success of Japanese management practices and the power of a Total Quality Management (TQM) approach. The program's emphasis on differentiation strategies, innovation, quality and service was favorably received by the managers.

It appeared that the subsidiary had the blueprint for its new strategy, in essence making money through superior quality and innovation rather than solely through cost reduction. It appeared that the company had the way to bring together operations ("men of the art") and finance, with a full set of principles, rules, methods and tools to guide its efforts. The basic goal was to restore the firm's dynamic energy in market place, improving innovation, quality and financial results. It was felt that the subsidiary was run by a highly professional team that now had a shared project and shared values, which would serve as the means to overcome an attitude of "Everything is fine! Why change?"

In early 1985, a strategic plan was developed that created a set of common goals and objectives. Reporting to a pilot group, several multidisciplinary strategic *focus groups* were created, project responsibilities with extended delegation were assigned, and budgetary resources were allocated. Change was introduced on a number of different levels, including organizational *structure* (moving from a functional to a product-centered divisional structure), values (enhancement of new mindsets, innovation *and* cost awareness, sense of common interests), behaviors (manager as coach, developer of talent), vertical (involving delegation and *evaluation* meetings) and lateral (task forces, multidisciplinary groups) relationships, and *strategic implementation* (e.g., introduction of a management by objective system).

Reflecting on the changes at the subsidiary during this period, the General Manager noted:

> The working climate is noticeably improving. Men of the art are smiling once again. Only the finance people are refusing to participate. Human resources people are in the spotlight. Creativity and initiatives are blossoming. The project portfolio is filling up once again. A stimulating and contagious enthusiasm is pushing evolutions forward.

Short-Lived Momentum

The subsidiary gained promising momentum. Launching a number of new products helped the company restore its position in the market place.

Unfortunately, the anticipated increase in earnings was not reached. Although improved, the earnings increase was below expectations and disappointing given the dose of energy injected. At the same time, there was growing enthusiasm among managers and the subsidiary's new participative management model was embraced by organizational members.

The organization, however, was increasingly being overwhelmed by the number of stacked-up projects and the managers' inability to run all of them efficiently. The management team began to become restless and tensions rose once again. In spite of enhanced communication efforts and the new MBO program, *strategic implementation* was weak. Managers began to complain that the majority of subsidiary's personnel were not supportive. Organizational members, in contrast, increasingly felt that the managers were unavailable and that they no longer listened to their ideas. The internal climate became very tense, and the new plant, which was still the most modern in Europe, was hit by a limited strike, the first in the company's history. The newly unleashed energy throughout the subsidiary seemed to be consumed in a short-lived blaze.

Problem Recognition

After nearly 2 years, the TQM project had mustered less than 100 champions. The project was largely shared among these champions, increasingly seen as "the group of the privileged few." This group was in charge of strategic projects, which afforded its members high *visibility*, in essence, a pathway to success. All their efforts and energy were directed toward the market place, with very few projects oriented toward internal operating matters, which contributed to increased delays and dissatisfaction. This majority of employees did not feel involved and they were not swayed by top management's "stratospheric objectives" when their daily problems, which constituted a potential source of savings, were not taken in consideration.

In the wake of this building discontent, four core issues were identified. First, managers had not followed through in enhancing their managerial skills and practices. It was felt that the example must come from the top of the hierarchy, an absolute necessity if management was to "pull" the personnel upwards, in terms of professional standards, both individually and collectively.

Second, the subsidiary's roughly 400 supervisors felt threatened. Objectives assigned by management, now materialized by the MBO program, were more demanding. In addition to the usual *productivity* demands, quality improvement and a mandatory "request" to change their management style overwhelmed the supervisors. They were also confronted by the possibility of organizational retrenchment, cutting the subsidiary's hierarchical levels (see DeMeuse & Marks, 2003). Questions abounded about their future with the company. Yet, despite these demands and stressors, the supervisors received very little personal development, training or additional *management tools* to support their efforts. As a result, thus supervisors were increasingly criticized by upper management, for being unable to "drive the troops," and by their employees, for being unwilling to act according to values as posted on the company premises. Both sides began to see the supervisors as "petty tyrants."

Third, the "new" management practices, particularly the MBO system, were deemed insufficient. Although MBO was supposed to be the solution gearing everyone to common strategic decisions and action plan implementation, it was not seen as a success. Despite a program-specific communication program, the objectives put forth by the system were seen as unrealistic and unfairly assigned, at times even triggering conflict among employees. The demands from the top created a backlog of projects, but without any priority assignment to the point where employees felt that to upper management "Nothing seemed impossible!" As one of these individuals noted,

> Their demands are broken down into objectives in a top/down procedure.
> The bottom/up validation process doesn't work. But no one dares to say no!
> This would be dangerous from a personal point of view.

The company had neglected introducing appropriate tools for calibrating, on a factual basis, the resources necessary to reach these objectives. While this was calculated on the basis of financial resources, it was never examined in the context of time resources. Assignment became a matter of authority and, as a consequence, objectives were either submissively accepted or else rejected in an attitude that was considered "rebellious."

Finally, the delegation granted to *project leaders* was insufficiently supported by *piloting* and appropriate information. With the exception of existing financial reporting, no other indicators were available to communicate a clear understanding a project's status. *Project leaders* were isolated, with neither warning nor assistance systems, and the projects began to slip in terms of delays, targets and final outcomes. In general, there was a gradual return to the financial approach, which literally choked the local "baronies" as each site tried to protect its own interest in "common" projects.

THE SEARCH FOR A TRULY INTEGRATIVE APPROACH

A collective decision was made in the spring of 1986. To avoid a quick return to the subsidiary's historical financial approach, it was critical that the organization had to pull together to reach the level of collective

efficiency it was striving for. A focus group was created with the mission of finding a new management method that would:

- provide significant improvement capabilities in strategic program implementation by developing true commitment on the part of all organizational members a decentralized, gratifying but also demanding working frame;
- include methodologies and the tools necessary to solve organizational weaknesses (e.g., controlling costs, coordinating actions, assigning realistic objectives);
- integrate investments made in previous approaches (e.g., the MBO program);
- maintain the same strategic direction of the firm, while filling in its weaknesses; and
- provide methods and tools that were rapidly transferable to the organization, maintaining the project's momentum.

In general, universities throughout France and Europe were of little help. Only a few of them envisioned quality as a subject matter. At that time, area consulting firms did not have many developed products in this field, considering quality as a utopian target. The subsidiary's focus group, however, began to encounter peer organizations on the same path. A series of meetings and working sessions were held with companies operating in a range of different fields (e.g., services, high technology, steel, food industries), with different sizes (from 200 to 55,000 people) and types of ownership (e.g., subsidiaries, corporate, privately owned).

Based on this exploration, a set of common conclusions emerged:

- Projects founded essentially on values, aimed at enhancing empowerment, experienced the same drawbacks as those identified in the subsidiary.
- Projects using a single tactic (e.g., quality circles) produced unsatisfactory results.
- Projects showing the best results were based on a number of related management tools focused on improving daily practice.

The Socio-Economic Management Method

The focus group highlighted the rapid evolution of the company in terms of its move toward participative management. The group proposed a "transformational" model that consisted of:

- introducing participation mode at the operating team level in daily activities;
- plugging these teams into their external and *internal environment* in order to generate a proactive approach to the company's strategy;
- finding ways, methods and tools to ensure the autonomy of these teams while maintaining a commitment to the firm's strategic action plan;
- creating a means for such delegation without loosing control; and
- reinforcing internal balance between line and staff who were expected to become experts.

The focus group recommended implementation of ISEOR's *socio-economic management* method (Savall, 2003b; Zardet, 2005). The approach was selected due to its conceptual and methodological content, the proposed intervention's characteristics, and the perspectives it offered in terms of the evolution of management style and techniques.

Based on its examination of the *SEAM* approach, the focus group concluded that the model's conceptual and methodological content:

- was based on total quality principles, but with an integrating and essential economic dimension, which had been verified through a number of successful projects;
- was consistent with the firm's previous approaches, with the added benefit of a coherent approach;
- provided substance and consistency through its link with the techniques previously utilized by the company (e.g., strategic planning, MBO, performance appraisal); and
- brought additional tools in the domain of vertical implementation strategy (e.g., *internal and external strategic plan action (IESAP)*, *priority action plans (PAP)*, *periodically negotiable activity contracts (PNAC)*) and diagonal implementation strategy (e.g., *synchronized decentralization*).

Through contracting with ISEOR, it was decided that (1) the methodology would be transferred through training the entire management line, (2) an internal intervener group would be established, and (3) project implementation would be undertaken by the company itself, not by external consultants (Savall, Zardet, Bonnet, & Moore, 2001).

The Intervention

The intervention was focused on the entire company. It unfolded in two stages: an initial phase of learning and implementing tools and methods, and a second phase of dissemination throughout the organization. Within the first year, the company planned to train 150 managers and involve approximately 800 employees in the project, with an additional 12 months set aside to integrate the rest of the 2,600 company members. At that point in time, each staff member was to have had an individual *PNAC*, negotiated with upper management with specific targets to be reached within the following 6 month period. These objectives were checked for feasibility, and were deemed compatible with colleagues' objectives. The objectives would also be derived from the *priority action plan* of the department, which had been integrated along the hierarchy and synchronized diagonally with suppliers and customer departments. The project was named "Tremplin" (meaning springboard) and the starting date was fixed for September 1987.

Phase I Implementation

The intervention was based on *Horivert* principles, addressing both top management and every department in the company. The *horizontal action* was focused on the first 150 top-ranking managers. Eight vertical interventions took place in different services and departments throughout the company.

Horizontal Action

The main objective of the horizontal action was to strengthen the top management teams, which were composed of managers from a variety of merged and acquired companies. This action involved 150 managers in three main processes.

The first step was to conduct a *mirror-effect* diagnostic aimed at taking an inventory of all dysfunctions. This process was carried out through individual interviews, approximately an hour and a half long, with each manager. The consultants took notes during these interviews and categorized the quotes with the assistance of *SEAMES® software*, the expert system software developed by ISEOR. As illustrated by the following quotes, the diagnostic enabled the intervener-researchers to identify the communication problems between the different components of the company:

> Contact between the production and the marketing department is very infrequent. I would like for them to visit us more often.

Headquarters' management control demands data that we don't understand or that doesn't make sense to us.

The relationship between sales reps and logistics management is tense. There is a complete absence of teamwork.

We know nothing about the other divisions. It is as though they were different companies alongside us.

What strikes me is the lack of *cooperation* when something goes wrong. People always blame each other instead of working things out together.

The diagnostic resulted in a 110-page report that served to create an awareness of the urgent need to solve the myriad *dysfunctions* that plagued the company. This awareness was enhanced by the explanation of organizational "taboos" identified by ISEOR's experts during the diagnostic (e.g., lack of awareness of business environment threats, lack of cost consciousness in spite of cost control, belief in the sustainability company's financial prosperity).

The second step involved project-group that held several meetings with groups of managers. The *project* themes that emerged from these meetings included the need to:

- redesign the department *structure* with sufficient resources to empower performance (e.g., skills, specialization and expertise);
- reflect on the coherence of wage policy and the need for performance incentives;
- define and clarify the rules of company operation, especially concerning the role played by the company's management hierarchy; and
- improve the company's *communication-coordination-cooperation*, notably between the production, sales and *engineering* departments.

Finally, the third stage involved training of the 150 managers who would be using key socio-economic tools, ranging from *time management, competency grids, priority action plans* and *strategic piloting indicators*, to *periodically negotiable activity contracts*. ISEOR management consultants in each of the 10 training *clusters* led nine training sessions of three hours each. In addition, *personal assistance* was provided to each manager in order to help them bridge the gap between theory and practice. For example, the CEO prepared his semester schedule using ISEOR *time management* techniques, which take into account the impact of the *priority action plan* on time allotment. The time budget synthesis for the training period is summarized in Table 12.1.

Table 12.1. Time Budget for First Phase Training Production Lines

Focus	Time (Hours)
Project for a new factory	200
Department redesign	90
Redefining communication policy	50
Improving operational efficiency	22
Designing strategic plans	26
Financial reporting to the parent company	110
Conducting communication improvement sessions	170
Human resource management	20
Personnel training	30
Participating in the *socio-economic intervention*	40
Total	190

Source: © ISEOR 1989.

The implementation of these *management tools* resulted in an improvement of the individual and group efficiency of the managers. This brought about several immediate impacts. First, the training improved the selection practice of priority actions and projects. Strategic projects such as new product development, for example, were not delayed due to poor *time management*. Second, enhanced delegation to subordinates was due to more informed *negotiation* of priority objectives and the means necessary to attain those objectives. Actions aimed at reducing delivery delays, for example, were broken down and attributed to the relevant departments (e.g., sales, operations management, production). Finally, there was enhanced communication of the company's strategy through the *priority action plans*, which helped to translate the overall strategy of the company into concrete actions that were understandable to all organizational members.

Vertical Actions

Diagnostics were also carried out in eight departments representing the company's different activities (e.g., production line, recently acquired sister companies, operational departments [computer, finance and accounting services], the sales force). These diagnostics were carried out through interviews with 443 people, which permitted both involving a large number of the company's key actors and re-explaining the objectives of the intervention.

A calculation of *hidden costs* revealed a total *dysfunction* cost of 11,000 Euros per person per year, which amounted to an annual total 28.6 million euros (extrapolating these calculations to the entire company). Examples of these hidden costs (in different departments) included: (1) the production line that was spending too much time training temporary help; (2) computer service unintentionally blocking applications due to a lack of respect for procedures; and (3) the commercial (sales) department with frequent *personnel turnover*, which reflected a poor image to clients.

Project Implementation

The vertical diagnostics gave rise, in each department, to a "*mirror-effect*" presentation, first to management, then to the entire group of people interviewed. Following these presentations, focus groups were set up in order to involve management in the search for dysfunction prevention solutions. In each department, four or five focus group meetings were held, in addition to work sessions in small groups destined to associate the largest possible number of actors to the definition of concrete solutions. Every action that was kept became the object of a report in the *priority action plan* of the concerned department, as well as becoming the object of an *economic balance*. For example, the report on the production line included the implementation of self-quality control, the economic balance of which is summarized in Table 12.2.

The example of *economic balance* in Table 12-2 demonstrates how an action can be reimbursed in less than five months, without necessarily assessing the numerous other related gains (e.g., reduction of customer complaints, stronger sense of responsibility on the part of operators, faster reactivity when faced with deviations).

This *foucs group* stage thus permitted training management in the methods of dysfunction prevention. In the past, the company's managers had been accustomed to dealing with *dysfunction* on a day-to-day basis, without ever addressing their root causes and without integrating solutions into an overall plan.

Table 12.2.   Provisional *Economic Balance* of the *Implementation* of Self-Quality Control on Production Lines

Costs		Performance	
Time spent training operators and supervisors	56,000 euros	Reduction of wastes	53,800 euros/year
		Reduction of defects	76,800 euros/year
Purchase of control equipment	18,500 euros	Reduction of production line stops	44,000 euros/year
Total	74,500 euros	Total	174,600 euro/year

Source: ©ISEOR 1989.

First-Year Outcomes

By the end of the first year of the intervention, the basic tools of socio-economic management were being utilized in the principle departments and approximately 800 employees were involved in completing *periodically negotiable activity contracts*. In addition, every team responsible for a department had created a *priority action plan* that was related to the company's *internal and external strategic action plans (IESAP)*. This effort permitted the organization to consolidate management teams and furnish them with appropriate tools for communicating with their entire team of personnel.

As an example, the *priority action plan* of one of the divisions included a number of specific objectives for the last semester of the intervention's first year: successfully launch a new product, improve the quality of current products, reduce delivery delays, improve manufacturing production, improve the *efficiency* of the "purchase" function, and enforce stricter evaluation criteria for collaborators. Each objective was broken down into concrete actions. Product quality *improvement* required the development of self-quality control, as was cited concerning production line focus groups, which made it necessary to purchase extra quality control equipment and to train operators in self-control.

Periodically negotiable activity contracts were also implemented at the management level of the company, as well as for employees in the vertical departments cited above. In all, more than 800 employees of the 2,600 in the company had an interview per semester with their direct superior to negotiate their *PNAC*. These contracts reiterated the principle objectives outlined in the *priority action plan* and examined the efforts accomplished, the indicators of *evaluation* and the means allotted for the project. The financial counterpart of the *PNAC* resulted in a 5% bonus of the semester's salary, paid during the month following the evaluation. An average of 80% of the objectives of the *PNACs* was achieved. This corresponded to an average of roughly a 24% salary supplement that was considered a nonnegotiable bonus. The *PNACs* dealt with levels to be attained, both in production and in quality for each team, as well as improvements on points outlined in the *priority action plan*. The *PNAC* of production-line operators, for instance, dealt with the preparation of the equipment, rigorous self-quality controls of ingredients, and the cleanliness and orderliness of their units.

Development Of Managerial Competence

The implementation of such tools as *priority action plans* and *periodically negotiable activity contracts* helped managers to become true team leaders. In order to prepare the *priority action plans*, they had to thoroughly explain the company's strategy to their team and translate it into concrete actions. They also had to involve their team in reflections on dysfunction prevention actions. The negotiation of *periodically negotiable activity contracts* led them to listen more attentively to their collaborators and to develop the capacity of negotiating the efforts necessary to implement *priority action plans*.

This development of managerial competence was accompanied by the formalization of the rules of the game regarding the role of managers. These "playing rules" were defined and set down by a work-group led by ISEOR experts as part of the Policy and Strategy Decisions Axis. The intervention involved four principle types of rules, which are illustrated in the following examples:

- *Synchronized Decentralization*: The company supported the creation of autonomous, flexible strategic operational units, specialized in lines of products or in types of clients. With respect to the functional departments of the company, they were meant to play a role of internal service for the other departments considered as clients.

- *Social Relationships*: Every member of the company renegotiated their objectives and allotted means with their supervisor as part of their *periodically negotiable activity contract*. The superior evaluated the fulfillment of this contract during each semester's interview.

- *Communication—Coordination—Cooperation*: Every supervisor of a service or a department was responsible for setting up a cooperation-unit, holding meetings and appointments with all of their collaborators. The project manager was responsible for designing a unit of *coordination* and *negotiation* with all of the functions concerned.

- *Cooperative Delegation*: The scope of the delegation (object, limits) was to be defined, negotiated and formalized in a manual of synchronized delegation. The time necessary to manage delegation was to be programmed, with regular appointments scheduled to ensure project progress.

Phase II Expansion and Development

The second and third years of the intervention focused on the application of new management methods in all departments of the company. Internal interveners were also trained to ensure the transmission of *socioeconomic intervention* methodology to the company.

Training and Internal Intervention

A team of eight *internal-interveners* was coached in an ISEOR training workshop on the socio-economic method. The team was composed of management controllers and human resource managers. The objectives of the composition of the team were to:

- Facilitate better *integration* of socio-economic methods within the company, based on the deeper understanding the participants had of the company (in contrast to external consultants). This involvement enables internal interveners to play an educational role at the heart of the company, explaining how the new management tools introduced by the intervention can enhance existing methods.

- Progressively take the relay of *external interveners*, who would not remain in the company over a long duration. Internal interveners, for example, took charge of carrying out diagnostics and *socio-economic projects* in the departments of the company that had not been touched during the first year.

- Play a permanent role of audit and guidance in the implementation of *socio-economic management tools*. Table 12.3 presents an example of the audit work. *Internal-interveners* employed this grid during their *personal assistance* with managers from each department of the company.

Implementation of Tools Throughout the Entire Company

By the end of the second year of the intervention, *PAPs* and *PNACs* were set up throughout the entire company. *Strategic piloting indicators* were used to compile a complete array of piloting indicators for each service, including the consolidation of data at department levels and at the level of the entire company. These indicators, in turn, became the object of monthly reports to headquarters, which enabled the organization to track the state of advance of the *IESAP*. These indicators were grouped according to *IESAP* categories:

- *Improvement of financial indicators* (e.g., operational margin by department and by product, reduction of the cost of *nonquality* (nonquality) by service)

- *Product innovation* (e.g., status on advance of 4 new product projects, status on respect of planning for new product launch)

- *New product performance of factories* (e.g., productivity gains by production lines and by product, ISO 9000 certification)

- *Logistics* (e.g., delivery delays respected, increase of orders by category of service)

**Table 12.3.   Audit Grid for Measuring the
Utilization of *Socio-Economic Management Tools***

Tools	*Actions*	*Remarks/ Action Plans*
Time management	**Semester activity review posing the following questions**: • Have we thought about our time organization? • Have we done the work of our collaborators? • Have we forgotten to pilot our services? • Have our meetings been prepared? • Have we respected the times and agendas of our meetings?	
Competency grid	**Semester competence review for each collaborator using:** • Security management (what is indispensable to the missions of the service?) • What would permit improving the efficiency of services? • Current know-how-capacities (utilized or not?) • Competence to be acquired to keep up with professional evolutions? • Measure of the flexiblity or weakness of services designing training plans?	
Priority action plan	*Evaluation* of *pap* **of the service including:** • Results of the previous pap • Priority actions of the superior level • Actions affecting the service (e.g., dysfunction, training, productivity) • Harmonization of pap with other services and directions • Communication of final pap to teams • Intermediate evaluation of pap recommedations (2 times per semester) • Final evaluation of team members	
Critical action contracts	• Interviews to define objectives for coming semester (the means to implement, the followup indicators, the consideration of different objectives) • Intermediate evaluation of pap (recommendation every 2 months during a.c.) • Interviews held to evaluate the attaining of objectives	
Strategic piloting logbook	• Search for indicators to follow used in the operation of the service and at a.c. • Design and follow up	
Activity check-point	• Formal individual meetings between level x and $x - 1$ destined to follow the evoluton of the service's activity • Recommendations (at least one A.C. per month, dedicate one every 2 months to intermediate *evaluation* of the *pap* and pnac	
Communication-coordination-cooperation	• Information by every appropriate means (meetings, memos) of communication at different levels of the company • 3-C communication strategy (pap unity, messages from general director [to be done every 24 hours])	

Source: ©ISEOR 1989.

- *Development of **human potential*** (e.g., status of training programs and progression of ***competence***, management use of socio-economic tools)

These indicators made it possible to complete existing ***strategic piloting logbooks***, which were mainly based on the calculation of costs, follow-up analysis of ***productivity***, and budget control. All managers in every department found that the piloting logbooks were quite useful, largely because they helped them better organize the gathering of qualitative, quantitative and financial information. Furthermore, these indicators made it possible to survey the results of the actions implemented through the ***PAPs*** and to evaluate the results of their collaborators' ***PNACs***.

Developing the Intervention

After 2 years of intervention, the principle ***socio-economic management tools*** were implemented throughout the company. This organization-wide application helped to transform the direction and role of the entire operational management team, while measurably contributing to improved ***economic performance***. It remained necessary, however, to consolidate the results of the intervention by helping functional services to reposition their roles in a context where operational service managers were increasingly assuming a dual focus on the development of ***human potential*** and tracking economic performance.

Three principle actions were carried out, applying a products portfolio approach to central functions, training in a methodology of coordinated development of products and technologies, and finally renewing the role of management controllers.

Applying A Product Portfolio to Central Functions

At the beginning of the intervention, central functions were considered as structural overhead assumed by operational divisions. The evolution of management methods induced by the intervention began questioning them, both because divisional management skills were on the rise, especially in human resource management, and an augmented awareness of costs and performance in the company led everyone to question their true ***value-added***. It thus became a question of repositioning them in the client/supplier relationship at the center of the company. This action led to clearly identifying the product portfolio of central functions, starting with computer services and the direction of human ***resources***.

As an example, within computer services, three categories of products were identified, each one corresponding to a different client:

- Design of new computer programs whose main clients were other central functions (e.g., logistics service).

- Maintenance of the company computer equipment whose clients were all the services of the company.

- Harmonization of computer methods and procedures across the entire company. In this case, the principle client was the top management.

In the same way, work within human **resources** revealed a long list of functions of advice, both to top management and to the different divisions. This identification of products made managers more aware of the need to develop this vital expertise and to reduce the portion of other products in direct competition with the new role of managers (e.g., recruitment, evaluation of personnel, control of time schedules, management of training programs).

Methodology of Coordinated Product and Technology Development

Before the intervention, the development of products and technologies was limited to engineering and technical development services. Vertical *socio-economic diagnostic*, as well as horizontal diagnostic, however, pointed to the costs of dysfunction related to this organization:

- Difficulties in putting new equipment into service, with numerous breakdowns following their introduction.

- Lack of training of production personnel in the use of new, automated production lines, resulting in the under-utilization of this equipment.

- *Overconsumption* of design induced by the lack of communication between production services and *engineering* services. In particular, the production management, feeling despised by the engineering service, were not inclined to transmit their knowledge that would have been useful in developing new industrial applications.

As part of the process of continuing to develop and institutionalize the intervention, it was necessary to teach the different services how to work in a coordinated fashion. To achieve this goal, ISEOR experts accompanied the creation of inter-service focus groups in the context of two cases of investment. The first case involved investment in a new bakery/factory. The *focus group* was composed of representatives of almost every service in the company in order to define together the specifications manual corresponding to the needs of each function. This process allowed the group to consider a range of variables usually neglected by engineering services, such as ease of cleaning, maintenance accessibility, conviviality of spaces designed for communication, and the flexibility of installations. Furthermore, the *focus group* painstakingly reviewed all dysfunctions identified in their *diagnostics*, with the goal of avoiding their reoccurrence in the new factory. To accomplish this, a dedicated database was constructed, grouping together more than a hundred key *dysfunctions*. Among other things, this database led to the designing of U-shaped production lines, rather than straight lines, in order to group together those occupied with the preparation of the dough, those in charge of the baking and the personnel responsible for packaging.

The second case involved investments in the renovation and automatization of production lines. *Focus groups* involved production managers and personnel in the design of the automated production lines. This collaboration enabled an appreciable *improvement* in the ergonomics, with synoptic logbooks of installations, designed with a graphic representation of the production line and with other significant variants. The initiative facilitated operator learning and limited sight fatigue.

Development of Socio-Economic Management Control

It became apparent, during the first year of the intervention, that financial and management control were reticent about the implementation of socio-economic tools. Indeed, during their calculation of *hidden costs* they questioned the relevance of traditional methods for calculating visible costs in the preparation of budgets. It was necessary, therefore, to help the management controllers transform their role by catering more to the needs of operational services. This was accomplished through a series of ten work sessions where ISEOR intervener-researchers played the role of instructor. Two main axes were adopted in the training:

- *Renewal of the methods of calculating costs.* The calculation of hidden costs (Savall & Zardet, 1992) fostered awareness of the need to go beyond calculations of overconsumption to include the cost of opportunity connected to *overtime* and *nonproduction*. For example, the cost of stopping production lines was largely underestimated by the customary methods of calculation. An effort to calculate the hourly contribution of timetables to the margin of variable costs thus permitted the harmonization of calculation practices throughout the company concerning the *conversion of hidden costs into creation of value-added*.

- *Development of the role of expertise of management controllers.* The intervention consisted of training management controllers in a role of assistance for operational personnel, so that *strategic piloting logbooks* included both *immediate results* and the *creation of potential*.

It was also necessary to adapt piloting logbooks to include data relative to productivity, quality and the management of human resources.

This evolution of management control practices resulted in two main positive impacts: (1) first line managers had more leeway to make decisions on visible costs (e.g., buying additional tools or equipment), and (2) management controllers could gather more data on intangible investments and the *creation of potential* gains.

SEAM'S IMPACT ON THE COMPANY

The following narration by the company's general manager reflects SEAM's influence on daily management practices and organizational efficiency. This first-year "diary" captures the rhythm of change in the intervention process and some of the key benefits, including enhanced company *cohesion*, increased *competitiveness*, and gains generated for major stakeholders, including shareholders, customers and organizational members.

*Our first objective is work hour recovery. One hundred and fifty executives and managers are overloaded, beyond the saturation point. There are many bottle necks that need to be removed. Time has to be made available for the TREMP-LIN project. Solution: **Time management.***

*This experience is the first contact with the SEAM training method: training, **coordination**, action. The entire group of 150 managers has been trained in a one-week period. They were divided in groups of 10-12 people: the boss and his direct reporters, an architecture that follows hierarchical lines like **clusters** of grapes on a grapevine. A resolution report concludes each training session, with immediate practice implications, auto-diagnostic preparation and self-analysis. Within a few weeks, individual coaching sessions were systematically programmed: tool handling assistance, solution finding and stimulation.*

*The next training session started with a collective **progress** review of the previous resolution report. Under this scheme, each group is locked in a collective move for mutual improvements. The tool is plugged in!*

Time analysis indicates that 32% of my own time is spent on matters that could be delegated (24%) or abandoned (8%). Our managers are showing similar results—more than 20% of their time is poorly employed. This collective disclosure was a first shock, but it created the conditions for real changes in daily practices, on an individual basis and collectively. Within a few weeks, everyone started working with a diary—meetings are programmed, timing respected.

Meeting are shorter, people are prepared, and they are always concluded with a resolution report issued on the spot, The participants become bonded to fulfill common resolutions for the next meeting, and the **cooperative delegation** *principle was introduced. Time has become a valuable resource of the company, enhanced thorough negotiation.*

The impact is rapidly visible on management. The atmosphere cooled down, and Many work hours are progressively saved, fueling the energy necessary for the success of TREMPLIN. **Time Management***'s* **effectiveness,** *from the beginning, gave strong credibility to TREMPLIN—Things are really moving.*

(Month Five). *At this point in time, a baseline has been created as part of our diagnostic. Many employees have been interviewed on* **dysfunctions** *observed in their work environment, but often with "mixed-feelings" … they expressed themselves openly and widely, but it could be dangerous. So it is important to proceed with a symbolic and effective move towards the baseline to give credibility to TREMPLIN and reduce anxiety and expectation levels. For this purpose, within the dysfunction diagnostic, top management selected a symbolic dysfunction which bothered a lot of people in many places, one that could be easily and rapidly solved—the toolboxes.*

Interviews with bakers revealed that 170,000€ per year are wasted in dysfunctions caused by the lack of small tools and parts in the workshop. These tools are necessary in the event of small breakdowns. In the past, a toolbox and set of parts were placed close to each key machine. Today, after a decade of budget cuts, only one box is available in the department. When the line is stopped, workers wait and production output falls down. A project was developed on this item, proposed and run by the production department: appropriation request—31,000€; pay-back period—3.5 months.

The reintroduction of these toolboxes becomes a symbol of the new direction for hundreds of floor workers: 'Finally, at the top, they are starting to understand! We are moving!' Several selected projects of this nature—fast, cheap, micro actions—have begun to generate momentum—no more speeches on values but actions based on facts and accountability for change. The social lever has started to unfreeze.

(Month Six). *The first part of the diagnostic was issued: the* **mirror-effect***. This was not the consultant's opinion, but rather the outcome of company member interviews, an auto-diagnostic. It pointed to 10,000€ of* **hidden performances/costs** *per year per employee. This represents 10% of sales, in a company already consistently reaching 12% income before tax, twice the biscuit industry average. The figure is a shock for everyone and a slap for top management, previously dubious about the* **hidden costs** *concept. This figure was initially rejected and strongly challenged. From the top to the base, everyone was requested to check the validity of these dysfunctions. After a month of investigation, the hidden cost computation was verified by controllers; only minor adjustments were made. A 27 M€ figure of hidden performances/ costs was accepted by the organization as being representative of its situation.*

*The **mirror-effect** began to break down internal barriers to the change process. An irreversible drive is in motion. Faced with the magnitude of this disclosure, the necessity to change became a strong and shared feeling at the top and on the line. At the same moment, the effect of the local micro-projects became visible in the field. The change was seen as possible for everyone.*

(Month Seven). *150 managers have been trained and are using common management tools—**competency grids**, **piloting logbooks**. Workers on the shop floor have started to unfreeze thanks to local micro actions. **IESAP** shows strategies and medium-term goals to reach. The **priority action plan (PAP)** was introduced, defining the objectives for the next six months. The PAP process included vertical negotiation—top-down and bottom-up. PAPs are estimated in terms of required resources ensuring their feasibility. Contribution necessary from other departments is being negotiated and programmed. Overall coherence is examined, trade offs made. The final company PAP has become a collective commitment for the next six months.*

*At this point, 150 **periodically negotiable activity contracts** (PNAC) have been introduced, as a first trial. The objective is to contractualize individual objectives, derived from the PAP, between each hierarchical rank. The virtuous loop is now in place, allowing the combination of top management strategies and local initiatives— the **synchronized decentralization principle**.*

The first PAP campaign has been a trial run. Management was over-ambitious, reluctant to assign priorities. It has been unrealistically built-up, and it is unlikely that it will be completed. The 150 PNACs are not enough to have a driving effect on the organization.

(Month Twelve). *The second PAP campaign has been more professional. Efforts have been made to really calibrate the resources required. **Project** feasibility was scrutinized; priorities were assigned. For the first time, projects were postponed and some even abandoned. Realism started to prevail. Planning and programming have become part of our vocabulary and into practice. **PNAC**s were extended to 800 employees.*

*The second campaign reached 80% of **PAP** achievement. Not perfect, by far, and yet it has provided a new feeling of a group maneuvering at the same pace and in the same direction. Local initiatives have blossomed, and they are becoming more coherent, placed within the context of a flexible, coordinated and stimulating process. Unfrozen energy is being correctly used within the group.*

*Each PAP campaign has gently and progressively improved the virtuous loop **efficiency**. It has become a sustainable process, satisfying all major stakeholders—shareholders, employees and customers.*

Synchronized Decentralization: A New Feeling of Cohesion

According to the general manager, the company was a "patchwork of different components" that were aggregated over time. There was a history of organization-wide conflict, especially between historical sites and the new headquarters, opposition between central functions and local managers, and resistance to the idea of synergies (Sirower, 1997) between sites. Through the **SEAM intervention**, in contrast, a new feeling of **cohesion** began to appear along a number of dimensions as illustrated by the general manager's comments:

- **(Vertically)** *"Throughout the line, managers and employees were working on the same* **PAP** *objectives, external as well as internal. A new communication line was established, with both top-down and bottom-up information flows. Supervision* **efficiency** *was enhanced by new management tools, especially PAPs,* **PNACs** *and* **competency grids***. Almost everyone on the line has gotten involved. They were all made accountable, but also compensated on the base of negotiated and challenging objectives, individual and collective. The social/human lever is in action, at all levels, in all places, oriented towards performance* **improvement***."*

- **(Transversally)** *"The PAP process introduced, within the line, negotiated resource allocation, particularly in terms of time. This was also done transversally, across the functions and the divisions. Commitments were taken, programmed quasi-contractually, and introduced in an internal customer-supplier relationship."*

- **(Among the functions)** *"A new balance of power has been progressively reached, beneficial to operational functions involved in the 'product-customer process.' New tasks have been assigned to staff functions: finance, quality, personnel. Monitoring the* **strategic piloting logbooks** *device has taken on a 'new controller' role: set up assistance, data-providing indicators, reliability control. We now have 'socio-economic management controllers.'"*

- **(Among the sites).** *"At the site level, SEAM has rapidly developed into a* **'survival/development** *capability.' Through the* **IESAP** *and* **PAP** *process, each site continually assesses its competitiveness within the marketplace. This focus has developed local economic* **vigilance***, stimulated innovation, increased local flexibility to demand, and accelerated reaction time to competitor moves. Stimulated by market conditions, each site has spontaneously sought* **cooperation** *with the other sites, as a way of increasing its own efficiency. The sites spontaneously look for synergies that they had previously blocked, when they were imposed by the head office. Sites have become a kind of confederation, in which they can and must build up their own future."*

- **(Within an acquisition context).** *"The company's industrial sites have been repositioned within the acquirer's organization. Their sharpened* **competitiveness** *allowed all major sites to win a desirable position in the new industrial configuration.* **SEAM intervention** *facilitated a managerial structuring of each site and each product line, strengthening their competi-*

tiveness, easing their integration and ultimately securing their future in the new configuration.

Performance Improvement

In general, the company's major stakeholders benefited from the intervention and *implementation* of the SEAM framework. Over a 3-year time period, for example, shareholders have benefited from greater investment in the company and increased earning per share. Potential creation spending (e.g., communication, R&D, training) increased by 80%, raising the investment in the company from 6.8% to more than 10% of sales. The margin ratio to sales is maintained in the range of 12%, but the investment made on immaterial side results in higher sales, increasing earnings per share. The 14% margin mark is passed two years later. It was clear to everyone involved that SEAM implementation, especially in terms of managerial skill development, *strategic implementation* capability enhancement, and *hidden cost* elimination, has been a decisive contributor to these results achieved—despite heavy turbulence throughout the industry.

Benefits from SEAM implementation were also tangible for the firm's workers, another important stakeholder. First, in terms of compensation, a 5% bonus linked to performance objectives was created. Second, SEAM has enhanced the sense of purpose and commitment experienced by organizational members (see Lewin, 1947). At first, workers were quite anxious about entering into individual contracts, especially swayed by trade union opposition. Their acceptance of the process, however, was visible during the second campaign as more than 90% of the employees signed PNACs. The internal climate at the firm also improved substantially as the working environment changed, including a more effective operational communication framework, individual merit recognition, and the elimination of many annoying *dysfunctions* (Bonnet, 1981, 1987). Managers were put into a position of true authority, supported by the new skills and competencies they developed during the intervention: "Everyone in the company has their own toolbox—the *same one for all*." This intervention also had positive impacts for the customers who benefited from high quality, inexpensive and environmentally-friendly products.

Overall, the implementation of the SEAM method not only enabled the company to increase its profit margin in the short run, but also helped to sustain its *economic performance* over the longer term, through a period that was characterized by a roller coaster environment (e.g., takeover bid, intrusion of new competitors, more and more demanding norms and standards in the field of product quality).

REFERENCES

Birkinshaw, J., Bresman, H., & Håkansson, L. (2000). Managing the post-acquisition integration process: How the human integration and task integration processes interact to foster value creation. *Journal of Management Studies, 37*(3), 395-425.

Bonnet, M. (1981). *Transformation du statut de la maîtrise au sein d'un processus de restructuration des emplois dans les ateliers en milieu industriel* [Transforming the status of supervisors as part of a job-restructuring process in industrial workshops]. Unpublished doctoral dissertation, University of Lyon, France.

Bonnet, M. (1987). *Liaisons entre organisation du travail et efficacité socio-économique. Analyse d'expérimentation dans des services de fabrication en milieu industriel* [Connections between work organization and socio-economic efficacy: Analysis of experimentation in industrial manufacturing services]. Unpublished doctoral dissertation, University of Lyon, France.

Buono, A. F. (1997). Technology transfer through acquisition. *Management Decision, 35*(3), 194-204.

Buono, A. F. (2005). Consulting to integrate mergers and acquisitions. In L. Greiner & F. Poulfelt (Eds.), *Handbook of management consulting: The contemporary consultant—Insights from world experts* (pp. 229-249). Cincinnati, OH: Southwestern/Thompson.

Buono, A. F., & Bowditch, J. L. (1989). *The human side of mergers and acquisitions: Managing collisions between people, cultures and organizations*. San Francisco: Jossey-Bass.

Buono, A. F., & Nurick, A. J. (1992). Intervening in the middle: Coping strategies in mergers and acquisitions. *Human Resource Planning, 15*(2), 19-33.

DeMeuse, K. P., & Marks, M. L. (Eds.). (2003). *Resizing the organization: Managing layoffs, divestitures, and closings*. San Francisco: Jossey-Bass.

Lewin, K. (1947). Frontiers in group dynamics. *Human Relations, 1*(1), 5-47.

Marks, M. L. (1997). Consulting in mergers and acquisitions: Interventions spawned by recent trends. *Journal of Organizational Change Management, 10*(3), 267-279.

Mirvis, P. H., & Marks, M. L. (1992). *Managing the merger: Making it work*. Englewood Cliffs, NJ: Prentice-Hall.

Savall, H. (1974, 1975). *Enrichir le travail humain dans les entreprises et les organisations* [Work and people: An economic evaluation of job enrichment]. Paris: Dunod.

Savall, H. (1987). Les coûts cachés et l'analyse socio-économique des organisations [Hidden costs and the socio-economic analysis of organizations]. *Encyclopédie du management* (pp. 599-628). Paris: Economica.

Savall, H. (2003a). An updated presentation of the socio-economic management model. *Journal of Organizational Change Management, 16*(1), 33-48.

Savall, H. (2003b). International dissemination of the socio-economic model. *Journal of Organizational Change Management, 16*(1), 107-115.

Savall, H., & Bonnet, M. (1988). Coûts sociaux, compétitivité et stratégie socio-économique [Social costs, competitiveness and socio-economic strategy]. *Encyclopédie de la Gestion* (pp. 742-757). Paris: Editions Vuibert.

Savall, H., & Peron, M. (2003) *Management research as an across-the-board source of knowledge: An attempt at conceptualization and contextualization via the socio-economic approach*. Paper presented at the Academy of Management Conference, Seattle.

Savall, H., & Zardet, V. (1987). *Maîtriser les coûts et les performances cachés: Le contrat d'activité périodiquement négociable* [Mastering hidden costs and performances: The periodically negotiable activity contract]. Paris: Economica

Savall, H., & Zardet, V. (1992). *Le nouveau contrôle de gestion: Méthode des coûts-performances cachés* [New management control: The hidden cost-performance method]. Paris: Éditions Comptables Malesherbes-Eyrolles.

Savall, H., & Zardet, V. (1995). *Ingénierie stratégique du roseau, souple et enracinée* [Strategic engineering of the reed, flexible and rooted]. Paris: Economica.

Savall, H., & Zardet, V. (2004). *Recherche en sciences de gestion: Approche qualimétrique. Observer l'objet complexe* Unpublished English translation: *Research in management sciences: The qualimetric approach. Observing the complex object*. Paris: Economica.

Savall, H., & Zardet, V. (2005). *Tétranormalisation: Défis et dynamiques* [Competitive challenges and dynamics of tetra-normalization]. Paris: Economica.

Savall, H., Zardet, V., & Bonnet, M. (2000). *Releasing the untapped potential of enterprises through socio-economic management*. Geneva, Switzerland: International Labor Office-ISEOR.

Savall, H., Zardet, V., Bonnet, M., & Moore, R. (2001). A system-wide, integrated methodology for intervening in organizations: The ISEOR approach. In A. F. Buono (Ed.), *Current trends in management consulting* (pp. 105-125). Greenwich, CT: Information Age.

Sirower, M. L. (1997). *The synergy trap*. New York: Free Press.

PART III

ISSUES AND CHALLENGES IN SEAM INTERVENTIONS

CHAPTER 13

SOCIO-ECONOMIC INTERVENTION AS INTEGRATED TRAINING ACTION FOR INTERMEDIATE SUPERVISORY STAFF

Marc Bonnet

Most management consultant interventions give little consideration to intermediate supervisory staff (International Labor Organization, 1998). Indeed, when interventions address the enterprise's general organization and strategy, the intermediate supervisory staff itself is typically not explicitly taken into consideration, but is merely expected to implement and adapt to decisions made at the summit (Dopson, 1992). Such is the case when, for example, a "process reconfiguration" intervention (or "reengineering") leads to grouping together activities from different divisions of the enterprise, provoking modifications in the number and composition of intermediate supervisory teams. In other instances, an intervention may concentrate on one of the enterprise's divisions, for example seeking to improve the ***productivity*** of a department or a project team (Kotter, 1996). The intervention substitutes itself for the intermediate supervisory staff, which has been either ignored or possibly even

Socio-Economic Intervention in Organizations: The Intervener-Researcher and the
SEAM Approach to Organizational Analysis, pp. 307–329
Copyright © 2007 by Information Age Publishing
All rights of reproduction in any form reserved.

deemed incompetent to conceive and implement organizational change (Mintzberg, 1989; Simon & March, 1958). It can also be observed that interventions destined to impact the entire enterprise, as is the case with total quality and quality certification initiatives, rely more on the enterprise's senior level management and its core staff than on its intermediate supervisory staff (Lawler, 1986), often considered to be more of an obstacle to improvement initiatives than a motor for change (Coch & French, 1949).

In the socio-economic method, in contrast, the intermediate supervisory staff plays an important role (Bonnet, 1987; Savall, 1974, 1975; Savall & Zardet, 1987). Indeed, an underlying premise of this approach is that the intermediate supervisory staff is critical to the enterprise's performance, but, as a group, it lacks change management training (Blake & Mouton, 1964). *Socio-economic intervention* constitutes a training method for the intermediate supervisory staff, helping to remedy this deficiency. The chapter will also attempt to show that excluding intermediate supervisory staff from change management programs can be detrimental to the *sustainable performance* of the enterprise (Savall, 1974, 1975, 1979, 2003a; Savall & Bonnet, 1988; Savall & Zardet, 1995, 2005), given the need for managers to guide and synchronize *development* actions (Drucker, 1973; Fayol, 1916). Based on an intervention in a 600-employee organization, the discussion will focus on how the socio-economic method can transform the role of intermediate supervisory staff.

The first section of the chapter describes the intervention context. The second part examines how the intermediate supervisory staff was weakened through repeated management consultant interventions in the enterprise. The third section will explicitly detail the manner in which the intervention process constituted an integrated management training method for the intermediate supervisory staff. The chapter then turns to a discussion of how socio-economic tools helped the intermediate supervisory staff to develop its managerial role and add value in the enterprise (Kets de Vries, 2001). Finally, the chapter concludes with an assessment of the role that consultants can play in improving *socio-economic performance* (Bonnet & Cristallini, 2003; Savall, Zardet, & Bonnet, 2000) through training the intermediate supervisory staff.

INTERVENTION CONTEXT

The intervention took place in an enterprise with roughly 600 employees specialized in the conception, production and distribution of insulation products. The enterprise is a subsidiary of a multinational group. It is composed of a commercial direction and five main departments: production, organized around three manufacturing lines, logistics, maintenance, engineering and technical services, and administrative services.

The enterprise's strategic situation (Cattin, 2002) can be characterized as follows:

1. *Cost reduction*: The firm was faced with the necessity to reduce the cost of traditional products by 10% per year in order to remain competitive in the international marketplace, particularly in the context of competitors benefiting from lower salary costs.

2. *Technological progress*: There was an increased demand for technological progress, especially the ability to offer customers products that were differentiated from those of competition, while ensuring production line productivity. Since the enterprise is a subsidiary of an international group, it was called upon to play a pilot role by refining certain new fusion technologies for the entire group.

3. *Customer satisfaction*: There was also a need to make progress in customer satisfaction in terms of quality and delivery delays, which includes rendering conditioning and logistic work more reliable in order to minimize **nonquality** in a just-in-time production system.

4. *Personnel costs*: Finally, there was pressure from the group's management to reduce personnel costs, both through progressive atomization and by reducing the number of hierarchical levels. However, the desired reduction in personnel costs could not negatively affect the firm's social climate as a strike could lead to a complete shutdown of deliveries and the loss of enterprise's credibility in the eyes of its clients.

The intervention was initiated by the director to simultaneously achieve a number of seemingly contradictory objectives, including the need to reconcile technological innovation with quality improvement, reduce personnel cost reduction without degrading the social climate, and improve reliability while reducing delays in product delivery. ISEOR's proposal thus created a framework for measuring the enterprise's global value added, attributing a central role to senior-level managers and the management staff in this process. Given the context, the enterprise's senior-level managers were conscious of the necessity to involve personnel, making them the artisans of this improvement and change process in the enterprise. They wished to carry out a preliminary pilot test of the method in the enterprise's logistics department composed of 63 people managed by a department head and seven control agents.

The Pilot Intervention

The pilot intervention was viewed as a success because it helped to create both an improvement in the social climate and an improvement in economic performance. This project consisted of carrying out an intervention in the logistics department of the enterprise, where supervisors played a central role in controlling forklift drivers. Special attention was devoted to controlling security rules, work discipline and application of procedures for product storage and loading trucks. Forklift drivers were unsatisfied with their working conditions and complained that managers didn't to listen to them, with the risk of creating a situation of conflict and a strike capable of paralyzing the entire enterprise.

The pilot intervention involved analyzing the current and potential hidden costs, induced by the absence of high-quality management. The *mirror-effect* showed, in particular, that costs due to **absenteeism**, **personnel turnover** and **work accidents** were much higher than the wage bill of the management sector, representing 1.9 million euros per year. The *focus group* had modified the enterprise's initial policy, which had been to eliminate a hierarchical level in order to economize on salaries and increase operator autonomy. Thus, the *project* set the stage for transforming the role of management, focusing on improvement actions and entrusting daily management and self-control to operators, while revalorizing remunerations.

The Horivert Intervention

The *Horivert* approach was chosen to involve the company's entire higher-level management staff, which was composed of 20 executive managers and most of the 50 members of the intermediate supervisory staff. The advantage of this process was to help synchronize improvement actions conducted at the top management level with development actions at the level of each department. Up to this point, the actions conducted with the consultants had been carried out in a disconnected manner, which led to some dispersal of the actions and loss of energy.

The resulting *Horivert* intervention included the following actions:

1. *Horizontal level*:

    - Carrying out a diagnostic of *strategic vigilance* and a diagnostic of dysfunctions.
    - Assisting in developing a strategic project and a dysfunction prevention project.
    - Training and coordinating 12 management directors.

2. *Vertical level:*

- Carrying out *socio-economic diagnostics* in two of the enterprise's departments (maintenance and production).
- Assisting in the conduct of a project in those two departments, as well as facilitating the *implementation* of the *project* in the logistics department.
- Training, coordinating and providing assistance to the 18 members of the intermediate supervisory staff of the two departments in the implementation of socio-economic tools.

Figure 13.1 shows the overall architecture of the intervention, including the *piloting group* composed of four senior-level managers. This group was in charge of making decisions about the implementation of socio-economic methods. The entire intervention took place over a 9-month period, following the experimental phase (pilot program) that lasted 6 months. Figure 13.2 presents the timetable schedule of the intervention.

The Initial State: The Precarious Position of Intermediate Supervisory Staff

At the outset of the intervention, it was clear that relatively little consideration had been given to the intermediate supervisory staff. In fact, it appeared that a number of political and organizational choices at the company level questioned the very relevance of the intermediate supervisory staff. For example:

- Training programs for production and maintenance operators were undertaken, but these activities did not include the intermediate supervisory staff.
- Production lines were reorganized with augmented operator tasks. This action was conducted by a management consultant specialized in socio-technical procedures. One of the consequences of increasing the operators' autonomy was a questioning of the role of the middle managers.
- There was an implementation of a semi-autonomous group organization on one of the production lines, completely eliminating intermediate supervisory staff on that line.
- A progressive reduction of control agents was undertaken over the years to reduce visible production costs. The increased responsibil-

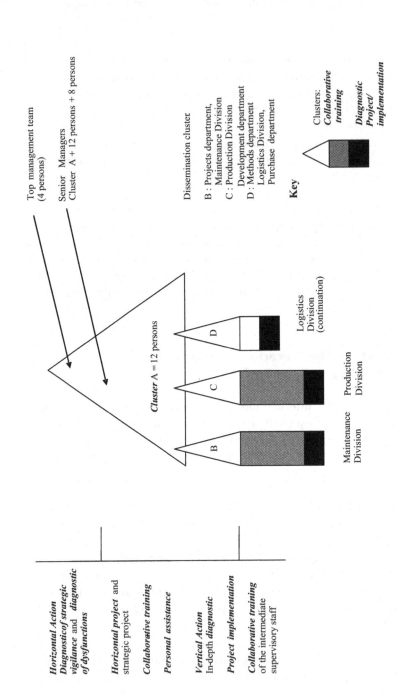

Horizontal Action
Diagnostic of strategic
vigilance and diagnostic
of dysfunctions

Horizontal project and
strategic project

Collaborative training

Personal assistance

Vertical Action
In-depth diagnostic

Project implementation

Collaborative training
of the intermediate
supervisory staff

Top management team
(4 persons)

Senior   Managers
Cluster   A + 12 persons + 8 persons

Dissemination cluster

B : Projects department,
      Maintenance Division
C : Production Division
      Development department
D : Methods department
      Logistics Division,
      Purchase department

Key

Clusters:
*Collaborative
training*

*Diagnostic
Project/
implementation*

Cluster A = 12 persons

D

Logistics
Division
(continuation)

C

Production
Division

B

Maintenance
Division

Source: ©ISEOR 2000.

Figure 13.1.   **Socio-economic intervention architecture.**

	Dec	Jan	Feb	Mar	Apr	May	June	July	Aug	Sept	Oct	Nov	Dec	Jan	Feb
Steering group	■ 1		2												
Horizontal Diagnostic		■													
Horizontal Project — Top management team				1		2				3		4			
Plant management team				1		2				3		4			
Training sessions focused on management tools — *Cluster A*	1		2	3		4	5								
Personal assistance			1	2	3					4					
Cluster B	1		2	3		4	5								
Personal assistance			1	2	3					4					
Personal assistance	1		2	3		4	5								
Cluster C			1	2	3	4	5								
Cluster D	1		2	3		4	5								
Personal assistance			1	2	3					4					
Training *internal interviewers*	1			2			3			4					
Vertical diagnostics — Maintenance division					■	■									
Production division					■	■									
Vertical projects — Maintenance division							1				2		3		4
Production division							1				2		3		4
Maintenance — Assistance to logistics division		1									2				

Source: ©ISEOR 2000.

Figure 13.2.   Intervention schedule.

ity of operators regarding control here again led to questioning about role of middle managers.

- A pivotal position responsible for training, communication and security was created as a means of developing initiatives in these domains. These needs had largely been neglected by the intermediate supervisory staff.

- Implementation of quality certification, relying more on higher supervisory staff than on the intermediate supervisory staff.

All of these actions constituted a strong, if not explicit, warning to members of the intermediate supervisory staff. They signified that company directors were questioning the role of this group and that interventions were meant to destabilize it even more. Nevertheless, it seemed to the directors that a minimum number of intermediate supervisory staff was necessary to maintain discipline, ensure a full complement of work hours, and apply security regulations. The one department where intermediate supervisory staff had been entirely eliminated was perceived as an exceptional situation, since the best-qualified operators had been grouped together there. Transposing that autonomous group organization to other company departments seemed like a desirable solution, but a distant one in terms of realistic possibilities for putting it into practice. The socio-economic intervention was thus charged with clarifying this issue by attempting to transform the role of the intermediate supervisory staff.

At the start of the intervention, the company's supervisory staff was composed of 12 senior-level managers, plus 8 other managers and 50 members of the intermediate supervisory staff, including team supervisors posted in five 8-hour shifts. The role of the intermediate supervisory staff, as analyzed by the *self-analysis of time* tool, was strongly focused on routine management activities (e.g., keeping station journals) and on the *regulation of dysfunctions* (e.g., absentee replacements, intervention on production and packaging incidents). The intermediate supervisory staff of the production division had acquired operator experience prior to its promotion to supervisory positions, and thus had greater affinity with the rank and file than with the senior-level managers. In the maintenance and new products departments, the intermediate supervisory staff was made up of trained technicians and played a role chiefly focused on project management and piloting.

The *diagnostic*, carried out through semi-directive interviews of the 59 members of the direction and management personnel, as well as 85 members of the core staff, revealed the following phenomena concerning the intermediate supervisory staff:

- Many supervisors complained about their *working conditions*, including stress related to frequent interruptions, the feeling of being mistrusted, lack of budgetary means for solving *dysfunctions* and lack of possibilities to affect remuneration in order to compensate the most deserving operators.
- Senior-level and higher management thought that the intermediate supervisory staff was not playing its role of interservice and interteam coordinator. The group was viewed as being too exclusively preoccupied with its team's and department's own interests than with the company's overall interests (e.g., supervisors in the logistics department complained that supervisors in production supplied them with poorly-conditioned products).
- The core personnel complained about the behavior of the intermediate supervisory staff (e.g., supervisors didn't listen to their problems concerning work conditions, behaved in an overly directive manner and did not behave like leaders). They also complained that the intermediate supervisory staff did not sufficiently inform them about market changes and company strategy.

In sum, the role of the intermediate supervisory staff appeared to be in question, and the intermediate supervisory staff itself was not satisfied with the functions that had been assigned to them.

SOCIO-ECONOMIC INTERVENTION AS TRAINING ACTION

The *socio-economic intervention* process was carried out in four stages: *diagnostic, project, implementation*, and results *evaluation*. Each of these stages produced effects on the evolution of intermediate supervisory staff competencies.

Socio-Economic Diagnostics

Each of the three *vertical diagnostics* carried out in the production and maintenance divisions involved the intermediate supervisory staff at different phases of the project:

- Every supervisor was individually interviewed and given the opportunity to express his or her difficulties, notably in terms of time management, characterized by sliding functions and frequent interruptions. These interviews provided the intermediate supervisory staff with the opportunity to reflect on the causes of dysfunction, becoming conscious, in particular, of the misbalance between time spent on the **regulation of dysfunction** and time dedicated to dysfunction prevention.

- A second series of interviews took place with each member of the intermediate supervisory staff in order to evaluate the *costs of dysfunction* based on information collected from every category of personnel. These interviews dealt with the calculation of dysfunction frequency (e.g., manipulation errors in lots of parts, time spent redoing work due to erroneous information). These calculations revealed a total hidden cost of 2.7 million euros per year, that is, a sum superior to the combined wages of the company's entire management staff. This analysis permitted the intermediate supervisory staff to become aware of the magnitude of **hidden costs**, and that it constituted a potential *budgetary leverage* to increase the means available to their departments. This analysis enabled management to realize that their failure to negotiate operators' responsibility for cleaning had incurred a significant cost.

- **Mirror-effect** sessions for the presentation of results took place with the intermediate supervisory staff and senior-level managers.

The next step involved the intermediate supervisory staff helping to present the diagnostic to the core personnel who had been interviewed. These reenacted issues helped the intermediate supervisory staff to reestablish a dialogue with senior-level managers as well as with the core personnel on subjects that had previously been "taboo." During these sessions, discussion touched on such delicate subjects as the lack of respect for equipment (which was causing hidden costs), and the overly directive management style of certain control agents. This exchange provided the opportunity for explicit expressions of concerns and served as a foundation for subsequent behavioral changes among intermediate supervisory staff members.

The Project Phase

In the course of the **project phase**, in every one of the enterprise's divisions, each intermediate supervisory staff member was assigned at least *one work issue (pool)*, destined to prevent dysfunction, both at the horizontal level as a participant, as well as at the vertical level as a group leader. Some examples of these work issues were:

1. The **horizontal project**:
   - harmonizing working methods for buying

- updated policies and procedures
- drawing up a simplified synthesis of management control results
- drawing up a charter on the role of intermediate supervisory staff

2. *The vertical project* of the production division:

  - implementing shared forms for conducting meetings
  - defining performance indicators
  - reporting logbook per area
  - improving team management procedure

3. The vertical project of the Maintenance Division:

  - improving work management procedures
  - improving maintenance rules for loading vehicles
  - developing preventive up keep

4. The vertical logistics project:

  - classifying product quality norms
  - developing operator *competence*
  - equipping and arranging office and storage space

These examples reflect concrete work issues. All of these issues concern *tasks poorly managed* by the intermediate supervisory staff, either because the staff did not have time for them or because they were customarily shunned by the enterprise's operational services. Thanks to the participation of the different focus groups, the intermediate supervisory staff developed competency in problem analysis and solution creation. As a group, they also learned to more effectively participate in transdepartmental communication, which was important given the variety of the divisions that took part in the *task groups*. Furthermore, the intermediate supervisory staff regained credibility in the eyes of its subordinates, as the latter saw that the supervisors were taking charge of solving the dysfunctions from which they suffered (Nonaka & Takeuchi, 1995).

At the end of the project stage for each division, the intermediate supervisory staff was charged with establishing the *economic balance* for all project actions. This process consisted of evaluating the project action costs and performance by measuring the impact on dysfunction cost reduction. For example, it was estimated that greater collaboration and *cooperation* between the maintenance and production departments helped, on average, to reduce waiting times from 5 to 10 minutes, which was evaluated annually at 724 hours. This time reduction was estimated to generate a savings of 6,112 euros per year. In all, the reduction of hidden costs budgeted in the economic balance of the production line project came to 259,000 euros per year, representing 32% of the 809,000 euros of *hidden costs* evaluated during the diagnostic process. These cost reductions gave the intermediate supervisory staff new leverage for negotiating complementary budgetary means. For example, the hidden cost reduction from improving cleaning and up-keep procedure of the loading vehicles was used to purchase cleaning equipment for the factory, a purchase that had been deferred for budgetary reasons.

Implementation

The implementation phase of the project also afforded an opportunity for the intermediate supervisory staff to increase its value added. Indeed, project actions consisted, for the most part, of carrying out *dysfunction* prevention. As an example, in the logistic department, time spent by the intermediate supervisory staff in training operators on security and quality procedures came to more than 100 hours per year. This time did not create the need to increase the number of intermediate supervisory staff, for it was compensated through a decrease in the time spent *regulating dysfunctions*. The control agents of the logistics department gained time because they had less of a need to supervise and verify the work of better trained and more responsible operators. Furthermore, the evolution of the role of the intermediate supervisory staff toward higher *value-added* activities created a better image for the control agents, both in the eyes of the core personnel as well as for senior-level managers. At the end of the *project implementation* phase, the issue of reducing the size of the intermediate supervisory staff, a question posed at the beginning of the intervention, was no longer being asked. The firm's management had thus evolved from a cost reduction logic to a strategy of *value-added development*.

Evaluation of Results

Eighteen months after the beginning of the intervention, senior-level and middle managers undertook a partial evaluation of project outcomes, which was carried out with the aide of internal (two executives from the personnel department and quality control department) and external (ISEOR intervener-researchers) interveners. This initiative resulted in a presentation, given to the entire top and middle management team that focused on qualitative, quantitative and financial results:

1. *Qualitative* results:

- ***Strategic development*** and implementation were accelerated both at the overall company level as well as at the individual department level.
- Greater awareness by the intermediate supervisory staff of economic constraints and hidden cost reduction that could be reconverted into value added.
- Improved quality management, as reflected by fewer manufacturing defects/rejects and fewer delays.
- Acceleration of competency progress, with the intermediate supervisory staff in particular playing more a role of instructor for their personnel.
- Improvement of security, largely due to the implementation of a greater number of prevention initiatives. For instance, the storage zone and the forklift truck traffic area were reorganized so as to cut down collision hazards.

2. ***Quantitative*** *results*:
   - Forty-four percent of the 310 dysfunctions identified in the diagnostic stage had been overcome.
   - Seventy managers and members of the intermediate supervisory staff had been trained and were using ***socio-economic management tools***, in particular ***priority action plans (PAP).***
   - Twenty-six project themes had been development and implemented

3. ***Financial*** *results*:
   - Every project theme was the object of an economic assessment. An example of this type of balance, taken from one of the 300 actions, is presented in Table 13.1, citing the case of setting up ***strategic piloting indicators*** (referred to as ***strategic piloting logbooks***) on one of the production lines.
   - Overall, the evaluated gains, after one year of intervention, represented 15% of the total sum of 2.7 million euros in hidden costs estimated during the diagnostic stage, or 405,000 euros.

Table 13.1. Economic Balance of the Implementation of a Production Line Strategic Piloting Logbook

Costs	*Performance*
Creation of the journal and keeping it up-dated: 48 hours × 16.90€/hour = 811.20€	Time saved by the intermediate supervisory staff conducting 6 meetings per year × 4 hours saved per meeting × 16.90€ /h = 2028€.
Total: 811, 20€	Total: 2,028€

These financial results represent only part of the qualitatively and quantitatively evaluated economic results.

RENEWAL OF THE INTERMEDIATE SUPERVISORY STAFF

Customarily, management tools are intended more for top management and operational supervisors than for intermediate supervisory staff. In the case of the firm discussed in this chapter, the *management tools* used by supervisors were limited to budget management of external expenses, follow-up measures of equipment *productivity*, and checks of personnel hours and absences for payroll preparation. *Socio-economic intervention* reinforced the use of such tools, training the intermediate supervisory staff, along with the rest of the company, in their use. This work was largely accomplished through three-hour *collaborative training* sessions, which were carried out over an 8-month period (with one session per month for top management). In the particular case of this firm, there were five sessions for the intermediate supervisory staff, conducted jointly by their supervisory staff superiors and internal interveners, all of whom were previously trained in the use of *socio-economic management tools*.

The *collaborative training* sessions mainly addressed four key points: *time management, competency grids, priority action plans (PAP)*, and *strategic piloting logbooks*. It should be noted that this company did not wish, at this early stage, to introduce *periodically negotiable activity contracts (PNAC)*, in order to give itself time to study the necessary changes and ensure coherent adaptation of remuneration and merit systems in all departments. ISEOR, nonetheless, clearly explained to senior-level managers what possible inconveniences could stem from this choice, particularly with regard to the performance and sustainability of *socio-economic management* approach.

Time Management

Training sessions focused on time management, including self-analysis of time usage and the implementation of common work methods (e.g., *resolution charts* for meetings, specification manuals for *cooperative delegation* of project management tasks). This work permitted the intermediate supervisory staff to better identify development actions, a task which had been sacrificed due to everyday management and emergency demands. It also provided them with the opportunity to more readily delegate tasks with moderate or low *value-added*. As an example, the daily control of equipment in the logistics department was delegated to opera-

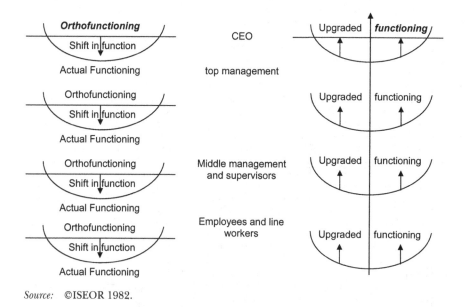

Source: ©ISEOR 1982.

Figure 13.3.   Raising levels of *value-added* in various personnel categories

tors, which permitted supervisory agents to gain time while simultaneously nurturing greater responsibility in the operators. This overall job
enrichment of tasks is illustrated in Figure 13.3, which presents the raising of levels of value added for the various categories of personnel, from
the rank and file to the top management, including the intermediate
supervisory staff.

Competency Grids

During the **collaborative training** and **personal assistance** sessions,
every supervisor created a "map" depicting his or her team's skill sets.
Figure 13.4 provides an example, drawing on the case of a production
line. This assessment allowed the intermediate supervisory staff to better
plan *skills management* actions, which facilitated their ability to deal with
ongoing organizational constraints (e.g., lack of all-around operators,
making replacements difficult in the event of absence, sickness or vacation leave; vulnerability of certain operations for which only a few persons
had been trained). For certain critical operations, supervisors created formal **integrated training manuals**[1] to train operators and make them more
autonomous, so as to develop **cooperative delegation**. This was the case,

Production Division Lines 1 and 2 : Team 1

WORK ORGANIZATION AT THE START OF THE PERIOD

ACTIVITIES

ACTORS

EXISTING OPERATIONS

SECURITY- MANAGEMENT

DEVELOPMENT-MANAGEMENT

EXISTING SPECIFIC KNOW-HOW

OBSERVATIONS

- Operating Equipment C
- Maintenance of Equipment C
- Operating Equipment D
- Tuning Equipment D
- Operating Station K
- Quality control of product
- Producing asphalturn panels
- Stacking product of line 1
- Palettizing heavy products
- Boxing products on line 1 and 2
- Rolling products on line 2
- Wrapping granulated wool
- Palettizing granulated products
- Stacking light products
- Palettizing light products
- Operating a lift elevator
- Intervening in an electrical cabinet
- Taking responsibility for an area
- Replacing the supervisor
- Analyzing findings
- Managing dysfunctions
- Daily team organization
- Proposing improvements for evolution.
- Intervening in the event of human accident
- Intervening in the event of fire in the plant
- Utilizing equipment D
- Utilizing the equipment F

KEY
- ■ Efficient command of daily operations
- ◪ Occasion operation or partially acquired efficiency
- □ Theoretical knowledge without practice
- — Neither theoretical nor practical knowledge
- ○ Training to be undertaken

Source: ©ISEOR 2001.

Figure 13.4. Example of a *competency grid*.

for example, with instruction manuals for loading vehicles, with the objective of attempting the identification of *multiskill dosage*. It should be noted that the **competency grid** approach goes beyond simply capturing the existing skill set in a particular area, facilitating the ability of the intermediate supervisory staff to better organize work activities and customize work content.

Priority Action Plans

The members of the intermediate supervisory staff were trained in preparing *priority action plans (PAP)* for their teams by programming actions derived from the company's **internal-external strategic action plan (IESAP)** linked to the **PAP** and the horizontal and vertical projects. These initiatives were introduced at the end of the **horizontal project** phase. This work, in its first phase, permitted the intermediate supervisory staff to become familiar with the company's **IESAP** and **PAP** in a detailed manner. Prior to this step, the intermediate supervisory staff had been informed only in a very distant manner about the company's strategic direction. It thus discovered the PAP facilitated their dialogue with senior-level managers, facilitating their ability to improve both economic, operational performance and **social performance** through more consideration and enhanced status.

An example is given in Figure 13.5, taken from the company's PAP, which explains the expansion of one of the IESAP lines relative to reinforcing **competitiveness**. This line is broken down into priority actions and the PAP explains the part of this action that is entrusted to production line supervisors—in this instance, they are called to participate in the energy-saving plan piloted by the production department. They are also tasked with developing AC products, piloted by the methods department in collaboration with the purchasing department. The intermediate supervisors were encouraged to optimize the use of work time, notably by using their **competency grids**. The PAP also clarifies the role the intermediate supervisors should play in converting **hidden costs** into **value-added** on production lines 1 and 2. Finally, it shows how supervision can cooperate with the logistics department in reducing **nonquality**.

These efforts facilitated the ability of the intermediate supervisory staff to break down their objectives into concrete actions, making them aware of their pivotal role in the implementation of company strategy. Labor optimization actions, for example, were expressed in terms of several precise actions focused on improving the use of work time, some of which had been suggested in the **socio-economic project** framework including: (1) implementing lists of work to be done by zone and by station; (2) training

Strategic Axes	Priority Objectives	Priority Actions	Department involved (in charge of actions)										Provisional Planning 1st semester						Observations
			Maintenance dept.	Production dept.	New Products dept.	Logistics dept.	Methods dept.	Purchase dept.	Projects dept.	Personnel dept.	Management Control	Top Management	January	February	March	April	May	June	
Axis #7 Reinforce our *competitiveness*	Accelerate production cost reduction and reduce *dysfunctions*	Energy saving on line 2		X					X						↕ March–May				
		Develop product C		X			X	X				↕					May		Quality of substitution products
		Streamline use of work hours on line 1 and line 2		X						X	X	↕					May		Time management *improvement*
		Cutting *hidden costs* in maintenance and production divisions	X	X								↕					May		Cutting *hidden costs* by 15%
		Reduce quality defects		X	X		X				X	↕	January–February						*Nonquality* cost limited to 1.3% of the budget

Source: © ISEOR 2001.

Figure 13.5. Excerpt from the company's *priority action plan (PAP)*.

operators in equipment cleaning and maintenance operations; (3) implementing follow-up of extra workdays; (4) implementing a common tool for temporary help management, shared by both the supervisory staff and the personnel department; (5) reorganizing teams and balancing skill distribution within each team; and (6) improving operator training on procedure conduct and implementing an operator evaluation form.

During the *evaluation phase* of the intervention, many members of the intermediate supervisory staff found this tool to be very useful, for it allowed them to better anticipate their team-level needs and actions, instead of seemingly always operating in an "emergency mode." They also found that the PAP piloting meetings, which took place every two months, permitted better *coordination* between departments.

The company's top management found that *breaking down the PAP* at the intermediate supervisory staff level had brought about three principal advancements in the company's management:

- A better global vision of development actions at the enterprise level, with more than 300 actions assigned to 6-month time frames.
- Better synchronization and *improved transversal action*, thanks to notably improved coordination among members of the intermediate supervisory staff in the different departments. At this point, the supervisors did not have to rely on senior-level managers for too-numerous decisions on details of inter-departmental actions.
- A better *respect for commitments* made by the group in terms of advancements on strategic actions. This last point is particularly important to stress for it shows that the intermediate supervisory staff participated in improving the enterprise's competitiveness by helping top management accelerate the implementation of strategic actions and the *creation of value-added*.

Management Strategic Piloting Indicators

At the start of the intervention, all members of the intermediate supervisory staff piloted a few management indicators, such as budgets and personnel follow-ups. However, training in the use of *strategic piloting indicators* led to enriching those indicators and placing them all side-by-side in one notebook for use in each department's weekly or monthly meetings. These piloting indicators included, among others, the PAP and the resolution charts cited above. In addition, two other important indicators were developed:

- *Immediate results indicators*, derived notably from dysfunction and hidden cost evaluations carried out during the socio-economic diagnostic (e.g., time lost transmitting orders, production line stoppage related to faulty maintenance, product *nonquality, absenteeism, work accidents* and identified incidents and *risks*).
- *Creation of potential* indicators, which included advancement on PAP actions and economic assessments (balance) of the different projects (e.g., rearranging space usage).

Furthermore, several of these indicators were displayed at work stations, in such a way as to instigate a daily dialogue between the intermediate supervisory staff and their subordinates on economic performance improvements. This produced a stimulating effect in terms of professional behavior (see the *HISOFIS effect*).

Strategic piloting logbooks thus turned the intermediate supervisory staff into actors of management control in a decentralized manner. During the evaluation, senior-level managers found that the involvement of the intermediate supervisory staff in the company's economic performance brought about three important advantages:

- Improvement in the piloting of economic *progress*, based on more precise indicators than those traditionally used in management supervision.[2]
- Assistance in following up the cost and performance of potential creation actions, supported by the progress indicators in the priority action plans.
- Greater reliability in budgets and results predictions, due to the more involvement of literally every actor playing a role in economic progress.

CONCLUSION

This example of an intervention in an insulation company illustrates the pioneering role played by intervener-researchers (Savall & Zardet, 2004) applying the socio-economic method. In this method, the consultant does not seek to substitute him or herself for the intermediate supervisory staff, but rather to help them transform their role of improving operations in their zones of responsibility. The overall intent is to contribute to their ability to convert *hidden costs* into *creation of value-added*.

The intervention presented here also shows how the socio-economic method can facilitate traditional management consulting techniques, while simultaneously integrating several types of action. Overall, the intervention:

- Played a role of integrated "coaching" of the intermediate supervisory staff members, who became aware of the necessity to change their role and behavior, notably through the diagnostic and *project phases*, as well as through *personal assistance* in the implementation of the various management tools. However, in contrast to coaching, which acts on the person in a fashion isolated from the enterprise and runs the risk of increasing desychronization between actors, *socio-economic intervention* focuses on all categories of actors.

- Contributed to the implementation of quality certifications (e.g., ISO9000), because the intermediate supervisory staff began to play an important role in formalizing procedures and contributed to ongoing quality *improvement*, due primarily to dysfunction prevention activities.

- *Integrated training* with personal development and human relations, prompting the intermediate supervisory staff to develop more of a participative management style.

- Contributed to knowledge management because the intermediate supervisory staff began to play an important role in formalizing and perfecting tacit knowledge, notably by implementing *integrated training manuals*.

- Incorporated an aspect of process reconfiguration (re-engineering), particularly at the project stage. These efforts helped the organization to eliminate activities (dysfunctions) that destroyed value added and improve inter-departmental *coordination* (e.g., preventing production line breakdowns, improving just-in-time delivery).

- Facilitated the implementation of change through participative methods. The socio-economic approach differentiates itself from traditional participative management techniques by integrating dysfunction cost calculation and implementing economic performance indicators for both immediate results and *creation of potential*.

In sum, as this chapter has indicated, socio-economic intervention can readily contribute to implementing an integrated methodology of management consulting for *creation of value-added* through the transformation of the intermediate supervisory staff's role.

NOTES

1. For additional information, see the work carried out by ISEOR between 1978 and 1981 on adequate on-the-job training on qualified job variables. As an example, see Beck (1978) and Bonnet and Beck (1980).

2. For deeper insight into the tensions between traditional approaches to management control and the concept of *socio-economic management control* see Savall and Zardet (1992).

REFERENCES

Beck, E. (1978). *Monograph on an enterprise in a branch of metallurgy* (ISEOR Research Report). Lyon, France: ISEOR.

Blake, R., & Mouton, J. (1964). *The new managerial grid*. Houston, TX: Gulf.

Bonnet, M. (1981). *Transformation du statut de la maîtrise au sein d'un processus de restructuration des emplois dans les ateliers en milieu industriel* [Transforming the status of supervisors as part of a job-restructuring process in industrial workshops]. Unpublished doctoral dissertation, University of Lyon, France.

Bonnet, M. (1986). Un nouveau métier: Contremaître [A new profession: Supervisor]. *Revue Gérer et comprendre, 5*, 18-25.

Bonnet, M. (1987). *Liaisons entre organisation du travail et efficacité socio-économique. Analyse d'expérimentation dans des services de fabrication en milieu industriel* [Connections between work organization and socio-economic efficacy: Analysis of experimentation in industrial manufacturing services]. Unpublished doctoral dissertation, University of Lyon, France.

Bonnet, M., & Beck, E. (1980). *Monograph of an integrated training action with the socio-economic approach in a glass factory* (ISEOR Research Report). Lyon, France: ISEOR.

Bonnet, M., & Cristallini, V. (2003). Enhancing the efficiency of networks in an urban area through socio-economic interventions. *Journal of Organizational Change Management, 16*(1), 72-82.

Cattin, G. (2002). Les enjeux du développement de la performance d'une entreprise implantée en milieu rural, face à accroissement de la compétitivité des marchés européen [The stakes of performance development in an enterprise in the rural milieu faced with the growing competitiveness of European markets]. In ISEOR (Ed.) *Le management des entreprises culturelles* (pp. 145-160). Paris: Economica.

Coch, L., & French, J. R. P. (1949). Overcoming resistance to change. *Human Relations, 1*, 512-532.

Dopson, S. (1992). Middle management's pivotal role. *Target Management Development Review, 5*(5), 8-11.

Drucker, P. (1973). *Management: Tasks, responsibilities*. New York: Harper & Row.

Fayol, H. (1916). *Administration industrielle et generale* [Industrial and general administration]. Paris: Dunod.

International Labor Organization. (1998). *Le conseil en management: Guide pour la profession* [Management consulting: A guide for professionals], (3rd ed.) Geneva, Switzerland: ILO Editions.

Kets de Vries, M. (2001). *The leadership mystique*. London: Reason Education.

Kotter, J. P. (1996). *Leading change*. New York: Free Press.

Lawler, E. E. (1986). *High-involvement management: Strategies for improving organizational performance*. San Francisco: Jossey-Bass.

Mintzberg, H. (1989). *Inside our strange world of organizations*. New York: Free Press.

Nonaka, I. & Takeuchi, H. (1995). *The knowledge creating company*. New York: Oxford University Press.

Savall, H. (1974, 1975). *Enrichir le travail humain dans les entreprises et les organisations* [Work and people: An economic evaluation of job enrichment]. Paris: Dunod.

Savall, H. (1979). *Reconstruire l'entreprise: Analyse socio-économique des conditions de travail* [Reconstructing the enterprise: Socio-economic analysis of working conditions]. Paris: Dunod.

Savall, H. (1987). Les coûts cachés et l'analyse socio-économique des organisations [Hidden costs and the socio-economic analysis of organizations]. *Encyclopédie du management* (pp. 599-658). Paris: Economica.

Savall, H. (2003a). An updated presentation of the socio-economic management model. *Journal of Organizational Change Management, 16*(1), 33-48.

Savall, H. (2003b). International dissemination of the socio-economic model. *Journal of Organizational Change Management, 16*(1), 107-115.

Savall, H., & Bonnet, M. (1988). Coûts sociaux, compétitivité et stratégie socio-économique [Social costs, competitiveness and socio-economic strategy], *Encyclopédie de la Gestion* (pp. 742-757). Paris: Editions Vuibert.

Savall, H., & Zardet, V. (1987). *Maîtriser les coûts et les performances cachés: Le contrat d'activité périodiquement négociable* [Mastering hidden costs and performances: The periodically negotiable activity contract]. Paris: Economica

Savall, H., & Zardet, V. (1992). *Le nouveau contrôle de gestion: Méthode des coûts-performances cachés* [New management control: The hidden cost-performance method]. Paris: Éditions Comptables Malesherbes-Eyrolles.

Savall, H., & Zardet, V. (1995). *Ingénierie stratégique du roseau, souple et enracinée* [Strategic engineering of the reed, flexible and rooted]. Paris: Economica.

Savall, H., & Zardet, V. (2004). *Recherche en sciences de gestion: Approche qualimétrique. Observer l'objet complexe*. Unpublished English translation: *Research in management sciences: The qualimetric approach. Observing the complex object*. Paris: Economica.

Savall, H., & Zardet, V. (2005). *Tétranormalisation: Défis et dynamiques* [Competitive challenges and dynamics of tetra-normalization]. Paris: Economica.

Savall, H., Zardet, V., & Bonnet, M. (2000). *Releasing the untapped potential of enterprises through socio-economic management*. Geneva, Switzerland: International Labor Office-ISEOR.

Simon, H., & March, J. G. (1958). *Organizations*. New York: John Wiley.

CHAPTER 14

INTERVENING IN SMALL PROFESSIONAL ENTERPRISES

Enhancing Management Quality in French Notary Publics

Laurent Cappelletti

In France, there are approximately 600,000 professional offices employing over 1 million people. All of these businesses are managed by someone holding a state-recognized qualification in a particular field of expertise and who bills fees for the services provided. Most of these companies are very small enterprises (VSEs) with less than 10 employees. These professional offices are also very diverse in terms of the expertise they offer, ranging from health care professionals (e.g., doctors, pharmacists, dentists) to legal representatives (e.g., lawyers, notary publics, bailiffs). Although these professions wield significant economic, social and political power in France and Europe, little academic or consulting-related research has been done on the management needs of these professional offices, beyond that required for any type of VSE (a notable exception is Maister, 1993, 1997). Similarly, university career paths leading to these professions offer little in the way of management training, as if it were unnecessary for a professional office to be managed at all (Altman & Weil, 1996; Boutall & Blackburn, 1998).

Socio-Economic Intervention in Organizations: The Intervener-Researcher and the SEAM Approach to Organizational Analysis, pp. 331–353

Nevertheless, these businesses have, in recent years, become subject to harsh strategic constraints, which require more than the professional expertise of their owners. Doctors, pharmacists and dentists, for example, must reduce their structural operating costs at the same time as guaranteeing the quality of patient care—an area in which their responsibility is increasingly being invoked in the courts. Lawyers must develop new legal products to survive in a climate of fierce competition. In short, today's professional offices must improve the quality of their management if they are to maintain their ability to survive and grow.

This chapter focuses on the challenge of improving the quality of management in very small professional offices. The guiding premise is that the introduction of *socio-economic management* (Savall, 1974, 1975, 1987, 2003a, 2003b; Savall & Bonnet, 1988; Savall & Zardet, 1987, Savall, Zardet, & Bonnet, 2000) principles will allow a professional office to improve the quality of its management, creating *value-added* in the process. In developing this central hypothesis, the chapter presents the results of a study of the use of ISEOR's scaled-down *implementation* method to introduce socio-economic management in 300 French notarial offices between 1998 and 2003. Notary publics constitute a representative group in which to examine the management problems of professional offices. In France and some other European countries, notary publics work as professional offices and, since the beginning of the 1990s, have been subject to new and exacting strategic constraints.

THE STRATEGIC CONTEXT OF NOTARIAL OFFICES

In France and some other European countries, notary publics represent a profession quite distinct from lawyers. Unlike the position in Great Britain or the United States, where legal professionals act as both lawyers and notary publics, in France notary publics provide legal security. They authenticate legal documents and provide advice, especially in real estate transactions (real estate purchase and sale) and family issues (e.g., inheritance, divorce). Since the French property crisis of the early 1990s, which resulted in the first bankruptcies ever seen among notarial offices, notary publics have had to come to terms with a new strategic environment, which, in turn, has obliged them to improve the quality of management in their businesses.

Notary Publics in France

France has approximately 4,600 notarial offices, managed by over 8,000 notary publics and employing 40,000 salaried staff. On the average, notary public offices employ eight staff members and are managed by a notary or several lawyers working as associates. Generally speaking, the notary deals with clients, and the clerks draft the legal documents. The accountant often plays a pivotal role as well, since he or she shares the notary's task of preparing client bills. The remaining employees are involved in administration, operating the telephone switchboard, arranging meetings and filing papers. Together, such offices turn over a total of some €3 billion annually. Approximately 80% of this revenue is generated from legal activities connected with family law (e.g., inheritance, marriage, divorce) and real estate law (e.g., real estate purchase and sale). Within these areas of activity, notary publics enjoy a state-regulated monopoly in which charged rates and fees are fixed by law. However, they are in competition with each other, since clients have a free choice of which notary they use. The remaining 20% or so of revenues is derived from nonmonopoly activities, where fees are unregulated (e.g., company law, asset management, real estate negotiation (Poli, 1997). In this market, notary publics are in competition not only with each other, but also with other professionals, such as lawyers and certified public accountants.

To ensure compliance with these regulations, notary publics are members of the regulatory organizations that control them, promote the profession and help it develop. Notary publics are also appointed by decree of the Department of Justice and belong to a chamber, a body containing all the notary publics in the same geographical *département* (there are 95 such chambers in France). The chamber is the basic unit of the profession, elects a notary as president every 2 years, and plays a disciplinary, promotional and management role within the profession. Each chamber forms part of a regional council. There are 33 regional councils that cover the same districts as courts of appeal and are constituted of notary publics elected for a 4-year term. This authority plays a role in regulating the profession by organizing inspections of notary publics by other notary publics. These bodies act within policy guidelines that are set and monitored by a national authority, the *Conseil Supérieur du Notariat* (CSN). Comprised of 80 permanent members and elected notary publics, the CSN plays an institutional role by setting policy and a single set of regulations for the profession. It also has management and significant business roles to play. As a result, the profession is organized in a unified and pyramidal fashion, as illustrated in Figure 14.1.

In summary, notary publics are organized into regulated professional offices: the notary is a business manager who is autonomous in the way he or she manages the business, but who is subject to regulation that has a direct effect on professional *ethics* (e.g., advertising is not permitted) and monopoly services. The notary is a public official, but works within a professional legal framework and receives his or her income from the business (Rymeyko, 2002).

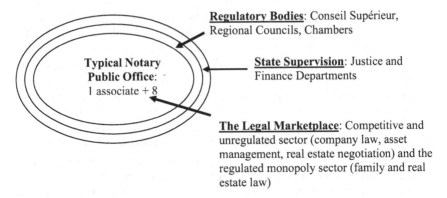

Figure 14.1.   The French notary public: A regulated profession.

The Problems and Strategic Constraints Faced by French Notary Publics

Since the French property crisis of the early 1990s, notarial businesses have had to cope with new strategic constraints that have required them to improve the quality of their management. This requirement, which is also being felt in other professions such as health care and law, is a recent development in a world where, for many years, the lawyer's expertise had been sufficient to ensure the *survival* and *development* of notarial offices. These new strategic constraints relate mainly to the need to create more value added in order to make monopoly activities more profitable, finance the development of new services, respond to the rising expectations of clients and staff, address the lack of corporate image, and cope with intense competition (see Figure 14.2). These constraints have been identified from the results of a study that examined 300 offices and the published personal accounts of approximately 30 notaries, chamber presidents, and regional council presidents attending ISEOR symposia (ISEOR, 2000, 2001, 2002, 2003).

The Financial Constraints on Creating Value Added

In terms of their monopoly activities, notary publics have tended to be rather somnolent when it comes to strategy: their working methods have evolved very little and have rarely focused on improving *effectiveness* and *efficiency*. ISEOR *intervention-research* (Savall & Zardet, 2004), for example, has demonstrated that very few notaries have introduced any form of management control to monitor the profitability of their monopoly activities, preferring to manage by focusing primarily on a single source of revenue. The time taken to provide a monopoly service varies from a few

Source: ©ISEOR 2000.

Figure 14.2.   The new strategic constraints faced by notary public offices.

hours to several days, while the price for that service remains fixed. Certain complex, contentious files can demand several months to process.

Focusing on monopoly work has diverted many notary publics from doing the research and development work that would allow them to offer non-monopoly services—estate management, for example. The growing reality is that notary offices need to exercise management control (Savall & Zardet, 1992) over both their monopoly and non-monopoly work if they are to improve the profitability of their offices (Cappelletti & Khouatra, 2004). They also need to apply innovative management methods to reduce their costs, increase their value-added initiatives, and invest in new product development. Otherwise, notaries will remain focused on routine monopoly activities, where margins are uncertain because of the high sensitivity to fluctuations in the real estate market and which are further threatened with eventual extinction as a result of liberalization in the European market for legal services. The harmonization of European regulations poses a genuine long-term *strategic threat* to notary publics because if that harmonization extends to the legal profession, notaries could easily lose their monopoly.

The Rising Expectations of Clients and the Need to Promote the Image of the Profession

As in other professions, notary publics have to cope with increasing competition and rising client expectations. Despite the discipline imposed by the profession, there is fierce competition between notaries for monopoly sector business. There is also intense interprofessional competition from lawyers, certified public accountants and realtors in the free market sector for services such as company law and real estate negotiation. This level of competition, which simply did not exist 20 years ago, is compounded by rising levels of client expectations. Clients are increasingly behaving as consumers of legal services. In the past, every family had its own preferred notary, but today's clients have no hesitation to change notaries according to the type of service offered, or even its cost.

Foreign clients of French notary publics are also contributing to this trend, because in some regions they make a significant contribution to notary revenues. This is the case, for example, with English and Dutch clients who are very keen on buying real estate in southwestern France. Notarial offices do not have the necessary client relationship management skills to respond to these expectations and, as a result, suffer from a poor public image. In fact, the professional image of notaries is regularly undermined by legal suits brought against them by clients and reported in the press. This helps maintain a certain level of aggressiveness among clients, despite CSN-funded campaigns to promote the profession in the media. ISEOR's intervention-research demonstrated that improving this image must go further than corporate advertising to address the quality of service actually delivered by individual offices (Cappelletti, 1998, 2006; Cappelletti & Khouatra, 2004). It seems that the reality of the notarial business is still a very long way away from an image that conveys any notion of effective management.

The Rising Expectations of Staff

Notary publics must also cope with the rising expectations of staff who are demanding to have a direct interest in financial results and to become more involved in the business. For a very long time, notary publics have been content to be paternalistic bosses, retaining the same staff throughout their professional lives. However, they must now become true managers with the ability to lead a team of staff members who are much more demanding in terms of training, promotion, career prospects and profit sharing. Notary publics are often confronted with poor staff motivation and commitment and are even finding it difficult to attract new skills. As businessmen, notary publics are literal beginners when it comes to management. As with other professions, they did not receive any formal management training during their time at the university and are typically ill-equipped to respond effectively to these expectations (Parsons, 2004).

Generic Problems of the Profession

In attempting to overcome these constraints, notary publics are handicapped by generic problems that affect their profession and the practical *functioning* of their own offices, problems which make it even more vital that they improve management quality. In terms of the profession, the hybrid legal and managerial status of the notarial office slows down or (over the short term) completely prevents any suitable response to the threat of losing their monopoly. State disciplinary and control regulation prevents some managerial or service-related innovations from being introduced. Moreover, the way in which the legal and financial professions work encourages a certain degree of corporatism. In terms of the way notarial offices function in general, governance and management are often handicapped by serious conflicts arising from disagreements or strategic differences between partner notary publics. ISEOR's intervention-research demonstrated that such conflicts can occur, for example, over the appointment of a new partner, the purchase of new premises, or branching out into new activities (especially in the unregulated sector). Where they exist, these conflicts are a powerful factor in breaking down *cohesion* inside the office, which in turn compromises external coherency and *strategic force* (Savall & Zardet, 1995).

INTRODUCING SOCIO-ECONOMIC MANAGEMENT IN 300 NOTARIAL OFFICES

Encouraged by the Superior Council of Notary Publics (Conseil Supérieur du Notariat), regional councils and chambers, notary publics requested the assistance of ISEOR in helping to improve management quality and accelerate office development. Between 1998 and 2003, ISEOR worked with 300 notarial offices in nine regions of France (Savall, Zardet, Cappelletti, Krief, Benollet, & Delattre, 1998-2003). *Socio-economic management* was introduced using a scaled-down implementation method adapted to suit the needs of VSEs and professional practices. This method is referred to as *multi-SB (small businesses) Horivert*. The aim of the method is to introduce the socio-economic management method within a company in a durable way that will improve the quality of management and ensure the *survival* and development of VSEs, as well as prevent any regression after those involved in the implementation have withdrawn (Savall, 2003a).

The Representative Nature of the 300-Office Sample

The socio-economic management method was introduced into 300 offices in nine regions of France, involving a total of 2,700 notary publics and salaried staff (see Table 14.1). This sample is representative of the total population of 4,600 French notarial offices in terms of size, geography and areas of business. The offices making up the sample have between 1 and 55 staff members, with an average of 8 employees (which reflects the national figure). The offices in the sample are both city-based (urban) and country-based (rural) offices. The sample contains equal numbers of traditional practices, focused primarily on monopoly business (mainly family law), and more innovative practices involved in significant levels of competitive business (mainly real estate negotiation).

Table 14.1.   The sample of 300 Notary Offices in 9 Regions of France

Region	Number of Offices	Office Size	Geography	Business Type	Period
1	34	4 to 25 people	Urban and rural	Monopoly and negotiation	1998
2	24	4 to 25 people	Urban and rural	Monopoly	1998
3	27	1 to 22 people	Urban and rural	Monopoly	1999
4	81	2 to 55 people	Urban and rural	Monopoly and negotiation	1999-2000
5	31	4 to 25 people	Urban and rural	Monopoly and negotiation	2001
6	4	1 to 41 people	Urban and rural	Monopoly and negotiation	2001
7	57	3 to 20 people	Urban and rural	Monopoly	2002-2003
8	14	3 to 22 people	Urban and rural	Monopoly	2002
9	28	4 to 25 people	Urban and rural	Monopoly	2002-2003
TOTALS:					
9	**300** (approx. 2,700 people)	**1 to 55 people**	**Urban and rural**	**Monopoly and non-monopoly**	**1998-2003**

Source: ©ISEOR 2003.

Objectives of the Multi-SB Horivert

The basic objectives of introducing the *multi-SB (small businesses) Horivert* approach in this intervention are illustrated in Table 14.2 through an example of the objectives set for the 57 offices in Region 7, which are representative of the goals pursued in the other sample regions. These objectives were formulated following **negotiation** between the offices concerned and staff from ISEOR, and summarized into the five general goals incorporated into the intervention-research contract. A critical goal was to use the *multi-SB Horivert* approach to actively involve all notary publics and staff in improving office management. In reality, it seems that teamwork and cohesion are very underdeveloped in many offices. A related objective was to transfer **management tools** and methods to notary publics and their staff in an effort to help them improve the service they provide to their clients. It also aims to help notary publics improve the relationship they have with their staff and the quality of their human **resources** management. Notarial offices are run by people who are experts in law, but absolute beginners when it comes to management and have received little or no training in how to manage or motivate their staff. The ultimate goal is to improve office *time management* and, more generally, to bring new impetus to internal office operation by improving financial and employment performance simultaneously.

Architecture of the Multi-SB Horivert Method

The offices in each region were brought together into groups of four. Each office within a group was involved in an *intracompany action plan*

Table 14.2.   Illustration of *Multi-SB Horivert* Objectives: Region 7

Intervention Objectives

1. To involve all notary publics and staff within the region in the process of improving the quality of office operation, management and service.
2. To transfer socio-economic management principles, methods and tools to all offices in order to improve professionalism and the quality of service delivered to clients.
3. To assist notary publics within the region to improve the relationship they have with their staff and their clients by highlighting directions for improvement and potential internal and external resources.
4. To support offices requesting assistance with work-time management and the implementation of the new Collective Notarial Agreement (*Convention Collective du Notariat*).
5. To bring new impetus to offices within the region by improving their internal organizational **structures**.

Source: ISEOR *intervention agreement.* ©ISEOR 2002.

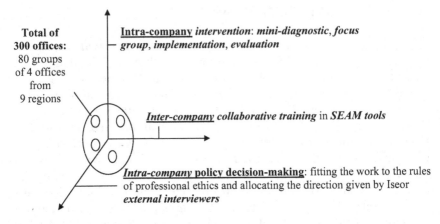

Total of
300 offices:
80 groups
of 4 offices
from
9 regions

Intra-company *intervention: mini-diagnostic, focus*
group, implementation, evaluation

Inter-company *collaborative training* in *SEAM tools*

Intra-company policy decision-making: fitting the work to the rules
of professional ethics and allocating the direction given by Iseor
external interviewers

Source: ©ISEOR 1997.

Figure 14.3. The three dimensions contained in the scaled-down *multi-SB Horivert* method.

(work done within the office) and an *intercompany action plan*. This was coordinated in each region by a *steering group* made up of notary publics elected to represent their region. The *multi-SB Horivert* approach comprises three dimensions: (1) bringing about change through intracompany action plans, (2) *collaborative training* in *socio-economic management tools* through intercompany action plans, and (3) the development of an overall synchronization policy (see Figure 14.3).

The *multi-SB Horivert* method follows the same principles as the *Horivert* method, but uses action plans that are scaled down to suit VSEs and professional offices. All the action plans were coordinated by twenty or so ISEOR consultants distributed across the 300 offices.

Intracompany Action Plans

The intracompany action plans were similar in all 300 offices. They consisted of carrying out a minidiagnostic focusing on the problems responsible for disrupting office effectiveness and efficiency. These problems were grouped into six themes that model the quality of management within a company (see chapter 1). In each office, notary publics and staff were interviewed separately about the problems relating to these six topics. A collective *evaluation* meeting was then held to evaluate the *hidden costs* of these problems (i.e., the amount of value-added lost). The qualitative, quantitative and financial results of these mini-diagnostic sessions were used as the basis for the work done by a two-tier focus group: a small group containing only the notary publics, and a larger group involving the notary publics and all office staff. The ***personal assistance*** sessions designed around the ***management tools*** introduced in the training/consultation sessions were combined with the intracompany action plan focus group sessions. This combining of ***focus group*** sessions with ***personal assistance sessions*** is a special feature of the scaled-down ***multi-SB Horivert method***. It introduces an intentional flexibility into the method that enables it to be adapted to the special nature of very small enterprises.

Intercompany Action Plans

Each of the 300 offices was also involved in an intercompany action plan built around groups of four offices of different sizes. The purpose was to organize training/consultation sessions focusing on the six basic tools of socio-economic management: ***time management, competency grid, internal and external strategic action plan, priority action plan, strategic piloting indicators***, and ***periodically negotiable activity contract***. Each office is represented by the notary public and one, two or three members of staff, depending on the size of the office.

Steering Group Action Plans

A ***steering group*** of between four and six chamber-appointed notary publics and one superior council (*Conseil Supérieur du Notariat*) representative was set up in each of the nine regions. The consultants were responsible for leading these steering groups and presenting anonymous assessments of the work accomplished in the offices. The steering groups for two of the nine regions (Regions 3 and 7) asked their consultants to help provide ***maintenance action plans*** to support the continuation of the initiative after completion of the ***project***. The result of this request was that the consultants led working days within offices 6 to 10 months after the original initiative ended in order to stimulate the use of these management tools, consolidate problem resolution processes, and increase the financial ***value-added*** created by the office.

Implementation Schedule

The method used to allocate action plans under the ***multi-SB Horivert*** method was designed to optimize the effectiveness and efficiency of these initiatives in each office. It was felt that for each office in a group of four, the involvement should be spread over a period of 8 months to allow for the ***integration*** of management and design tools and the implementation and evaluation of the selected improvement initiatives (see Table 14.3). For each group of offices, four intercompany ***collaborative training*** sessions on ***socio-economic management tools*** were held every 2 months, alternating with five intracompany diagnostic sessions, followed by focus group and tool implementation sessions. Each office in every group of

Table 14.3.   Typical *Multi-SB Horivert* Schedule for a Group of Four Offices

Action Plans	Month							
	1	*2*	*3*	*4*	*5*	*6*	*7*	*8*
Policy-decision axis								
• ***Steering group*** (3 sessions)	1		2					3
Intercompany intervention								
• Collaborative training (4 intercompany sessions totaling 2½ days)	1 (1 day)		2 (½ day)		3 (½ day)		4 (½ day)	
Intracompany intervention								
• Minidiagnostic (2 intracompany sessions totaling 1 day)	1 (½ day)	2 (½ day)						
• ***Personal assistance*** and ***focus group*** (3 intracompany sessions totaling 1½ days)				1 (½ day)		2 (½ day)		3 (½ day)

Source: ©ISEOR 2000.

four received the same number of intercompany action plan sessions as intracompany sessions (2½ days). Three steering group sessions provided the opportunity to monitor how work was progressing in the various groups of offices within the region.

The coordinating role played by the ISEOR consultants is similar to that played in the ***Horivert*** method. Its main objective is to implement the skills and ***behavior*** patterns inherent to the role of the socio-economic management consultant as ***methodologist*** (to transfer the socio-economic management method), ***mediator*** (to manage the conflicts inherent in all human groups) and ***therapist*** (to provide a degree of psychological support to those involved in the process of change) (Savall & Zardet, 2004).

Results of Introducing Socio-Economic Management in the 300 Notarial Offices

The results of the study will be reported in two sections. First, the outcome of the 300 diagnostic sessions that were held to identify the main operational and management problems in the office, and their related *hidden costs* (value-added lost) will be discussed. Second, the results of the *evaluation* of the effects of introducing socio-economic management on office management quality, and the consequent recovery of value-added, will be explored. The study makes a distinction between *immediate results* (those which have had an effect on practice performance during the current year) and the *creation of potential* (the investments—most of them intangible—that will have an effect on future office performance).

Results of the 300 Diagnostic Sessions

The study identified the qualitative, quantitative and financial aspects of the main problems affecting the quality of operation and management seen both in the offices themselves and in the services they offer. The results of the socio-economic management projects described in this chapter demonstrate improved performance at all three *integral quality* levels: quality of operation, quality of management, and quality of products and services (see chapter 1).

Qualitative results obtained from diagnostic sessions. ISEOR's *intervention-research* demonstrated that the managerial problems affecting over 80% of the offices in the sample can be divided into four categories: personnel management, business management, client relationship management and strategic action.

- **Personnel management** is characterized by poor *integrated training* and human resources management. Notary publics devote too little time to developing career plans for their staff and evaluating their skills. Collective meetings and individual staff management appraisals do not exist in most offices and there is no management of the way work is distributed. Staff members are very often left to their own devices, which can lead to a lack of involvement and motivation.

- **Business management** was handicapped by poor filing and classification procedures, despite the fact that these activities are priorities in any notary public office. There is no time management associated with the composition of files and case notes, which prevented the introduction of management controls, made every task an urgent one, and resulted in *poorly executed tasks*. The methods used by the profession to compile files and case notes varied widely between notary publics and clerks, even in the same office. Offices also suffered from an absence of any *synchronization* between the notary and the staff in respect to current cases. These business management problems led to quality failures in the legal documents themselves.

- **The customer relationship** suffered chiefly as a result of the strains imposed by the poor reception clients receive, both on the telephone and in person (clients have to wait for long periods at reception or on the telephone). This relationship was also upset by fee quotation procedures that are sometimes less than reliable, resulting in differences between the quotation given to the client and the final bill received. Similarly, clients were often given too little information about their cases at every stage of the process. These problems resulted in clients being lost and made it difficult to retain the loyalty of existing clients.

- **The strategic action** taken by notary publics was compromised by many failures. The level of collaboration between notary publics working in the same office was too low or nonexistent, and was determined to be the cause of many hidden costs. Almost all of the offices did not have any new product development strategies or any strategy for implementing initiatives aimed at collectively pushing the business forward.

- The *root causes* of all four categories of problems lie partly in the fact that no time was scheduled or programmed for personnel management, business management, client management (other than client meetings with the notary) or strategy setting. They also stemmed from the fact that neither the notary nor the staff received any training in management principles and tools. They also emerged from the notary's unwillingness to depart from his or her main activity of meeting clients and dealing with cases in order to devote time to purely managerial issues. These root causes lead to a lack of *direction*, *synchronization*, *clean-up* and communication, all of which simply perpetuated the problems.

Quantitative and financial results obtained from diagnostic sessions. The intervention-research demonstrated that the financial impact of these problems translated into **hidden costs** (or loss of value-added) equivalent on average to roughly €13,000 per person, per year. On the average, this loss of value-added represents 20% of the office's *variable cost margin* or variable **value-added** margin (i.e., the difference between the revenue generated and the office costs), which vary automatically according to activity level. In order to evaluate the hidden costs present in each of the 300 offices in the sample, the impact of problems was evaluated qualitatively, quantitatively and financially. As an illustration, the results of a **minidiagnostic** session conducted in an office in Region 4 are presented in the form of key points in Table 14.4. Presented in this form, results can be used as a working platform for internal office **projects**.

**Table 14.4. Example of an _Internal_
Minidiagnostic Session Conducted in a Notary Public Office**

Dysfunctions		Hidden Costs (= Destruction of Value-Added)
Working conditions	Poor _working conditions_	€4,300
	Obsolete office equipment	€24,000
	Unreliable filing system	€12,000
	Lack of small items: telephone, etc.	€60,500
	IT problems	€2,500
	Lack of client parking	Not evaluated
	Total	**€103,300**
Work organization and _communication-coordina-tion-cooperation_	Poor links between head office and local site	€3,600
	Poor supervision of clients and cases	€89,200
	No definition of expected service quality	€82,900
	Poor time and deadline management	€67,800
	Lack of meeting and information communication resources	€3,200
	Total	**€246,700**
	Grand Total	**€350,000**

Source: © ISEOR 2001.

ISEOR's intervention-research in these 300 offices demonstrated that _hidden costs_ average €104,000 per office, per year (see Table 14.5). Most of the value-added lost through the problems identified in the offices were a result of _overtime_ (extra time spent recreating poor quality files and case notes), _nonproduction_ (the loss of dissatisfied clients, work left unfinished due to _absenteeism_ and underproductivity resulting directly from lack of motivation) and _excess salaries_ (a situation the notary could delegate, but refuses to, such as opening mail).

Effects of Introducing Socio-Economic Management

In nearly 70% of the offices in the sample, the introduction of socio-economic management led to positive effects that significantly improved the quality of management, operation, products and services. Table 14.6 provides an example of the positive effects observed in an office in Region 2 in terms of the three _integral quality levels_ (management, operation,

**Table 14.5.   Summary of *Hidden Costs*
as Evaluated in the 300 Notary Public Offices**

Region	Number of Offices	Hidden Costs (*Destruction of Value-Added*) per Person, per Year	Hidden Costs (Destruction of Value-Added) per Office, per Year
1	34	€10,000	€80,000
2	24	€12,000	€96,000
3	27	€10,000	€80,000
4	81	€11,000	€88,000
5	31	Not evaluated*	Not evaluated*
6	4	€16,000	€128,000
7	57	€12,000	€96,000
8	14	€15,000	€120,000
9	28	€10,000	€80,000
Total	**9**	**300**	**€13,000 on average**	**€104,000 on average**

Source: © ISEOR 2003.
Note: *Not evaluated due to the limited amount of time available for the study.

products and services), and the qualitative, quantitative and financial impact of these effects.

These positive effects had an ***immediate result*** in terms of performance, as well as a deferred result in terms of the ***creation of potential***. Analysis of the results illustrated that the number of positive effects was greater in those regions where the partnership between the offices and the ISEOR team was the closest. Analysis also showed a difference between those offices in regions where the Chambers decided on compulsory involvement for all offices (Regions 2, 3, 4 and 9), and offices in regions where involvement was voluntary (Regions 1, 5, 6, 7 and 8). Where a region decided on compulsory collective involvement, the partnership between the offices and the ISEOR team was very close in one-third of cases, average in one-third of cases ,and very poor in a final one-third of cases. This final third related to those offices that did not enter into the improvement project voluntarily and had the decision imposed upon them by the Departmental Chamber or Regional Council. In those regions where involvement was not compulsory, the partnership between the offices and the ISEOR team was close or very close and the positive effects were more pronounced than in other regions.

Immediate results. Significant levels of success were achieved with the introduction of ***socio-economic management tools*** into the 300 offices studied. The ***time management*** and ***competency grid*** tools were implemented in over 80% of offices. The management indicators, internal and external

Table 14.6.   *qQfi Evaluation* of *Socio-Economic Management* Methods on *Integral Quality* Levels and Financial Performance in One Office

Positive Effects Observed Within 1 Year	Socio-Economic Results		
	Qualitative	Quantitative	Financial
Quality of service for *External Client*			
• Extending telephone switchboard opening times to 09:15-18:30 every day	• Increased revenue	+13%	+€90,000
• Installing a telephone answering machine for when the switchboard is closed	• Increase in no. of legal documents	+9%	(included in revenue)
• *Scheduling* client meetings hourly instead of half-hourly	• Improved client meetings	2/3 fall in hidden costs	+€3,000
• Renovating the reception area to provide a better welcome for clients			
Quality of functioning			
• Changing the accounting and document processing software	• Less time wasted in the office	2/3 fall in hidden costs	+€7,000
• Buying an additional photocopier			
• Installing a phone for each person	• Better working atmosphere	Not evaluated*	Not evaluated*
• Changing the alarm and surveillance system			
• Introducing a loan procedure			
Quality of management			
• Organizing regular meetings between notary publics and staff to discuss and resolve problems	• Better cohesion within the office	Not evaluated*	Not evaluated*
• Introduction of improved salary packages for all staff			

Source: © ISEOR 2002.
Note: *Not evaluated due to the limited amount of time available for the study.

strategic action plan and *priority action plans* were implemented in 60% of offices.

Table 14.7 shows an example of management indicators and their architecture. This example is taken from a notary public office in Region 7. *The periodically negotiable activity contract (PNAC)* was implemented in only 10% of offices, due to the limited time available (short period of 8 months and limited number of sessions as a result of *scaling down the HORIVERT method*) and significant resistance among the notaries to address the problem of the salary gap between full-time staff and notary publics in many offices.

In nearly 70% of the offices, the *focus groups* implemented management quality improvement initiatives, which emerged during the first month and were developed throughout the 8-month project period. These activities have considerably strengthened the offices' ability to sur-

**Table 14.7.   Architecture of
Notary Public Office *Strategic Piloting Indicators***

Sections (Names and Content)	Indicators		
	Qualitative	*Quantitative*	*Financial*
1. General management of activities			
• List of urgent tasks to be carried out	X		
• General scheduling of legal documents	X		
• Allocation of cases to staff	X		
• *Scheduling* of signing meetings	X		
2. Financial results			
• Revenue and monthly results (monthly variances)			X
• Number of legal documents signed (during the month)		X	
• Number of inspections made (during the month)		X	
3. Communication with clients			
• Telephone messages (client call-backs)	X		
• Business cards	X		
4. Quality policy			
• List of problems and hidden costs	X	X	X
• Number of poorly handled cases to be repeated (with reasons)		X	
• Quality meeting action sheets	X		
5. Office strategy			
• Priority action plan for the current half year	X		
• Internal and external strategic action plan	X		
6. Communication with colleagues			
• List of chamber lawyers offices (notary public) (with phone numbers)	X		
• Notification to attend the chamber	X		
7. Legal and accounting references			
• Fast-access scale of fees, estate duty charging rates	X		
• Attrition ratio	X		
• Written references	X		
• Tables: contract of sale	X		
8. Personnel management			
• Annual evaluation meeting schedule	X		
• Staff holiday and school holiday planner	X		
• Skills matrix	X		
• Office training plan	X		

Source: © ISEOR 1999.

vive and develop. However, in approximately 30% of the offices, management quality improvements, although real, were less well-established. In these cases, the introduction of management tools, the reduction of ***dysfunctions*** and loss of value-added, and the solutions developed by the focus groups did not result in a lasting ***improvement*** of management quality. Analysis demonstrates that the offices concerned are those that did not volunteer for the initiative and simply took the passive route of following the policy instructions issued by their professional Chamber, which had decided to make the initiative compulsory for all offices within their region. In these cases, the lack of involvement of the notary managing the office impeded the introduction of the management tools, restricted the creativity of the focus group, and caused considerable disappointment among the staff. ISEOR's *intervention-research* has shown how important it is that the managing notary sets an example by welcoming the initiative and that this type of role modeling behavior has a direct effect on the qualitative and financial results obtained.

The positive effects observed in the study have been seen within the four categories of *dysfunctions*:

- **Personnel management**:   Changes included the development of training plans and career plans for staff, the introduction of monthly office meetings and biannual meetings between individual staff members and the notary, and notary public offices setting individual targets for staff members.

- **Business management**:   Initiatives encompassed introducing mini-management controls, drafting and monitoring of quality procedures, and creating action plans to address the need for synchronization between the notary and those staff responsible for legal drafting in complex cases.

- **Client relationship management**:   Actions involved the reorganization of client reception areas and telephone answering procedures, the introduction of personalized client relationships, clients receiving regular updates on the *progress* of their cases, and greater accuracy in the fee quotation process.

- **Strategic actions**:   Efforts focused on the definition and implementation of strategies for new activity development (in areas such as company law) and the definition and implementation of strategies to upgrade office IT systems.

The *intracompany focus groups* had a very positive effect on cohesion and teamwork, with a positive follow-up effect on the external coherency and strategic *efficiency* of the office. Analysis demonstrates that internal office *cohesion* has a major effect on external coherency as a result of the relationships created between notary public offices inside the focus groups and between the notary public and his or her staff.

These positive effects have resulted in improved financial performance in the offices (*immediate results*). These *improvements* were evaluated

**Table 14.8.   Financial Effects of
Introducing *Socio-Economic Management* in the 300 Offices**

Region	Number of Offices	Hidden Costs (Destruction of Value-Added) per Person, per Year	Reduction in Hidden Costs (= Conversion Into Value-Added) After 8 months
1	34	€10,000	36%
2	24	€12,000	29%
3	27	€10,000	27%
4	81	€11,000	37%
5	31	Not evaluated*	Not evaluated*
6	4	€16,000	45%
7	57	€12,000	38%
8	14	€15,000	36%
9	28	€10,000	47%
Total 9	**300**	**€13,000 on average**	**37%, i.e., €40,000 per office, on average**

Source: © ISEOR 2003.
Note: *Not evaluated due to the limited amount of time available for the study.

financially at the end of the process in each office, by measuring the reduction in hidden costs, i.e. the growth in value-added (see Table 14.8). The study demonstrates that, on the average, the positive effects of introducing *socio-economic management* have led to a 37% reduction in lost value-added, resulting in value-added gains of some €40,000 per office, or approximately 10% of the variable cost margin. Our intervention-research reveals that most offices contain the ability to conduct *proactive endogenous strategies* to cope with an environment that has become highly competitive.

Creation of potential by the offices. The notary offices have used part of these value-added gains to fund initiatives aimed at creating potential, which should have an eventual financial impact but will require real investment in the short term. The potential-creating initiatives observed in the offices include the computerized storage of files and case notes, the creation of office Web sites, and computerized data sharing between the notary and his or her colleagues. These initiatives also contribute to restoring the image of the office and the profession as a whole. Socio-economic initiatives also encouraged the development of a sales mentality and an awareness of client service quality in the offices, although for many of the offices in the sample diagnostic sessions showed a marked reluctance among notary publics and staff to accept the socio-economic management principle that "everyone involved has a greater or lesser sales role to play as part of their normal jobs" (see chapter 1). Generally speaking, notary publics and their collaborators do not consider themselves merchants conducting business, even though this is the case. Indeed, most of them have studied law, which develops a pejorative vision of sales and commerce. In essence, law is noble, commerce is not.

The *intervention-research* in this project, however, demonstrates that everyone in the office *does* play a role in sales initiatives and, in his or her own way, contributes to revenue generation—from the secretaries welcoming clients and the clerks in preparing quality case assessments, to the quality of information supplied to clients and notary publics in meeting and advising clients. ISEOR's intervention-research framework demonstrates that developing the "everyone is a salesperson" concept in professional practices can be a powerful influence on achieving sustainable improvements in *socio-economic performance* (Savall & Zardet, 2005).

CONCLUSION

The chapter has documented a long-term *socio-economic intervention-*research project (1998 to 2003) with a sample of 300 notary public offices involving approximately 2,700 people. The results demonstrate that in 70% of the cases, the introduction of socio-economic management using a scaled-down *method*, specially adapted to meet the needs of professionals and very small enterprises (Nobre, 1993), facilitates lasting improvement in the quality of management, operation and service provision in regulated professional offices. This improvement further translates into significant enhancements in personnel management, business management, client relationship management, and strategic initiatives. As reflected in other ISEOR projects, the study underscores the critical role that involved and energized managers (in this instance the managing notaries) play in creating a foundation for success. It is imperative that such involvement be encouraged and monitored very closely in the professions and VSEs during interventions aimed at improving management quality.

The *improvement* in management quality allows notary public offices to come to terms with the new strategic constraints imposed by the need to create more value-added, the rising expectations of clients and staff, a poor corporate image and intense competition. It appears to increase the ability of these offices to survive and develop on a long-term basis, while improving profitability. However, it is still necessary to conduct further long-term study into the effects of introducing *socio-economic management* using the scaled-down *multi-SB (small businesses) Horivert* method to assess the extent to which this management philosophy remains a permanent part of the business, years after the consultant has withdrawn.

REFERENCES

Altman, M. A., & Weil, R. (1996). *How to manage your law office*. New York: Matthew Bender.

Boutall, T., & Blackburn, B. (1998). *The solicitors' guide to good management, practical checklists for the management of law firms*. London: The Law Society.

Cappelletti, L. (1998). *L'ingénierie d'audit d'activité d'une entreprise: La production d'intelligence socio-économique. Cas d'expérimentation* [Engineering the activity audit of an enterprise: The production of socio-economic intelligence. Experimental cases]. Unpublished doctoral dissertation, University of Lyon, France.

Cappelletti, L. (2006). *Contribution à une épistémologie de l'audit et du contrôle de l'activité* [Contribution to an epistemology of activity audit and control] (French National Authorization to Supervise Research), University of Lyon 3, France.

Cappelletti, L., & Khouatra, D. (2004). Concepts et mesure de la création de valeur organisationnelle [Concepts and measures of the creation of organizational value]. *Comptabilité-Contrôle-Audit, 1*(10), 127-146.

Daudé, X. (2006). Démarche de changement dans la profession notariale [Change procedure in the notary public profession]. In ISEOR (Ed.), *Le management du développement des territoires* (pp. 165-182). Paris: Economica.

ISEOR. (2000). *Le Notariat Nouveau* [The new notary public office]. Proceedings of ISEOR Colloquium (pp. 13-83). Lyon, France: ISEOR.

ISEOR. (2001). Table ronde sur les perspectives d'évolution de la profession notariale [Round table on the perspectives of evolution in the notary public profession]. *Recherche-Intervention et Création d'Entreprises (accompagnement et évaluation)* [Proceedings of ISEOR Colloquium] (pp. 183-206). Lyon, France: ISEOR.

ISEOR. (2002). Les effets d'une démarche qualité dans la profession notariale [The effects of quality procedure in the notary public profession]. *Le management des entreprises culturelles* [Proceedings of ISEOR Colloquium] (pp. 179-192). Lyon, France: ISEOR.

ISEOR. (2003). Développement stratégique et management socio-économique dans le notariat [Strategic development and socio-economic management in notary public offices]. *L'Université Citoyenne, progrès, modernisation, exemplarité* [Proceedings of ISEOR Colloquium] (pp. 291-302). Lyon, France: ISEOR.

Maister, D.H. (1993). *Managing the professional services firm*. New York: Free Press.

Maister, D.H. (1997). *True professionalism: The courage to care about your people, your clients, and your career*. New York: Free Press.

Nobre, T. (1993). *La structuration des entreprises en phase post-création et de pré-développement. L'apport de l'intervention socio- économique. Cas d'expérimentations* [Structuring enterprises during the post-creation and pre-development phase. The contribution of socio-economic intervention. Cases of experimentation]. Unpublished doctoral dissertation, University of Lyon, France.

Parsons, M. (2004). *Effective knowledge management for law firms*. New York: Oxford University Press.

Poli, N. (1997). *L'influence de l'Union Européenne sur le Notariat, bilan et perspectives* [The influence of the European Union on notary public offices: Assessment and perspectives]. Unpublished doctoral dissertation, University of Montpellier, France.

Rymeyko, K. (2002). *Enjeux stratégiques des professions libérales réglementées, mutations des pratiques de management et impacts sur la performance. Cas des offices de notaires* [The strategic stakes of liberal, self-regulated professions: Transformation of management practices and impacts on performance. Cases of notary public offices]. France: University of Lyon.

Savall, H. (1974, 1975). *Enrichir le travail humain dans les entreprises et les organisations* [*Work and people: An economic evaluation of job enrichment*]. Paris: Dunod.

Savall, H. (1987). Les coûts cachés et l'analyse socio-économique des organisations [Hidden costs and the socio-economic analysis of organizations]. *Encyclopédie du management* (pp. 599-628). Paris: Economica.

Savall, H. (2003a). An updated presentation of the socio-economic management model. *Journal of Organizational Change Management, 16*(1), 33-48.

Savall, H. (2003b). International dissemination of the socio-economic model. *Journal of Organizational Change Management, 16*(1), 107-115.

Savall, H., & Bonnet, M. (1988). Coûts sociaux, compétitivité et stratégie socio-économique [Social costs, competitiveness and socio-economic strategy]. *Encyclopédie de la Gestion* (pp. 742-757). Paris: Editions Vuibert.

Savall, H., & Zardet, V. (1987). *Maîtriser les coûts et les performances cachés: Le contrat d'activité périodiquement négociable* [Mastering hidden costs and performances: The periodically negotiable activity contract]. Paris: Economica

Savall, H., & Zardet, V. (1992). *Le nouveau contrôle de gestion: Méthode des coûts-performances cachés* [New management control: The hidden cost-performance method]. Paris: Éditions Comptables Malesherbes-Eyrolles.

Savall, H., & Zardet, V. (1995). *Ingénierie stratégique du roseau, souple et enracinée* [Strategic engineering of the reed, flexible and rooted]. Paris: Economica.

Savall, H., & Zardet, V. (2004). *Recherche en sciences de gestion: Approche qualimétrique. Observer l'objet complexe.* Unpublished English translation: *Research in management sciences: The qualimetric approach. Observing the complex object.* Paris: Economica.

Savall, H., & Zardet, V. (2005). *Tétranormalisation: Défis et dynamiques* [Competitive challenges and dynamics of tetra-normalization]. Paris: Economica.

Savall, H., Zardet, V., & Bonnet, M. (2000). *Releasing the untapped potential of enterprises through socio-economic management*. Geneva, Switzerland: International Labor Office-ISEOR.

Savall H., Zardet V., Cappelletti, L., Krief, N., Benollet, P., Delattre, M., et al. (1998-2003). *Etudes Notariales* (1-9) [Notary public studies]. ISEOR Research Report. Lyon, France: ISEOR.

CHAPTER 15

MASTERING COMPUTER TECHNOLOGIES

Contributing to Research-Experimentation With Users and Computer Specialists

Véronique Zardet and Nouria Harbi

At the beginning of the twenty-first century, information and communication computer networks are omnipresent in professional life, playing an ever increasing role in our personal and extraprofessional life as well. Computer technologies have contributed to significantly increasing the global economic *productivity* of enterprises, especially in comparison to noncomputerized work. Nonetheless, close observation of businesses and organizations reveals major, frequent and costly *dysfunctions* in the operation of such computerized activity, specialized service, information and communications technology (ICT) support and computer service companies. These dysfunctions considerably reduce the contribution of computer networks to the actual economic productivity of businesses and organizations.

Based on numerous *socio-economic diagnostics* (Savall & Zardet, 2004), carried out both with ICT users and ICT service departments (e.g., Bou-

Socio-Economic Intervention in Organizations: The Intervener-Researcher and the
SEAM Approach to Organizational Analysis, pp. 355–372
Copyright © 2007 by Information Age Publishing

langer, 1997; Datry & Payre, 2004; Delattre, 2002; Demissy, 2002; ISEOR, 1998; Plane, 1991; Saint-Léger, 2005; Savall & Zardet, 1985; Zardet & Harbi, 2002; Zardet & Srajek, 2000), we have established a typology of dysfunctions related to computer network development, maintenance and use. The chapter examines the costly economic stakes these dysfunctions represent, owing to the hidden costs they generate. The discussion then turns to how a socio-economic perspective can increase the *value-added* of specialized computing, while improving ICT operations.

COMPUTER AND ICT TECHNOLOGIES

Computer and ICT technologies—including intranet, Internet, e-mail and groupware—constitute an important sector of activity, appreciable by the sales figures they generate, the number of jobs they create and their growth rate. For example, the *Institut National de la Statistique et des Études Économiques* (INSEE) estimated the number of employees working in computer technologies at 668,000 in 2002, but this figure does not include government employees in computer technologies, nor those working in the field of insurance for banks and liberal professions. Inside enterprises, ICT technologies play an essential role in activity processes, and in certain professions and business functions, this role is of vital importance. These technologies support internal and external information systems, as well as production and management systems, from raw materials and final product stocks, to production launch and organization, order and billing management, delivery and accounting, and human resource management.

Rather than attempting to identify the dysfunctions specific to each technology and application, a basic goal of the chapter is to analyze, in generic fashion, the various types of dysfunctions relative to ICT support activities and the utilization of these technologies. For businesses, the stakes are very high. This can be appreciated as much by the number of computer technicians in businesses, as by investment budgets for the ICT hardware and software utilized by companies. In 2004, for example, the American magazine *Information Week* conducted a study of 500 American businesses considered to be the most innovative in terms of technological practices. It found that these firms dedicated between 2% (automobile, chemical) and 9% (financial services) of their total budget to ICT hardware and software, depending on their branch, with specialized professional software representing one forth of that budget.

Despite the on-going tendency toward lowered hardware and software prices over the past 20 to 30 years, computer park renewal markets remain significant. As for information and communication technologies, many businesses, especially in France and Europe, are still in the phase of acquiring their first equipment (which also explains why so many small- to medium-sized enterprises are still not equipped with Internet access). The renewal market represents all potential clients already possessing one or several computers that they change approximately every 3 years. This market is as big as the market for initial purchase of computer equipment.

HIDDEN COSTS AND DYSFUNCTIONS LINKED TO ICT TECHNOLOGIES: USER AND TECHNICIAN PERSPECTIVES

ISEOR's knowledgebase, progressively developed over the past 30 years in more than a thousand enterprises and organizations (Savall, 2003a), contains numerous findings on dysfunctions linked to computer technologies. Today, virtually all organizations, regardless of their size, status or activity, make use of computer technologies in performing their work. Our findings are drawn from this enormous body of data on dysfunctions linked to the utilization of computer technologies. Furthermore, this database also contains data concerning computer service departments and companies in which in-depth *socio-economic diagnostics* (Savall, 1974, 1975, 1987, 2003b; Savall & Bonnet, 1988; Savall & Zardet, 1987) were performed. The cases have been extracted to serve as the basis for an in-depth study of these *dysfunctions* and *hidden costs* (see Savall, Zardet, & Harbi, 2004). The major characteristics of that study are presented in Table 15.1.

Lack of Rigor at the Conception Phase

The practice of depending on computer specialists for the design of computer programs and applications has significant repercussions on the cost of running and using those programs. One often finds, in particular, programs that are poorly adapted to the needs of users, or programs that cannot "communicate" with one another for lack of a common interface between the computer systems. This problem results in numerous disturbances for users and creates delays in carrying out their tasks. These disturbances include excessive time spent operating applications, multiple re-entries of data, errors that later require manual modification, and so forth. But hidden costs attributed to users are also caused, in a "*boomerang effect*," by computer specialists themselves who are called in to "rescue" users who require support and repair.

Diagnostics performed inside development teams have helped to identify the underlying causes of these dysfunctions. In addition to the lack of structured working methods among development teams, the lack of procedures and norms was observed in virtually every situation, both in internal team *interactions* as well as in the interfaces with users. The lack of

**Table 15.1.   Five In-Depth Cases on
ICT Service Departments and Companies**

	Case A	Case B	Case C	Case D	Case E
Date of study	1983	1988	1991	2002	2004
Company	Bank	Agro-food firm	Insurance and banking firm	Public agency for employment	Computer service company
Size	420 p.	3 000 p.	49,000 p.	3,700 p.	5 p.
Sector of *diagnostic*	Data processing service	Data processing	Production computing	Information systems management	Production computing
Size of sector of diagnostic	7 p.	45 p.	144 p.	50 p.	5 p.
Hidden costs/ per year*	8,600 €**	16,200 €	3,800 €***	68,500 €	34,000 €
Absenteeism	NE	200 €	NE	2,000 €	500 €
Work accidents	4,800 €	NE	NE	NE	NE
Personnel turnover	2,500 €	NE	NE	2,000 €	NE
Nonquality	NE	400 €	NE	44,000 €	2,000 €
Direct productivity gaps	1,300 €	15,600 €	3,800 €	20,500 €	31,500 €

Source: © ISEOR 2004.
Notes: *All hidden cost amounts have been revaluated on the 2004 base. **Partial estimation based on three indicators. ***Partial estimation based on one indicator. NE: Not evaluated, given the time allotted to the study.

know-how in computer project management and the absence of time norms for development contributed to increasing delays for application deliveries to users. Faced with these chronic delays, both computer technicians and users sacrificed the test phase, preferring to deliver the application as soon as possible. Inside computer support services, a frequent imbalance was observed between time dedicated to *security-management* and time dedicated to *development-management*. Even though they were considered less valorizing than development management, security management activities dominated and monopolized a large portion of computer support teams' time *resources*, whether for assistance or repair. Yet, these services were often demanded by users, thus entailing reorganization of development activities, which in turn were even further delayed.

Beyond the absence of a relevant policy for budgetary issues and computer investment choices, the predominant *root-cause* resided in the strategic choices of allotment and distribution of human and financial *resources*: What part should be dedicated to development and maintenance? What part of development should be externalized, as opposed to development by employees? Finally, an additional *root-cause* was the frequently observed temptation to flee in a "blind race forward" toward new technologies, succumbing to the powerful trends of fashion and novelty that characterize that sector of activity.

Investing in Hardware Networks

Equipping offices with computers, servers, printers and other hardware usually isn't a problem. However, offices that are poorly-equipped for computer networking activity can cause the same effects as those discussed above. Insufficient or defective cables and networks that lack broadband transmission can cause communication difficulties, and can even make it impossible to successfully use the hardware. One still encounters, from time to time, the absence of hardware (or out-dated hardware) in an organization, which clearly complicates users' work (e.g., some computers are slow, certain software programs are installed only on the most powerful computers).

Software Programs as Risk

All software applications developed for the enterprise's own needs must be readjusted every few months or years to the changing needs of that enterprise. Thus, access to "source programs" becomes indispensable, without which the enterprise has no other recourse than to develop a new application, which is much more costly in terms of development costs, adaptation and user training. However, in many of the computer service companies we observed, there were major *risks* in terms of loss or non-access to the program sources. For example, one company called in an external development company, and the source-program remained the possession of the latter. The enterprise was thus forced to contract anew with the computer service company, without any guarantee that the company would continue to exist, or that it would be able to preserve the source-programs. If the development was done internally, in contrast, there would have been less risk of dependence a priori. However, in practice, capitalization is poorly guaranteed particularly—but not exclusively—in computer service departments. In addition, computer services are characterized by high *personnel turnover*, thus their computer engineers leave, taking their programs with them. This loss of know-how is particularly serious in the computer sector, where technician are especially in demand. Also, dishonest practices were observed among some employees who deliberately left the enterprise with their "know-how."

This lack of capitalization in the domain of ICT technologies is harmful to businesses and computer support teams, as well as to users. For example, in Enterprise E, the IT manager failed to sufficiently capitalize on computer support to users. A result was delayed user training, which in turn made it impossible for users to become sufficiently empowered, leading to a cycle of mutual dependence between ICT technicians and users.

Isolation and Compartmentalization of Computer Support Teams

Socio-economic diagnostics, conducted both in computer service departments and among users, revealed a particularly high degree of isolation experienced by computer experts, accentuated by a feeling of extreme dependence on the part of non-experts. This is why ICT departments and directors often serve as scapegoats, as though they were entirely responsible for all dysfunctions linked to computer technology usage. These departments are often perceived as fortified castles, smug with the certainty of their superiority over users, who are typically reputed to be incompetent novices. However, in-depth analysis of internal client-supplier relationships in the computer technologies domain reveals many weaknesses. Relationships are either absent all together, or hindered by *misunderstandings* over what roles pertain to which service (e.g., computer technologies-logistics, equipment-cable networks). Relationships are also impaired by a lack of commitment and instruction on the part of computer support teams concerning strategic and technical options engaged by the enterprise regarding information and communication technologies. In fact, when users and ICT technicians mutually accuse one another for dysfunctions that are connected to computer technologies, it becomes evident that there is corespondibility between technicians and users. The poor quality of the internal client-supplier relationship is often at the heart of these *dysfunctions*.

In certain businesses, this problem is made even more serious by a "captive" market position where users have no other choice than to rely on the company's internal computer service department. This situation also occurs when an external service agency wields a local monopoly or oligopoly. For example, in the notary public sector in France, the specific software applications for the profession are offered by only two or three suppliers in the French market. In these small enterprises, computer competency is often limited to the utilization of these professional programs. The many dysfunctions connection to this utilization engender wasted time, estimated at 1-hour per person per day. Yet, since the suppliers of these software applications enjoy a dominant position, they typically provide their service only after long delays (several days after being called), which hinders the efficient operation of the client businesses. This situation makes it necessary, as we shall see later on, to reinforce the contracting process as it relates to intervention modes, **cooperation** levels and final deliverables from the supplier.

Finally, certain computer service departments in large firms suffer from a symptom of "never enough"—more hardware, more software, more technicians, more budgets. This "blind race forward" hits its limit when its strategies are to the detriment of other company departments, and when it aggravates difficulty for users, who are faced with constantly readapting to new technologies.

Lack of Precision in the Distribution of Roles and Responsibilities

Computer service departments can be conceptualized as part of the operational services of the enterprise, and, as such, they are affected by dysfunctions in every other operational service. In essence, they answer to a multitude of order-givers, which can make it difficult for them to prioritize. They struggle to assume a set of ambivalent roles—supervision, support and expertise—toward their client services. Indeed, users prefer the support function to the supervision function, but the latter is typically a better fit with top management expectations. Yet in practice, one observes that the technical assistance function and the training function are poorly assumed since the supervision role is placed first. One of the consequences, as has already been mentioned, is a vicious cycle in which the computer department finds itself trapped. The less time it has for user training to empower them, the more users solicit computer experts to help solve their problems. Time spent *regulating dysfunctions* begins to significantly overshadow time spent training users.

Fear of decentralization, on the part of operational units, especially those operations formerly assumed by the computer service department and management, is another cause of tension between technicians and users. Moreover, when the *risk* of *externalization* often makes itself felt, tension is aggravated. For example, some enterprises typically question the need to maintain central computer buying procedures (hardware or software), which extend delays and do not necessarily correspond to the very precise demands of service users.

We have indicated the fear on the part of computer technicians faced with behavior they judge "autonomous" from users, for example, rearranging a computer configuration, or not respecting procedures, adding software without the knowledge of the computer department, developing "microprograms" by users to resolve utilization difficulties and so forth. These fears and tensions often create stressful relationships between computer specialists and users, aggravated by a feeling of hopelessness—on both sides—regarding the possibility of improving the situation.

Lack of Integrated Training

All businesses and organizations are faced with a need for in-depth, integrated user training. Indeed, the purchase of hardware and software is often accompanied by basic training by the supplier. However, the first use of these new technologies typically provokes user-awareness of the inadequacy of this basic training—either the transferred knowledge to vague or superficial, which in turn engenders much wasted time in the use of these technologies), or training programs were simply inappropriate for first users. If offered too early, such programs do not allow for the application of the content of the training sessions, and everything is forgotten within a few weeks; if offered too late, they inhibit the use of software that was already installed.

Today, the major difficulty in implementing information and communication technologies is not related to material investments; their price has been in constant decline. Rather, the critical issue is the amount of thought and budget dedicated to *immaterial investments*, which should be planned for whenever any material investment is made. Material investments alone, the visible part of the iceberg in an investment decision, do not generate *economic performance*; neglecting immaterial investments is extremely detrimental to decision-making quality. Indeed, if the latter is omitted or under-estimated in the preparation of a strategic decision, the estimated cost of investment necessary to attain the desired performance can be misleading. Based on our experiences in the field, in today's environment, the amount of *immaterial investment* should be evaluated at five times more than the material investment—if the latter is to actually be profitable. Indeed, when businesses are too insensitive to the importance of the immaterial investments that are necessary to ensure the profitability of investments in information and communication technologies, they suffer serious dysfunctions as costs cancel out the potential benefits of the material investment. The final result is a loss of *effectiveness and efficiency* in comparison with the former situation. Coupling immaterial investment with material investment means: adapting material *working conditions* to facilitate the use of new equipment, modifying work processes and procedures when a formerly manual activity is computerized, adapting the distribution of tasks among employees, conducting user training sessions based on the precise operations and activities they will be called to perform, ensuring close *communication-coordination-cooperation* between computer specialists and users during the initial utilization periods for new programs and applications, and finally, readapting programs to fit users' needs.

Hidden Costs

ISEOR's database findings estimate *hidden costs* between 10,000€ and 50,000€ per person per year. Computer service departments are included in this range, in fact one of them (Table 15.1, Case D) even attains the level of 68,500€ per person per year. *Nonquality* and *direct productivity gaps* merit special attention, for they make it possible to establish a typology of *hidden costs*, as presented in Table 15.2. Every case of dysfunction requires *regulation*, which in turn leads to supplementary time and prolonged

Table 15.2.   Typology of *Hidden Costs* in Computer Service Departments

	Indicator: *Nonquality*
Type of Hidden Cost	*Examples*
Technical documents and instruction manuals not appropriately handled and filed	Numerous incidents during computer conception and operation
Development tasks reperformed in certain projects due to lack of autonomy	Estimated at 20% of total development activity
Unreliable new software	Incomplete documentation leading to emergency interventions to users
Lack of technical mastery of certain hardware and software	Insufficient instruction manuals or manuals not supplied, computer technician training for new products not done
Difficulties experienced by users during software installation	Lack of instruction manuals and of formal installation protocol
Central computer breakdowns	Caused by power shortage and power cuts
Machine breakdowns (printing, forwarding, unwinding, etc.)	Worn out or poorly maintained machines requiring repair and recovery
Information system breakdowns during certain transactions	Time wasted in processing these transactions
Central processing incidents	• Abnormal stoppage of programs • Job control language errors • Beyond-capacity files causing time loss in recovery work
Program errors	Demanding recovery work by operators

(Table continues on next page)

Table 15.2. Continued

Indicator: Direct Productivity Gaps

Type of **Hidden Cost**	Examples
Excessive maintenance time	20% of maintenance could be avoided through better user training
Program development interrupted	Frequent change in development priorities by top management
Re-edition of reports and lists	Reports lost or discarded by error (address error)
Frequent interruptions by users	Lack of user training resulting in "unnecessary" interruptions
Director of computer service department frequently interrupted by team technicians	Lack of training and autonomy of team members
Lack of coordination between conception and operation	Lack of disk space at the time of first operation
Poor estimation of development and intervention times with internal customers	Allotted time insufficient making it necessary to reschedule work
Supplier files do not interface with each other	**Overtime** spent on manual reporting
Accountants frequently interrupted by users of the accounting software	Poorly designed tool, without user-friendly interface, lack of user training
One printer shared by an entire department	**Overtime** spent sorting printed documents (errors when different printing supports used
Overload of calls to the computer service department	Users are unfamiliar with the operation of software programs (one third of the calls are unnecessary)

Source: © ISEOR 2004.

delays, sometime effecting clients. Figure 15.1 presents a graph of the regulation of a software program crash in Enterprise A (see Table 15.1).

In addition to *overtime* costs (referred to as **overtime**), **nonquality** in customer service can be another consequence of this situation. For example, in Bank A, a file manipulation error during the calculation of bank premiums was not detected before being sent to the customer, a mistake that had two consequences: certain clients were overcharged and others were overcredited. This error caused an initial loss for the bank of 70,000€, not counting the added the cost of regulating the situation: time for re-processing data, trouble-shooting, responding to customer complaints at the agency, and refunding clients that were overcharged for an overall supplementary cost of 50,000€, not to mention the impact on the bank's image and **risk** of losing clientele. In all, this incident generated a minimum of 120,000€ in **hidden costs**, not counting the many unpaid hours which

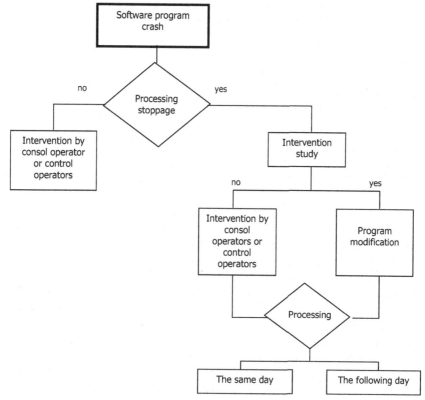

Source: © ISEOR 1988.

Figure 15.1.   Diagram of *regulation of dysfunctions* after a software program crash.

numerous employees dedicated to managing this incident, all the way up to the board of directors.

Finally, it should be noted that the cost of *personnel turnover* is particularly high in computer service departments. In addition to the usual recruitment costs, the cost of training and on-the-job learning of new recruits generate significant costs. In Enterprise B, for example, this was estimated to be roughly 18,000€ for a consol operator (see Table 15.3). It should be pointed out that the cost of on-the-job training accounts for 50% of the overall cost of integrating a new employee. Figure 15.2 presents a *competency grid* that reveals certain weak spots in the computer service team. "Network administration," for example, is mastered today by one single person out of the 13 employees in the department, which could put the department and the organization at risk.

**Table 15.3.   Estimated Cost of Integrating
a Consol Operator in Enterprise B**

Factor	Cost	Percentage
Cost of recruiting	3,000 €	16
Cost of external and *integrated training*	6,000 €	33
Cost of on-the-job learning • underproductivity of 60% the 1st of the month • of 20% the following 4 months	9,000 €	50
Errors: not evaluated	n/a	n/a
Total cost of integration	18,000 €	100

Soure: © ISEOR 1988.

Hidden Costs Involved in Adopting New Technology

Installing new information and communication technology is usually considered to be a profitable investment, while in reality the decision to invest is rarely accompanied with an economic study calculating the relevant provisional profitability. As we have seen, the lack of *immaterial investment* associated with material investment generates new dysfunctions, decreasing any performance gains generated by the material investment. As a further illustration, we evaluated the economic impact of setting up a site on the Internet, creating a company intranet, and setting up email in an airport with 425 employees. Table 15.4 represents the *economic balance* of this investment, not counting the *hidden costs* incurred by the utilization of new technologies, which are provided in Table 15.5.

Note that the amount of hidden costs incurred exceeds the additional operating costs that are visibly perceived by the enterprise. Thus the actual operating costs had been underestimated by 123%, which illus-

**Table 15.4.   *Economic Balance* of Installing New Technologies
(in Millions of Euros)**

Investment costs	0.57 M€	Performance attributed to the investment	63 M€
Additional operating costs per year	3.15 M€		
Total amount costs	3.72 M€	Total amount performance	63 M€
	Net annual result = 59.28 M€		

Source: © ISEOR 2004.
Note: Not including *hidden costs* incurred.

ACTIVITIES / ACTORS	IN AN ORGANIZATION AT THE BEGINNING OF THE EXERCISE																											OBSER-VATIONS
	EXISTING OPERATIONS																											
	SECURITY- MANAGEMENT																							DEVELOP MENT - MANAGEMENT		EXISTING SPECIAL KNOW-HOW		
	Secretary	Accounting	Exploitation	User training and support	Inventory	Handling	Installing workstations	Maintenance	Network administration	Receptionist	Server administration	Systems administration	Administration	Project analysis	Development	Software application maintenance	Conducting meetings	Supplier relations	Leadership	Specifications	Project follow-up	Strategic alert	Work follow-up	Management	Methods	Help desk	Client relations	
P1	■	■	■	◪	■	■	—	—	—	—	—	—	■	—	—	—	◪	■	—	—	—	—	—	—	—	—	◪	
P2	—	—	■	■	◪	◪	◪	◪	—	—	—	◪	□	■	■	■	◪	◪	—	—	—	○	—	—	○	—	◪	
P3	□	—	◪	■	□	◪	□	□	—	—	—	—	□	■	■	■	◪	◪	—	—	◪	○	—	—	○	—	◪	
P4	□	—	◪	■	—	—	□	□	—	—	—	—	□	■	■	■	◪	◪	□	□	■	○	—	◪	○	—	—	
P5	□	□	◪	■	—	◪	■	◪	□	—	—	■	◪	■	■	■	◪	◪	◪	□	■	○	—	◪	○	—	◪	
P6	□	—	◪	■	—	—	◪	◪	□	—	—	■	◪	■	■	■	◪	◪	◪	□	■	○	—	◪	○	—	◪	
P7	□	—	◪	■	□	■	■	■	■	◪	○	—	■	—	—	—	◪	■	○	□	◪	○	—	□	○	□	◪	
P8	◪	—	◪	■	□	■	■	■	■	◪	■	—	◪	◪	◪	◪	◪	■	○	◪	■	◪	—	◪	○	□	◪	
P9	—	—	—	■	■	—	■	■	—	■	—	—	■	—	—	—	◪	■	○	◪	■	◪	■	—	—	—	■	
P10	◪	◪	◪	◪	□	◪	□	◪	□	□	□	□	■	◪	□	□	■	■	■	■	■	■	◪	■	○	○	■	
P11	—	—	—	■	—	■	■	■	—	—	—	—	—	—	■	□	◪	—	—	□	—	—	—	—	—	—	—	
P12	◪	—	—	◪	—	—	◪	◪	◪	□	□	□	■	■	◪	—	■	■	■	■	■	■	—	—	○	—	—	
P13	—	—	—	—	—	—	■	—	—	—	—	—	—	—	—	—	—	—	—	—	—	—	—	—	—	—	—	

LEGEND : ■ Daily use mastery; ◪ Occasional use or incomplete mastery; □ Basic knowledge without practice; — Neither theoretical knowledge nor practice; ○ To be acquired

Source: © ISEOR 1988.

Figure 15.2. ***Competency grid*** of a computer service department (Case B).

Table 15.5.   *Economic Balance* Including the *Hidden Costs* Incurred by New Technology Investments (Millions of Euros)

Investment costs	0.57 M€	Performance attributed to the investment	63 M€
Additional operating costs per year	3.15 M€		
Hidden costs incurred by the utilization of information and communication technologies	3.86 M€		
Total amount costs	7.58 M€	Total amount performance	63 M€
	Net annual result = 55.42 M€		

Source: © ISEOR 2004.

Table 15.6.   *Economic Balance* of the Intranet Installation (in Thousands of Euros)

Investment costs	346 K€	Performance attributed to the intranet	1,535 K€
Additional operating costs per year	871 K€		
New *hidden costs* incurred	2,495 K€		
Total amount costs	3,712 K€	Total amount performance	1,535 K€
	Net annual result = 2,177 K€		

Source: © ISEOR 2004

trates the lack of precision in provisional cost estimates involving new technologies. Nevertheless, in this case, the impact on the net annual result remains relatively small (7% lower), given the higher level of performance attributed to the investment. The new *hidden costs* stemmed from such *dysfunctions* as support from ICT personnel (33%), creating documents that were not productive for the activity (e.g., email at 16%), loss of documents following computer breakdown (24%), information saturation resulting in time wasted dealing with unnecessary or poorly-formulated messages (10%), and so forth.

The case of the intranet installation in the same company was even more striking. In this instance, new hidden costs linked to the utilization of the intranet canceled out the anticipated *economic performances*, resulting in a net deficit of more than two million Euros (see Table 15.6).

PATHWAYS TOWARD IMPROVEMENT

With reference to the work process that characterizes computer departments, the major actions that improved *socio-economic performance* (Savall & Zardet, 1995; Savall, Zardet, & Bonnet, 2000) linked to ICT utilization are situated at the conception/design and utilization stages. At the application design stage, or during the preparation of investment projects, interactivity between computer specialists and users constitutes can be extremely fruitful. Users seldom know how to express their needs and computer specialists are rarely designated to assist them. Thus, the conception stage for the specification manual and the test stage that follows, both need to be transformed, integrating and involving users in a more intense and regular manner than is currently practiced in ICT design methodology. Process design should be oriented toward the *coconstruction* of best practices. In terms of technical considerations, ICT tool conception methodologies could be greatly improved by designing tools with the capacity to evolve, in keeping with the inevitable strategic, commercial and organizational evolution of businesses.

At the set-up and utilization stages, weaving structured relationships between computer specialists and users can improve relationships and reduce tensions. By regularly listening to users, by designating ICT reporters, and by meticulously following-up every reported dysfunction actually experienced by users, the origins of these difficulties can be identified. Making certain tools available to users, which include user guides and *integrated training manuals*, can also reduce user dependence on computer specialists and thus reduce interruptions in the computer department.

Management actions can further contribute to improving *socio-economic performance*. *Root-causes* point to the veritable management of computer experts and their activities within the computer service department (e.g., development, exploitation, maintenance). They touch on the organization of team work, especially by reducing distinctions between "noble" tasks and those considered less important by computer teams. Root causes also allow for a better prioritization of computer projects by top management and more rigorous methods of knowledge management, the development of a veritable client-supplier relationship between the computer service department and its internal clients.

Elaborating a *product portfolio* can further enable the IT manager to draw up a typology of the department's internal clients, identifying methods and services the department should offer to its various *internal-clients* targets. For example, by drawing up the product portfolio for the computer department of a firm employing 3,000 persons (see Table 15.1, Case B), the IT manager was able to better identify the unit's types of clients and to differentiate the products offered to them. This portfolio approach clarified the activities to be performed (or potentially performed) by the department, and identified the portion of time-resources assigned to each type of product, even the time allotment requested for the coming year. These time references were progressively established for every type of product, which contributed to reducing delivery delays thanks to a better estimation of the time needed to complete activities, especially time for development activities.

Such *improvements* have their source in acceptance of the fact that computer departments do not benefit a priori from an exceptional status in the enterprise, which would make it exempt from time and budgetary constraints, from *management tools* and methods, nor exonerate it from contributing to overall *effectiveness* and *efficiency*.

CONCLUSION

Utilization of information and communication technologies in businesses and organizations does not constitute a specific use of technology. Indeed, as was the case with production technologies, ISEOR's work on *dysfunctions* relative to computer technologies reveals a number of lessons. First, it shows that material investment does not produce positive economic results, except when they are intertwined with appropriate *immaterial investments* aimed at the human actors who design and/or utilize the technological equipment. Second, the wide-spread, tenacious belief in economic profitability from material investment was shattered by the multiple findings our team obtained, in all kinds of enterprises and organizations. As this chapter has attempted to demonstrate, technology constitutes, more often than not, a factor of *hidden cost* aggravation, when it is insufficiently mastered.

A third lesson is that information and communication technologies produce multiple dysfunctions and hidden costs, both real and potential (risks), which greatly reduce the global economic *productivity* of the enterprise. Finally, pathways toward mastering these dysfunctions and hidden costs essentially point to more advanced management of ICT teams, where "bugs" would not be perceived as abnormal, but rather considered symptomatic dysfunctions to be dealt with and taken seriously. ISEOR's socio-economic approach is clearly a step in the right direction as a way of assisting firms in mastering their technology investments.

REFERENCES

Boulanger, G. (1997). *L'usage des systèmes informatiques dans la performance économique et sociale des organisations* [Usage of computer networks to enhance economical and social performance of organizations]. Unpublished doctoral dissertation, University of Lyon-ISEOR, France.

Datry, F., & Payre, S. (2004). *Diagnostic socio-économique d'une TPE de services informatiques* [Socio-economic diagnostic of a very small SME computer service company]. Lyon, France: ISEOR.

Delattre, M. (2002). *Diagnostic socio-économique d'une direction informatique* [Socio-economic diagnostic of an IT management team]. Lyon, France: ISEOR.

Demissy, B. (2002). *Étude des impacts des nouvelles technologies d'information et de communication sur les performances* [Study of the impact of information and communication technologies on performances]. Unpublished doctoral dissertation, University of Lyon-ISEOR, France.

ISEOR. (1998). *Diagnostic socio-économique du service informatique central d'une entreprise agro-alimentaire* [Socio-economic diagnostic in the central computer department of an agri-food firm]. ISEOR Research Report. Lyon, France: ISEOR.

Plane, J. M. (1991). *Les difficultés de restructuration du département production informatique d'un grand groupe d'assurances* [Difficulties encountered in restructuring the production computing department in a large insurance group]. ISEOR Research Report. Lyon, France: ISEOR.

Saint-Léger, G. (2005). *Quel processus de changement peut permettre une mise en œuvre et une utilisation efficace et efficiente d'un système d'information de type ERP dans les moyennes structures de production de biens et de services?* [What change management permits implementation of an effective and efficient ERP information system in small goods and service businesses]. Unpublished doctoral dissertation, University of Lyon-ISEOR, France.

Savall, H. (1974, 1975). *Enrichir le travail humain dans les entreprises et les organisations* [Work and people: An economic evaluation of job enrichment]. Paris: Dunod.

Savall, H. (1987). Les coûts cachés et l'analyse socio-économique des organisations [Hidden costs and the socio-economic analysis of organizations]. *Encyclopédie du management* (pp. 599-628). Paris: Economica.

Savall, H. (2003a). An updated presentation of the socio-economic management model. *Journal of Organizational Change Management, 16*(1), 33-48.

Savall, H. (2003b). International dissemination of the socio-economic model. *Journal of Organizational Change Management, 16*(1), 107-115.

Savall, H., & Bonnet, M. (1988). Coûts sociaux, compétitivité et stratégie socio-économique [Social costs, competitiveness and socio-economic strategy]. *Encyclopédie de la Gestion* (pp. 742-757). Paris: Editions Vuibert.

Savall, H. & Zardet, V. (1985). *Diagnostic et projet socio-économiques dans la perspective de fermeture prochaine d'un service de traitements informatiques d'une banque régionale* [Socio-economic diagnostic and project in light of the forthcoming closure of a data-processing service in a local bank]. ISEOR Research Report. Lyon, France: ISEOR.

Savall, H., & Zardet, V. (1987). *Maîtriser les coûts et les performances cachés: Le contrat d'activité périodiquement négociable* [Mastering hidden costs and performances: The periodically negotiable activity contract]. Paris: Economica

Savall, H., & Zardet, V. (1995). *Ingénierie stratégique du roseau, souple et enracinée* [Strategic engineering of the reed, flexible and rooted]. Paris: Economica.

Savall, H., & Zardet, V. (2004). *Recherche en sciences de gestion: Approche qualimétrique. Observer l'objet complexe.* Unpublished English translation: Research in management sciences: The qualimetric approach. Observing the complex object. Paris: Economica.

Savall, H., Zardet, V. & Bonnet, M. (2000). *Releasing the untapped potential of enterprises through socio-economic management.* Geneva, Switzerland: International Labor Office-ISEOR.

Savall, H., & Zardet, V., & Harbi, N. (2004). *Analyse spectrale de diagnostics socio-économiques : traitement qualimétrique de données qualitatives* [Spectral analysis of socio-economic diagnostics: Qualimetric processing of quantitative data]. Paper presented at the International Conference AOM-Research Methods Division, Lyon, France.

Zardet, V., & Harbi, N. (2002). Processus de découverte de connaissances dans une application de diagnostic en réseau [Knowledge discovery process using a networked diagnostic application]. Paper presented at the 7th Information and Management Systems Association Conference.

Zardet, V., & Srajek, B. (2000). *Diagnostic socio-économique d'une direction administrative et financière d'un organisme consulaire* [Socio-economic diagnostic of administrative and financial management teams in a consular organization]. Lyon, France: ISEOR.

CHAPTER 16

SEAM, CHANGE, AND ORGANIZATIONAL PERFORMANCE

The Importance of Incorporating Quantitative and Qualitative Assessment

Rickie Moore and Michel Péron

Organizations can be conceptualized as organs of change. As biological entities within a constantly evolving environment, organizations are challenged to adapt to their new environment, new employees, new technologies, new laws, new processes, new skills, new *resources*, new markets, and new opportunities. A quick review of the buzz words and concepts used in current corporate vocabulary—including the need to *re*-orient, *re*-organize, *re*-engineer, *re*-structure, *re*-align, and *re*-allocate—highlights our pervasive focus on constantly *re*evaluating the ways in which our firms work.

Managing change is a complex challenge and many firms continue to struggle in their attempts to cope with organizational and employee reactions (Tichy, 1983). According to conventional wisdom, two out of every

Socio-Economic Intervention in Organizations: The Intervener-Researcher and the SEAM Approach to Organizational Analysis, pp. 373–383
Copyright © 2007 by Information Age Publishing

three change efforts fail (e.g. Beer, Eisenstat & Spector, 1990; Beer & Nohria, 2000a, 200b; Kerber & Buono, 2005). Indeed, even as our understanding of the change process increases, firms are still experiencing great difficulties in implementing organizational change. Although there is no shortage of literature on how to implement these initiatives, firms continue to grapple with the application of models and strategies (see Argyris & Schön, 1996)due to a lack of appropriate methodological tools (Moore, 2005).

With the increasing complexity of organizations, organizational change also becomes increasingly complex, which creates even more of a dilemma (Péron & Péron, 2003). As the stakes for the *survival* and viability of firms increase, the challenges of deciding how to change effectively and how to manage the change process become critical. It is therefore not surprising that all too frequently the political stakes involved in the change effort surpass the technical elements of the change itself. More often than not, organizations are reshaped and reorganized as part of the change process.

Fundamentally, there are three key issues in organization change: (1) how to implement the change effectively (***implementation*** tools), (2) how to manage the change (change method), and (3) how to institutionalize the change (long-term reinforcement and sustainability). One integrative approach, as illustrated throughout this volume, is the Socio-Economic Approach to Management (***SEAM***), a method that has been experimented and applied in numerous countries worldwide (Péron, 2002; Savall, 1974, 1975, 1987; Savall & Zardet, 1987; Savall, Zardet, & Bonnet, 2000).

SEAM AND ORGANIZATIONAL PERFORMANCE

This chapter illustrates the applicability of ***SEAM*** in the context of a large-scale change initiative in a major multinational firm, referred to as ComCorp, headquartered in the United States. ComCorp is the American branch of a leading global human resources company that is publicly held with private equity. Incorporated in the early 1970s, ComCorp employs approximately 3,300 employees. Specialized in offering innovative solutions to client organizations' staffing needs, and in finding full-time, contract and temporary work for job seekers (applicants), the company prides itself on "connecting and finding the right people for the right job." With three principal lines of products and 16 areas (sectors) of specialization in its staffing branch and traded on several stock exchanges internationally, ComCorp generated in excess of $900 million in revenue in 2004.

Dividing the country into a number of divisions, states, zones, and regions, ComCorp operated each agency or branch as an individual profit center. Several profit centers were then grouped together in a region, which was considered as a consolidated profit center. Regions, in turn, were grouped together to form zones within each state, and several states were grouped together in a division. Five years ago, ComCorp bought one of its rivals and its financial performance had improved significantly. In one of its Western regions (Region L), sales had increased 35% in 1998 compared to 1997, and by the end of May 2000, sales were already at 34% of the company's total sales volume in 1998.

One of ComCorp's global senior executive vice presidents (GSEVP), who had discovered *SEAM*, decided to commission a *socio-economic diagnostic* within the firm. Intrigued by the results of various SEAM interventions around the world, and given the challenges that ComCorp was facing in terms of sustaining its sales performance, the GSEVP felt that a *SEAM intervention* in ComCorp would help to identify hidden sources of profitability and areas for change. ComCorp had already utilized a number of improvement methods, including Six Sigma and ISO 9000, but the firm was not fully satisfied with the results to date. The GSEVP requested the local senior vice presidents (SVP) for national (US) operations to identify a region in which the SEAM diagnostic could be conducted. The SVPs identified the Southwest as the experimental region, as it was geographically located in close proximity to the SEAM *intervener-researcher* who had a visiting academic appointment in the United States. The Southwest Region was comprised of nine outlets (branches).

From the start, it was necessary to explain the SEAM approach to the VP of national operations, who was suspicious of the motives of the GSEVP and why her area was chosen as the target of the experiment. The suspicion of the VP is perhaps understandable as she had not been informed of the intervention by the head office. Naturally, this suspicion rippled through the region and the branches as everyone was wondering why their branch and region was chosen for the experiment, beyond their geographical convenience.

The Intervention

The basic objective of the intervention was to identify opportunities of improvement for the firm. The intervention was conducted in accordance with the *Horivert* method (see Figure 16.1) and the *communication-coordination-cooperation* approach, and was applied to all levels within the experimental sites. Semi-directive and informal interviews and discussions were carried out with the Regional VP, the outlet managers and their deputies, their operations managers, a selection of 18 supervisors, and more than 20 clients. The interviews and discussions focused on the

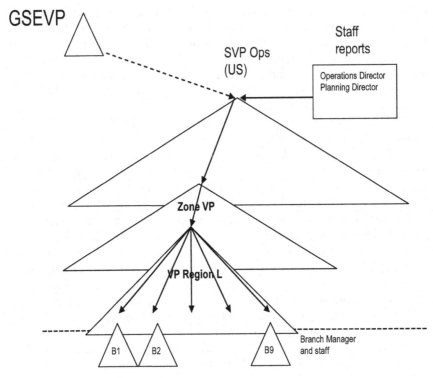

Figure 16.1.   *Horivert intervention* framework.

operations of the branches within the region and challenges that employees were facing, from such operational concerns as invoicing, expenses
and remuneration to broader sales-related and strategy considerations.
The interviews also sought to address *dysfunctions* related to *working conditions, work organization, time management*, recruiting difficulties, job
matching, communication-coordination-cooperation and operations.

   Direct observation of the way the nine outlets were operating and analysis of quantitative and financial data also accounted for a large part of
the intervention. The analysis found that new staff were performing
duties for which they had not been certified to perform, a low level of utilization of key recruiting tools, and new applicant interview locations and
conditions that were often suboptimal (e.g. lack of current software, lack
of comparable technology equivalent to that used by the clients, branch
staff in the field concentrating primarily on generating new clients and
orders without careful analysis of being in a position to fill the orders).
Quantitative and financial data focused on the number of hours sold per
month and the breakdown between permanent and temporary positions,

jobs completed, associates placed, revenue generated, settlements paid, and recruiting expenses. In fact, ComCorp was overloaded with quantitative data, and beyond the general objectives of keeping costs down and generating more revenue, the data was simply being collected and reported.

Diagnostic

Based on the initial analysis, the most appropriate tools for the intervention were the: (1) *competency grid* (the key instrument to prioritize the hiring of new employees); (2) *priority action plan* (which helps to differentiate high and low value-added tasks); (3) *time management* (focusing on the time spent on unprofitable tasks and how such down time could be eliminated); (4) *strategic piloting indicators* (these indicators are particularly useful when firms tend to focus too much on immediate results and the short-term and ignore the benefit which could occur from long-term potential gains; and (5) *periodically negotiable activity contracts (PNAC)* (having identified dysfunctions within the organization, the *PNAC* implements and ensures a continuous dialog among all interested parties within the firm). Considered as the backbone of *SEAM* (Savall, 2003a), the PNAC clarifies the objectives that are established for the various actors and suggests improvement actions.

Through the analysis of the data from the interviews, questionnaires, observation and financial analysis, it appeared that the organization was experiencing a lack of harmonization and linkage between its strategy, system assessment, and practices and procedures.

Strategy

The firm's current strategy, which was aimed at cornering the market, indiscriminately accepted all applicants and clients. In essence, this meant that (1) ComCorp's staff spent valuable time recruiting persons that they would not likely be able to place (time which could have been used elsewhere), (2) ComCorp's database was full of applicants that ComCorp could not possibly utilize in the short or medium term (ComCorps's placing statistics would show a relatively important underplacement ratio, because of nonplacement and underplacement), and (3) ComCorp's image was tarnished as an effective employer that got jobs for its applicants very quickly. These difficulties, in turn, led to massive applicant defection, especially among the most qualified people. With large amounts of applicants without the required competencies and abilities, ComCorp could potentially send one or more of them to a client, and would risk dissatisfying their client, which in turn would lead to client defection and loss of revenues. In short, this practice was synonymous with the absence or lack of a real recruiting strategy.

System Assessment

Information pertaining to all clients, employees, and applicants was provided by a number of purely quantitative indicators (e.g. invoice amount, length of contract, gross margin to be applied or sustained, number of applicants per day). Although one could follow the evolution of these indicators, it was more difficult to analyze their impact on the strategy due to a lack of a performance *evaluation* system. The indicators were solely quantitative and the qualitative aspects of the system (e.g. organization of recruiting events, response to client complaints, reference checking, resolution of associate pay and billing issues, service to special and high maintenance accounts) were not captured.

Practices and Procedures

The current specialization among the different outlets was loosely enforced and gave the impression of a blurring of the market segments. As illustrated in Figure 16.2, while some branches were specialized in certain sectors—industrial, construction, services, secretarial—and had client orders in these areas, most branches would also recruit talent for sectors for which they are not fully trained or equipped.

Concerned with not having a sufficient labor force to respond to incoming orders, the firm escalated its hiring practices, to the point of even reducing its hiring requirements in certain sectors, which ultimately contributed to a decline in service quality. This type of "take-all" strategy

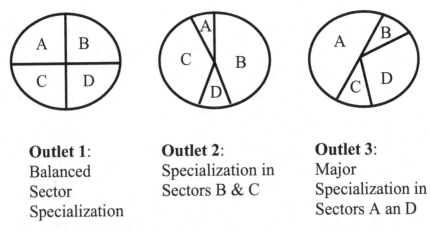

Outlet 1:
Balanced
Sector
Specialization

Outlet 2:
Specialization in
Sectors B & C

Outlet 3:
Major
Specialization in
Sectors A an D

Figure 16.2.   Outlet specializations.

led to a number of dysfunctions, ranging from unnecessary expenses and *hidden costs* in indiscriminatingly recruiting applicants who could not be reasonably placed and the acceptance of orders the company could not reasonably fill. In essence, ComCorp began to blur its image, both with its applicants and clients, even though the firm's sales appeared to be healthy. There was also a problem with a basic lack of *communication-coordination-cooperation* between the company's clients and the HR department. Deadlines were often not met, which suggested insufficient *time management* procedures.

Since these factors are interdependent, it is necessary to consider them as a whole, which helps to ensure greater coherence and cohesion in the management of the various operations of the company. The intervention revealed that there was no actual strategic plan that could clarify the causal relationship between the company's objectives and its proposed actions. Interviews and group discussions showed that the firm was decidedly focused on the short term. The *strategic piloting indicators* remained largely focused on sales and gross margins, and were excessively quantitatively oriented, even when client satisfaction was being recorded. It was decided that the *competency grid* would be useful in *piloting* the *development* of associate talent skills. Similarly, since bonuses and incentives were oriented to sales and gross margin generated, PNACs could help focus on the underlying *dysfunctions* rather than just on sales. None of the *strategic piloting indicators* included qualitative information, such as career prospects or job enrichment for its applicants. In terms of the *internal-external strategic action plan (IESAP)*, ComCorp was concentrating on securing market share rather than value for the company.

The diagnostic established during the *mirror-effect* stage showed that the firm's strategy needed to be revised. The essence of the company's strategy was summed up as a "Yes sir" strategy, that is, the firm sends the following message to potential clients and employees: "It doesn't really matter what you want, we will always say yes." Such an attitude is uniquely consistent with a quantitative approach that is aimed at undercutting the competition.

Overview of Hidden Costs

The financial results over a 30-month period demonstrated a potential loss of earnings amounting to roughly about 25% of the total financial value of its orders. As illustrated in Table 16.1, the firm was sitting on a potential income of $114.43 M, which was justified as follows: Incomplete orders were those client orders that had only been partially filled and for which the clients were expecting replacements. Cancelled orders comprised both the orders cancelled by the client and to a lesser extent client orders that were cancelled by ComCorp because of the client's track

Table 16.1. Operational *Dysfunctions* and Economic Losses

Situation: *Dysfunctions*	Economic Losses USD $	Explanation
Incomplete orders	52.50 M	Firm-related
Cancelled orders	40.00 M	Client initiated
Unfilled orders	21.30 M	Firm-related
Unsuccessful client prospecting	0.34 M	Approximately $6.5 K per week
Untapped earnings potential	0.29 M	Nonplacement /underplacement*
Total *hidden costs*	$ 114.43 M	

Note: *Nonplacement refers to people who are registered in the database as persons able to work but who were not placed on a mission/work assignment. Underplacement refers to people who were assigned to positions that did not utilize their full capabilities.

record. Unfilled orders were those for which ComCorp was unable to find appropriate associates to be sent on the mission/work assignment. Unsuccessful client prospecting consisted of the orders received from clients for permanent positions for which ComCorp incurred costs that were not reimbursed by the client. Untapped potential is an approximation of the potential revenue that could be generated if underplaced and non-placed associates were fully employed.

It was clear that ComCorp needed to address these issues if the company was to improve its performance. Based on the *SEAM* analysis, the company needed to improve the way in which it completed, delivered and filled orders, secured new clients, and maximized its earnings potential. The fact that as much of 45% of its employees were under- or non-placed (i.e., large numbers of hired applicants not placed in a job, an equally large number not having significant work or infrequent short-term assignments) suggested that there was a lack of alignment between associate skills sets and market needs. Even though ComCorp had taken steps in the past to reduce its hiring requirements in some sectors (e.g. over abundance of available employees in selected areas, lack of client orders in these sectors), under- and nonplaced employees remained a problem.

Possible Solutions

The problems entailed by the *dysfunctions* noted above could have been partly solved, or at least alleviated, through longer-term *SEAM intervention* (Savall, 2003b). Such application, however, would have required more time for the intervention than was initially negotiated. It was suggested, for example, that ComCorp's "take-all" strategy should transition to a more focused, market–minded approach, which would provide for a better control of the flow between client needs and organizational capabilities. By clarifying the criteria on which associates were to be contracted, the firm would be able to target its customers more precisely, thus minimizing the risk of either being swept away or overwhelmed by its current all-out attack on its competitors.

Another solution appeared to lie in the training function, where ComCorp could have invested in developing associates to be able to better to respond to client needs. Such an initiative could have resulted in a major market opportunity for ComCorp, as its unproductive "down time" could have been used to enhance the earning potential of the firm. Using the *competency grid* in the recruitment process and comparing the orders it received from its clients, ComCorp could have ensured that its applicants were able to meet client requirements. While this would incur expense for ComCorp, the expense can be considered as an investment that will be recovered through the successful completion and satisfaction of client orders. ComCorp could also have negotiated a contractual period with the associate for which they will agree to work only for ComCorp. ComCorp would thus be more highly valued by its associates and applicants, as it would be seen to be investing in the development of its personnel. ComCorp's clients would be more satisfied because ComCorp would have sufficient competent *resources* to fill their orders.

To offset the cost of training, ComCorp could harness the profits captured from the underlying hidden performances (i.e. ComCorp was able to place more associates in the open positions with the required skills). For example, a simple action on providing higher associate salaries alone would leave this potential untapped, not to mention the perverse effects which typically ensue (e.g. associates can claim to have the necessary *competences* but if they are not properly evaluated nor screened, and are sent on a mission, the risk of having to pay client settlement fees increases if the order is unfilled or needs refilling). As illustrated in Table 16.1, it is important to utilize *hidden cost* calculations that include the costs of *non-quality* and the non-realization of potential. We can thus estimate the waste of wealth and the loss of value-added potential that result from the current state of operations of the firm.

CONCLUSION

This case illustrates the applicability of *SEAM* and underscores one of its major tenets—the importance of looking at both qualitative and quantitative data in guiding organizational change and assessing organizational performance. Performance assessments, of course, should be undertaken with quantitative tools, but an equal emphasis should also be placed on qualitative data and insights. A purely statistical treatment does not allow the firm to account for the fairly high number of contractual obligations which are not fulfilled. In the ComCorp case, effectiveness was largely dependent on the qualitative aspects of the service provided rather than on the sheer number of employees involved, although both factors must be taken into consideration. The concept of *qualimetrics* (Savall & Zardet, 2004) aims to reconcile these two approaches.

The structure of the outlets as shown on Figure 16.2 created a number of dysfunctions, but this could have been resolved through a focused, marketing strategy, explaining to the various interested parties in the minor specializations that a particular outlet might not be the most appropriate for their market. The configuration of the outlets and the difficulties and blurring that ensued resulted from the dispersion of the different outlets (each outlet will attempt to be holistic in its response to their customers and clients) while serving specific areas in accordance with their market segmentation. The extreme complexity of this structure could have been made clearer and even contributed to greater efficiency if every employee was considered as a profit center, both psychologically and financially. This change could have led to positive consequences by ending unnecessary recruitments and rendering operational profit management more transparent.

In this firm, the outlet managers were among the persons responsible for recruiting new employees. Interviews were conducted according to identical lines, sometimes without updated recruitment guides. Better cross-training of outlet managers would have avoid these recruitment dysfunctions and allow the managers to better target appropriate applicants. The influx in applications, which can account for a marked shift in the outlet managers' responsibilities, does not allow them to keep the necessary free-time slots for high valued-added actions (e.g. setting up of specialized "recruitment centers" that would enhance operational performance and increase the company's dynamism).

When an organization decides to implement SEAM, it means that its priority should be to act on human behavior and cost control factors. To simply resort to downsizing, without considering other possibilities, would not solve any problem in the long term. The loss of millions of dollars from cancelled, unfilled or incomplete orders suggests that a noticeable improvement in the organization's performance could be accomplished through behavioral changes. The ComCorp case illustrates a fundamental principle of SEAM, which postulates that value is created by developing a company's **human potential** rather than by cutting it down. This company, like many others, was obsessed with its immediate and short-term financial results instead of focusing on its short-, medium- and longer-term **human potential**. Such potential remains hidden—thus untapped.

REFERENCES

Argyris, C. & Schön, D. A. (1996). *Organizational learning II: Theory, method and practice.* Reading, MA: Addison-Wesley.

Beckhard, R., & Harris, R. T. (1977). *Organizatinal transitons: Managing complex change.* Reading, MA: Addison-Wesley.

Beer, M., Eisenstat, R., & Spector, B. (1990). Why change programs don't produce change. *Harvard Business Review, 68*(6), 158-167.

Beer, M., & Nohria, N. (Eds.). (2000a). *Breaking the code of change.* Boston: Harvard University Press.

Beer, M,. & Nohria, N. (2000b). Cracking the code of change. *Harvard Business Review, 78*(3), 133-141.

Kerber, K., & Buono, A. F. (2005). Rethinking organizational change: Reframing the challenge of change management. *Organization Development Journal, 23*(3), 23-38.

Moore, R. (2005). *Evaluation de la performance économique durable des entreprises: Méthodes et pratiques américaines et européennes* [Evaluation of sustainable economic performance in enterprises: American and European methods and practices]. Doctoral dissertation in management sciences, University Jean Moulin Lyon 3, France.

Péron, M. (2002). *Transdisciplinarité: Fondement de la pensée managériale anglosaxonne?* [Transdisciplinarity: Foundation of Anglo-Saxon management thought?]. Paris: Economica.

Péron, M., & Péron, M. (2003). Postmodernism and the socio-economic approach to organizations. *Organizational Change Management, 16*(1), 49-55.

Savall, H. (1974, 1975). *Enrichir le travail humain dans les entreprises et les organisations* [Work and people: An economic evaluation of job enrichment]. Paris: Dunod.

Savall, H. (1987). Les coûts cachés et l'analyse socio-économique des organisations [Hidden costs and the socio-economic analysis of organizations]. *Encyclopédie du management* (pp. 599-628). Paris: Economica.

Savall, H. (2003a). An updated presentation of the socio-economic management model. *Journal of Organizational Change Management, 16*(1), 33-48.

Savall, H. (2003b). International dissemination of the socio-economic model. *Journal of Organizational Change Management, 16*(1), 107-115.

Savall, H., & Zardet, V. (1987). *Maîtriser les coûts et les performances cachés: Le contrat d'activité périodiquement négociable* [Mastering hidden costs and performances: The periodically negotiable activity contract]. Paris: Economica

Savall, H., & Zardet, V. (2004). *Recherche en sciences de gestion: Approche qualimétrique. Observer l'objet complexe.* Unpublished English translation: *Research in management sciences: The qualimetric approach. Observing the complex object.* Paris: Economica.

Savall, H., Zardet, V., & Bonnet, M. (2000). *Releasing the untapped potential of enterprises through socio-economic management.* Geneva, Switzerland: International Labor Office-ISEOR.

Tichy, N. M. (1983). *Managing strategic change: Technical, political and cultural dynamics.* New York: John Wiley.

CHAPTER 17

SEAM IN THE CONTEXT OF MERGER AND ACQUISITION INTEGRATION

Anthony F. Buono and Henri Savall

The overarching reason why firms enter into a merger or decide to acquire another company is the belief that the combination will allow the new entity to attain its strategic goals more quickly and less expensively than if the firm attempted to do it on its own (Haspeslagh & Jemison, 1991). The poor performance of combined firms, however, continues to raise questions about the efficacy of this strategy, as it appears that less than one third of mergers and acquisitions actually achieve the operational, financial and *strategic objectives* suggested in precombination feasibility studies (cf. Coff, 2002; Elsass & Veiga, 1994; Lubatkin, 1983). Most merger and acquisition (M&A) strategies are still dominated by financial analyses, legal considerations and power plays by dominant groups as individuals jockey for position and influence. Rather than focusing on the inherent dysfunctions that can emerge in the combined organization due to the informal power held by organizational members—low *productivity*, poor quality, reduced commitment, voluntary turnover, and related *hidden costs* and untapped potential—far too many companies seem to meander through the postcombination integration process.

Socio-Economic Intervention in Organizations: The Intervener-Researcher and the SEAM Approach to Organizational Analysis, pp. 385–399

THE SOCIO-ECONOMIC APPROACH TO
MANAGEMENT IN MERGERS AND ACQUISITIONS

The *SEAM* approach to organizational analysis underscores that there is an inherent difference between what may be intended in a particular situation and the actual experience of key stakeholders—including employees, managers, customers, shareholders and suppliers (Savall, 1974, 1975, 2003a; Savall & Zardet, 1987, 1995; Savall, Zardet, & Bonnet, 2000; Savall, Zardet, Bonnet, & Moore, 2001). The realization that organizational members readily draw on and exercise their informal powers to accelerate or thwart the pace and direction of change raises a number of challenges for M&A-related integration efforts. Of course, the idea that the human side of mergers and acquisitions must be attended to—from timely and informative communications, to helping organizational members deal with the concomitant stress and anxieties associated with the combination, to sensitizing them to the culture clashes that inevitably emerge when two autonomous firms come together—is not a novel idea. A growing number of researchers and practitioners have been raising such concerns for the past 20 years. Yet, far too many organizations continue to treat the merger and acquisition process as an engineering exercise, as a series of rational decisions rather than a far more chaotic set of events that readily affect people's lives and future prospects (Ashkenas & Francis, 2000). Precombination transition planning teams continue to be disbanded too early, many of the insights that are generated through systematic due diligence assessments of acquisition targets or merger partners fall into a literal interorganizational void due to time pressures and internal politics, and postcombination integration orchestration falls well short of expected and needed efforts (cf. Buono & Bowditch, 1989; Buono & Nurick, 1992; Haspeslagh & Jemison, 1991; Jemison & Sitkin, 1986; Marks & Mirvis, 1992, 1998).

Most observers of the M&A process readily agree that the personal, interpersonal, group and intergroup dynamics that follow the combination of two firms are significant determinants of merger success or failure. Yet, while an organization development (OD) approach to postcombination planning and integration emphasizes the human dynamics associated with such strategies, this focus is often at the expense of the economic realities accompanying the change. The hybrid nature of the *SEAM* approach—the integration of social *and* economic factors—provides a unique and needed approach to post-M&A integration.

As illustrated in Figure 17.1, a growing body of research and consulting experience documents the reality that such combinations—even those that are suggested to be friendly combinations—have far-reaching and often dysfunctional effects on those involved (Buono, 2005). Based on more than 2 decades of experience in studying and working with companies going through the merger and acquisition process, we have found that while initial reactions are frequently characterized by good will and a cooperative spirit, and senior management talks about the promise of the combined entity, the reality is often in stark contrast to such lofty promises. As suggested by the **SEAM** framework, the experience of organizational members is often far removed from what is initially intended by senior management. In fact, a merger or acquisition can sufficiently transform the **structures**, cultures and employment prospects of one or both the organizations that they cause organizational members to feel stressed, angry, disoriented, frustrated, confused, and even frightened (Buono & Bowditch, 1989; Buono, Bowditch, & Lewis, 1985; Buono & Nurick, 1992). Referred to by some observers as the *merger syndrome* (see Mirvis & Marks, 1992, 1998), these reactions fester under the surface of the combination and reflect high levels of anxiety and stressful reactions, heightened self-interest and preoccupation with the combination, cultural clashes, restricted communication and crisis management orientations, creating problems at both the individual and organizational levels. SEAM analysis can help managers focus on the resultant **hidden costs** and "postmerger drift" (see Pritchett, 1987) due to these **dysfunctions**.

Most early M&A planning, however, emphasizes **horizontal actions**, focused on the upper echelons of the organization and basic changes in organizational structure. While such initiatives are important for the overall success of the combined entity, **vertical actions**—focused on key individuals and work units—tend to be guided by shortsighted decisions. Clearly, an integrated emphasis on both dimensions—the **Horivert method** suggested by SEAM analysis—would go a long way toward ameliorating many of the resulting dysfunctions and hidden costs associated with such large-scale change. A key dimension of the Horivert approach in M&As lies in the **mirror-effect** that is produced by the reporting of detailed field notes, juxtaposing top-level executive quotes and perspectives with those from middle managers and employees. **Horivert** facilitates the analysis of what might be thought of as the "hierarchy gap," when the perceptions and foci of senior-level management are often quite different from the attention and focus of middle managers, supervisors and employees. While senior-level executives may have previously worked through the stress and conflicts associated with such change, turning their attention to the next strategic moves, middle- and lower-level organizational members are still mired in the anxiety, confusion and anger that typically accompanies the initial stages of a merger or acquisition (see Marks & Mirvis, 1992).

As an illustration of these dynamics, a U.S.-based company we will refer to as SteelCo acquired a petrochemical company (Petro) as part of

MANIFESTATIONS

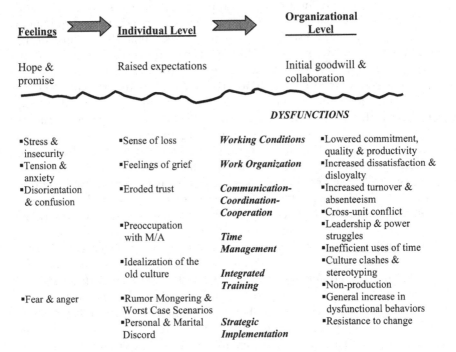

Figure 17.1.   Illustrative human reactions and organizational *dysfunctions* in mergers and acquisitions.

its diversification strategy and focused its initial energies on capturing short-term, financial gains (see Buono, 2003). Based on initial diversification planning, the positions of the technical experts, engineers and scientists at Petro were not threatened. In fact, SteelCo had largely acquired the firm to secure the expertise of these highly skilled, technical employees. During the postcombination aftermath, however, SteelCo's senior management focused on attaining immediate costs savings, consolidating basic functional areas and support groups—human resources, finance, legal—*without* communicating their intended strategy to the organization or reaching out to the target's technical core. In **SEAM** parlance, in their effort to attain short-term cost savings the executives ignored the long-term **hidden costs** inherent in the acquisition. Petro's technical staff interpreted the changes and terminations in other sections of the company as "a sign of things to come" and began bailing out of the company. Even the scientists who were willing to "give SteelCo's management a chance" found themselves under significant pressure from colleagues and coworkers to exit the firm. By the time SteelCo's executives realized what was happening, they found themselves in control of the petrochemical company but without the core of technical professionals that made Petro a desirable acquisition target.

A similar outcome has been found in the recent acquisitions of investment banks by the large commercial banks in the United States. After paying peak prices for these houses, the bigger banks failed to retain the key investment bankers themselves, in essence the individuals whose talents made the acquisitions enticing in the first place (Atlas, 2002). As **SEAM** suggests, **economic performance** has both a short-term, immediate set of outcomes and longer-term possibilities and *potential* gains. If organizations become too focused on capturing **immediate results**, they can inadvertently undermine their ability to capture longer-term possibilities. Given the stress and anxiety associated with these large-scale changes, however, most managers think about getting them over with rather than attempting to understand how to do them better (Ashkenas & Francis, 2000).

In comparison to the SteelCo-Petro case, the Brioche Pasquier Group (BPG, the French bakery referred to in earlier chapters in this volume), structured its acquisition strategy around the **SEAM** methodology, exploring the **hidden costs** that are inevitably associated with a combination strategy (e.g., cultural differences, learning curves, resistance to change). As Serge Pasquier noted, guided by ISEOR's philosophy, the company's strategy was "creating rather than merging," acquiring smaller firms with the intent of transforming them through **socio-economic management**. Using **mirror-effect** analysis to surface issues and uncover areas of potential discontent and untapped potential, BPG structured is preacquisition planning and postacquisition integration efforts to maximize employee input and participation through socio-economic analysis.

As firms attempt to capture the "softer" synergies related to their M&A plans, SEAM analysis becomes increasingly important. As reflected in Figure 17.2, managers face increased challenges as they move from financially-based, "hard synergies"—the visible costs and savings involved in the combination—to capture the "softer" synergies involving the transfer of core competencies and best practices and investing in their human resources (see Coff, 2002; Eccles, Lanes, & Wilson, 1998). Yet, as Marks (2003) underscores, even when companies look to the "softer" synergies illustrated in Figure 17.2 corporate staffers from acquiring organizations, armed with their charts of accounts, reporting cycles and planning approaches, tend to impose their systems on target firms rather than engage in a true joint diagnosis and analysis of the situation. **SEAM** analysis forces recognition of the resultant hidden costs inherent in these tendencies, revealing the **dysfunctions** that underlie traditional accounting methods and financial assessments. The untapped potential and **hidden costs** in

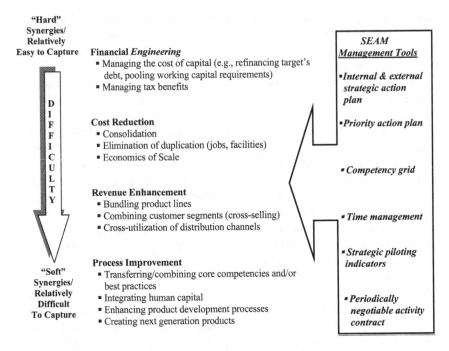

Figure 17.2.   Illustrative merger and acquisition synergies.

mergers and acquisitions (*nonproduction, noncreation of potential* gains, wasted time, *excess salaries* in SEAM parlance) will continue to linger unless senior-level management emphasizes the need to minimize the *dysfunctions* associated with the types of dynamics highlighted in Figure 17.1.

SEAM INTERVENTION AND ACQUISITION SUCCESS: THE GIFA-COLLET COMPANY

In 1999, the GIFA Company, an independent medium-sized enterprise that equips and markets emergency vehicles (e.g., ambulances, fire trucks, hearses) found the opportunity to take control of its principal national competitor that was on the verge of bankruptcy. After completion of a precombination feasibility study on the potential acquisition, legal formalities were completed in the second half of 1999.

In 2000, following the acquisition, an operational study was conducted that focused on drawing up an inventory and identifying the differences between the two GIFA plants (the acquiring company) and the Collet plant (the acquired company) in industrial, commercial, managerial and behavioral terms. The overall conclusion of the Chief Executive Officer, Jean Caghassi (2002), was that GIFA was a high-performance company, while Collet showed numerous difficulties and handicaps. In order to face the amplitude of these discrepancies, and to avoid creating a "winner-versus-loser" mentality, the CEO decided to set up socio-economic management as a way of establishing a common language and common set of management tools in the companies. Thus, both companies found themselves on the same terrain that of apprenticeship. The resulting intervention created a basis for drawing the two former "corporate cultures" closer to each other, with a new identity unified around *socio-economic management*.

Background

Preparation for the acquisition was done with the support of the key financial shareholders (venture capitalists). Jean Caghassi, the CEO, was owner of one third of the capital. The acquisition negotiations were carried out with the aid of his senior managers, a lawyer's office specialized in acquisitions, and the firm's CPA. The behavior of the acquiring CEO during the negotiations was already characterized by *SEAM ethics*, even before he became acquainted with the ISEOR. Thus, he made a point of preserving the interests of the former owners, the Collet family, by avoiding the sale of company's real assets, which would have forced the company into bankruptcy, but would have lead to a more profitable acquisition for the buyer. The former owner-CEO's retirement capital was thus preserved. In the same spirit, the GIFA Company kept all the employees of the acquired company, including the son of the former CEO, who became sales director several years later.

During 2000, following the acquisition, both companies experienced a crisis due to an inexorable resistance to change and the deplorable behavior of a manager hired to develop the acquired company, and individual whose strategy did not conform itself to the *ethics* of the acquiring CEO. The manager left the enterprise in 2002.

SEAM Intervention

Jean Caghassi, the CEO, met Louis-Marie Pasquier, general manager of the Brioche Pasquier Company (see chapter 2 in this volume), who introduced him to socio-economic management and advised him to contact the ISEOR. The preliminary *set-up phase* of the socio-economic intervention took place from 2001-2002 simultaneously in the GIFA Company and the Collet Company. The *development phase* and the *maintenance phase* were carried out from 2003-2006, during which *socio-economic performance improvements* were evaluated. This *evaluation* revealed that the success of the acquisition was attributable to the simultaneous acquisition by both companies of a common language, management philosophy and management tools.

The **hidden cost** method demonstrated that the high-performance company (GIFA) could harbor as many hidden costs as the company on the verge of bankruptcy (Collet), because the level of hidden costs points out the existence of a potential for self-financed performance improvements. This is an exemplary case since diagnostics revealed the same level of hidden costs in both companies, which incited modesty on the part of GIFA company actors, even thought they were already leaders of the French market. After the ISEOR had carried in-depth diagnostics in both plants, improvement actions and change dynamics were launched in both companies, including two actions common to both plants and others specific to each one of them (see Table 17.1).

Table 17.1.   Socio-Economic Innovation Decisions and Actions to Facilitate Integration

Domains	Common Actions	Specific Actions	
		GIFA Company	*Collet Company*
Working conditions	• material investments including machines, tools and computers; • redesigning workshops; • creating a Hygiene and Security Committee at both plants.	• creation of an assemblage zone.	• reorganization of vehicle stocking practices; • creation of an employee parking lot.
Work organization	• setting up *itemized operation charts* for production and manufacturing; • computerizing procedures (estimates and stock management with bar codes).	• creation of the position intermediary supervisory staff(production assistants); • redesign of job descriptions (workshop).	• reorganization of workshop teams by products; • creation and hiring of a production head; • creation of technical files and nomenclatures by products and by orders; • replacement of the stock management software program; • setting up self-controlling and quality-controlling tools and procedures.

(*Table continues on next page*)

Table 17.1.   Continued

| Domains | Common Actions | Specific Actions | |
		GIFA Company	Collet Company
Communication-coordination-cooperation	• creating regular coordination meetings: planning, quality, security and management team meetings in both companies.		
Time management	• more rigorous time management practices, both for individuals and teams.		
Integrated training	• setting up **competency grids**; • developing technical training programs in security and management.		
Strategic implementation	• -socio-economic intervention and improvement of the quality of management; • setting up **periodically negotiable activity contracts (PNAC)** with negotiated objectives and remuneration incentives; • ISO 9001 certification; • harmonization of selling prices; • one single stand at professional fairs representing both trademarks GIFA and Collet.	Recruitment of personnel.	Creation of a labor-management committee.

Source: © ISEOR 2004.

The most important instrument of this unification was **collaborative training** and the change process (axis 1 and axis 2 of the **SEAM trihedral**). The construction of this new identity was pursued concomitantly with deployment of the **development strategy** (axis 3 of the **SEAM trihedral**) designed by capitalizing on the strengths of both companies in terms of market share, commercial equipment, trademarks and the commercial advantages of each company's distinct products. Thus, the initial post-acquisition phase, which focused on **coordination** of the two companies, came to an end in 2002.

This initial success provided foundation for the *integration phase* (2003), involving the *socio-economic management* of commercial, industrial, financial and organizational processes, and culminating with the official, statutory merger of the two companies, GIFA and Collet, in December 2005. In 2006, the final stage of the *integration* of socio-economic management concepts and tools got off the ground with the setting-up of a *socio-economic management control system*.

Outcomes

These actions constitute *intangible investments* whose total cost, including the value of time spent by *internal actors*, came to 187,000€. This investment's profitability is considerable, since it represents a very small portion of the sales figure (0.6%), the margin on variable costs (2%), company wages (3.8%), and potential hidden cost recuperation (11.7%). Indeed, the speed of return on investments (4 months) and the rate of profitability (500% = 935,000€/187,000€) were extremely high. Table 17.2 illustrated the evolution of *economic performance* indicators during the period of postacquisition, following the simultaneous introduction of SEAM.

These merger cases demonstrate that the socio-economic approach based on the participation of actors and the construction of a cooperative system produces higher economic results than the traditional approach founded on the exploitation of the "victory" over the target company, evident in the war metaphors with which classic theories of economic competition are rife.

Implications

Most companies simply talk about their commitment to and the importance of "human capital." SEAM, in contrast, emphasizes *human potential* because human beings cannot be considered as other *resources*—organizational members are free to spend or withhold (e.g., nonproduction, noncreation of potential gains) their energy according to the quality of the informal and formal contract they have with the company (see Savall, 1979; 1987). This distinction is important since a true commitment to developing this key resource is often more talk than effort, a reality that is influenced by the inaccuracy of traditional accounting reports. This is where *SEAM* analysis, with its focus on developing a set of practices that integrates the different disciplines—management, marketing, accounting, finance—into a coherent and interactive whole is clearly needed.

Table 17.2. Evolution of Economic Performance Indicators (Postacquisition)

Indicators	GIFA Group Total Annually	GIFA Group Per Person/per Year	GIFA Company Total Annually	GIFA Company Per Person/per Year	Collet Company Total Annually	Collet Company Per Person/per Year
Hidden costs						
- At beginning of socio-economic intervention (2000);		20,650€		20,500€		20,900€
- At the end of the first post acquisition phase (2002);		7,150€		7,000€		7,400€
- Conversion of hidden costs (reduction) into value-added between 2000 and 2002:						
• Total	1,871,100€	13,500€		13,500€		13,500€
• Itemized earnings on						
° *Absenteeism*	93,000€		91,000€		2,000€	
° *Work accidents*	58,000€		1,900€		3,900€	
° *Personnel turnover*	18,300€		—		18,300€	
° *Nonquality*	450,100€		162,300€		287,800€	
° *Direct productivity gaps*	1,304,000€		972,600€		882,800€	

Indicators	GIFA Group Total	GIFA Group Percentage	GIFA Company Total	GIFA Company Percentage	Collet Company Total	Collet Company Percentage
Variation of performances between 2000 and 2002:						
• Sales figure	+7,034,000€	29%	+6,144,000€	+30%	+890,000€	+9%
• *Margin (or value-added) on variable costs*	+2,416,000€	36.3%	+1,823,000€	+49%	+594,000€	+22%
• Work hours	+23,900h	8%	+37,770h	+24%	-13,870h	10%
• *Hourly contribution to margin (value-added) on variable costs (HCMVC)*	+5€	20%	+4€	+11%	+6€	+25%
Number of employees						
• Variation in 2002	181 persons	+27 (+30%)	97 persons	+29 (+45%)	88 persons	0 (+0%)

Source: © ISEOR 2004.

As part of the acquisition strategy between two computer network companies (see Buono, 1997), for example, key objectives included integrating product lines, assimilating technology across the two companies, and consolidating a joint vendor base. This operational focus was supported by a concerted effort to (1) retain key talent, (2) involve organizational members from both companies in the planning and implementation process through a series of interlocking transition teams (*priority action plan*), (3) use focus groups to uncover key concerns and provide training and insight into acquisition-related dynamics, (4) support the combination with postacquisition integration ceremonies and team-building activities, and (5) use postacquisition training workshops focused on the lessons learned from the acquisition to enhance the core competencies of the management team. The SEAM method's diagnostic orientation (*diagnostic*, *project* planning, *implementation*), cost analysis and *mirror-effect* reporting process readily facilitates this type of integrated strategy. Management in the acquiring company quickly realized that the nature of the desired synergies *required* truly collaborative and cooperative relationships across the two organizations. Rather than focusing solely on short-term reductions in costs and achieving the "hard" synergies in acquiring another company, SEAM-based analyses provide a clearer indication of the strategies and tactics necessary to capture the longer-term, "softer" synergies inherent in true operational *integration* efforts.

Reflecting on this acquisition through the lens provided by the *SEAM* approach, the strategy included: (1) efforts to ensure that organizational members from both firms were aware of the activities and tasks (and related costs for *nonperformance*) necessary for the *strategic implementation*; (2) regularly scheduled meetings that focused on *piloting* the *progress* of postacquisition integration initiatives (*strategic piloting indicators*); (3) interviews and focus groups that allowed a fuller analysis of the gap between executive and manager/supervisory/employee perspectives on the change (*mirror-effect reporting process*); (4) *priority action plan* that attempted to integrate short-term objectives (e.g., coordinate product specifications across product lines) and longer-term gains (e.g., integrated next generation products) while minimizing potential dysfunctions (e.g., "we" versus "they" tensions); and (5) *Horivert interventions* (Savall, 1979, 2003b) that attempted to integrate actions at the top management team level with initiatives focused on lower-level work units.

CONCLUSION

Mergers and acquisitions by their very nature create significant upheaval in the lives of organizational members. The disruption caused by combination-related stress and anxiety, culture shocks and tensions, and job loss, relocation or *realignment*, among a host of other difficulties, obviously entails a number of dysfunctions and human costs that can trigger a series of negative reactions on the part of organizational members. A SEAM-based analysis during M&A planning would focus attention on these problems and the hidden dynamics and costs reflected in Figure 17.1, uncovering potential problems in *working conditions, work organization, time management, communication-coordination-cooperation* dynamics, *training* needs and *strategy implementation*. By focusing on these factors, managers can begin to ameliorate the inherent dysfunctions festering under the surface, creating a stronger foundation for the combined organization to achieve the type of critical operational synergies summarized in Figure 17.2.

The good news is that a growing number of companies are beginning to focus on the post-M&A integration period much more systematically, appointing highly visible and well respected integration managers to shepherd the two firms through what is often turbulent and unchartered territory as they attempt to function as a single entity (see Ashkenas & Francis, 2000). As suggested by the SEAM approach, today's mergers and acquisitions demand an overarching management style based on teamwork, involvement and empowerment, a training and development orientation, and high levels of communication and *negotiation*. Since these interorganizational strategies require a level of consultation and collaboration that goes well beyond typical patterns of management and organization (see Buono, 1991; Kanter, 1988), the time has come for managers (1) to understand the *hidden costs* involved in combining organizations, (2) to assess the gap between the experience of organizational members and the strategic plans of senior-level managers, and (3) to balance short- and longer-term dimensions of performance.

There are, of course, a number of approaches that can facilitate this type of analysis—ranging from action research models and sociotechnical systems analyses to appreciative inquiry, learning organizations and stakeholder management (Savall & Zardet, 2004). Yet, while these methods focus on social and technical interfaces and dialogue interventions across different groups, they lack the attention to the financial-oriented disciplines associated with organizational change in general and M&A integration strategies in particular. By integrating the social and economic realities inherent in mergers and acquisitions, *SEAM* provides a highly useful basis for intervention planning and *implementation*.

ACKNOWLEDGMENTS

This chapter draws heavily from an earlier article published in the *Journal of Organizational Change Management*, titled "Seam-less Post-merger Integration Strategies: A Cause for Concern" (Buono, 2003).

REFERENCES

Ashkenas, R. N., & Francis, S. C. (2000). Integration managers: Special leaders for special times. *Harvard Business Review, 78*(6), 108-116.

Atlas, R. A. (2002, May 26). How banks chased a mirage. *New York Times*, pp. 3-1, 3-12.

Buono, A. F. (1991). Managing strategic alliances: Organizational and human resource considerations. *Business & the Contemporary World, 3*(4), 92-101.

Buono, A. F. (1997). Technology transfer through acquisition. *Management Decision, 35*(3), 194-204.

Buono, A. F. (2003). SEAM-less post-merger integration strategies: A cause for concern. *Journal of Organizational Change Management, 16*(1), 90-98.

Buono, A. F. (2005). Consulting to integrate mergers and acquisitions. In L. Greiner, & F. Poulfelt (Eds.), *Handbook of management consulting: The Contemporary Consultant—Insights from World Experts* (pp. 229-249). Cincinnati, OH: Southwestern/Thompson.

Buono, A. F., & Bowditch, J. L. (1989). *The human side of mergers and acquisitions: Managing collisions between people, cultures, and organizations.* San Francisco: Jossey-Bass.

Buono, A. F., Bowditch, J. L., & Lewis, J. W. (1985). When cultures collide: The anatomy of a merger. *Human Relations, 38*(5), 477-500.

Buono, A. F., & Nurick, A. J. (1992). Intervening in the middle: Coping strategies in mergers and acquisitions. *Human Resource Planning, 15*(2), 19-33.

Caghassi, J. (2002). Démarche de changement dans un contexte de fusion de deux sociétés ou l'émergence du groupe [Change procedure in the context of merger of two companies or the emergence of the group]. In ISEOR (Ed.), *Le management des l'entreprises culturelles* (pp. 161-178). Paris: Economica.

Coff, R. W. (2002). Human capital, shared expertise and the likelihood of impasse in corporate acquisitions. *Journal of Management, 28*(1), 115-137.

Eccles, R. G., Lanes, K. L., & Wilson, T. C. (1999). Are you paying too much for that acquisition? *Harvard Business Review, 77*(4), 136-146.

Elsass, P. M., & Veiga, J. F. (1994). Acculturation in acquired organizations: A force-field perspective. *Human Relations, 47*(4), 431-453.

Haspeslagh, P., & Jemison, D. (1991). *Managing acquisitions: Creating value through corporate renewal.* New York: Free Press.

Jemison, D. B., & Sitkin, S. B. (1986). Acquisitions: The process can be a problem. *Harvard Business Review, 64*(2): 107-116.

Kanter, R. M. (1988). The new alliances: How strategic partnerships are reshaping American business. In H. L. Sawyer (Ed.), *Business in the contemporary world* (pp. 59-82). Landham, MD: University Press of America.

Lubatkin, M. (1983). Mergers and the performance of the acquiring firm. *Academy of Management Review, 8*(2), 218-225.

Marks, M. L. (2003). Making mergers and acquisitions work: A guide to consulting interventions. In A. F. Buono (Ed.), *Enhancing Inter-firm Networks and Interorganizational Strategies* (pp. 3-30). Greenwich, CT: Information Age.

Marks, M. L. & Mirvis, P. H. (1992). Rebuilding after the merger: Dealing with 'survivor sickness.' *Organizational Dynamics, 21*(2), 18-32.

Marks, M. L., & Mirvis, P. H. (1998). *Joining forces: Making one plus one equal three in mergers, acquisitions, and alliances.* San Francisco: Jossey-Bass.

Pritchett, P. (1987). *Making mergers work: A guide to managing mergers and acquisitions.* Homewood, IL: Dow Jones-Irwin.

Savall, H. (1974, 1975). *Enrichir le travail humain dans les entreprises et les organisations* [Work and people: An economic evaluation of job enrichment]. Paris: Dunod.

Savall, H. (1979). *Reconstruire l'entreprise: Analyse socio-économique des conditions de travail* [Reconstructing the enterprise : Socio-Economic Analysis of Working Conditions]. Paris: Dunod.

Savall, H. (1987). Les coûts cachés et l'analyse socio-économique des organisations [Hidden costs and the socio-economic analysis of organizations]. *Encyclopédie du management* (pp. 599-628). Paris: Economica.

Savall, H. (2003a). An updated presentation of the socio-economic management model. *Journal of Organizational Change Management, 16*(1), 33-48.

Savall, H. (2003b). International dissemination of the socio-economic model. *Journal of Organizational Change Management, 16*(1), 107-115.

Savall, H., & Zardet, V. (1987). *Maîtriser les coûts et les performances cachés: Le contrat d'activité périodiquement négociable* [Mastering hidden costs and performances: The periodically negotiable activity contract]. Paris: Economica

Savall, H., & Zardet, V. (1995). *Ingénierie stratégique du roseau, souple et enracinée* [Strategic engineering of the reed, flexible and rooted]. Paris: Economica.

Savall, H. & Zardet, V. (2004). *Recherche en sciences de gestion: Approche qualimétrique. Observer l'objet complexe.* Unpublished English translation: *Research in management sciences: The qualimetric approach. Observing the complex object.* Paris: Economica.

Savall, H., Zardet, V., & Bonnet, M. (2000). *Releasing the untapped potential of enterprises through socio-economic management.* Geneva, Switzerland: International Labor Office-ISEOR.

Savall, H., Zardet, V., Bonnet, M., & Moore, R. (2001). A system-wide, integrated methodology for intervening in organizations: The ISEOR approach. In A. F. Buono (Ed.), *Current trends in management consulting* (pp. 105-125). Greenwich, CT: Information Age.

CHAPTER 18

THE ROLE OF SOCIO-ECONOMIC MANAGEMENT IN ENHANCING SALES AND MARKETING ACTIVITIES

A Comparative Case Study

Isabelle Barth

Performance is a central concern of business enterprises and management research, and it extends to the performance of the sales representative (Rich, Bommer, McKenzie, Podsakoff & Johnson, 1999). For perhaps the past 80 years,[1] practitioners and researchers alike have been trying to better understand the dynamics surrounding sales force performance, attempting to determine the factors that stimulate and develop that performance. This focus continues to be a subject of considerable topical interest—enterprises are rethinking their sales strategies, reorganizing their sales forces, redefining sales-related jobs, and restructuring remu-

Socio-Economic Intervention in Organizations: The Intervener-Researcher and the SEAM Approach to Organizational Analysis, pp. 401–420
Copyright © 2007 by Information Age Publishing

neration plans to meet current social and economic constraints. Companies have to deal with consumers who are more and more discriminating, demanding faster delivery, better quality and wider selection at increasingly lower prices, without any guarantee of loyalty. The globalization of markets clearly strengthens this competition and leads companies, whatever their size and *resources*, to adopt an international approach, with the concomitant need to incorporate ever more complex parameters into their relationships and activities. Within the enterprises themselves, these *interactions* increase and become more extensive and more complex, whether these involve financial flows or human and logistic flows in outsourcing (e.g., production in factories, service by call centers).

Faced with all these problems, innovation may appear to be a key to performance, but attempts at innovative sales practices are often little more than a mad dash that becomes increasingly frantic and disorganized, generating vast management costs due to poor quality or failure to comply with market requirements. The spread of information and communication technology can also give rise to many operating problems, especially when systems are designed merely to computerize existing procedures rather than enabling firms to more effectively manage their interactions. It can also contribute to a lack of continuity in customer relations. The compartmentalization of customer relations—from presales and sales, to sales follow-up and after sales service—creates situations where the customer no longer deals with one single contact. Buy-outs, mergers, and takeovers further result in enterprise governance that is increasingly financially oriented. More than ever, emphasis is placed on strategies for maximizing return on investment and short-term *economic performance*, a focus that is often to the detriment of creating intangible assets, investing in non-material resources, and enhancing *social performance* (Savall, 1974, 1975, 1987; Savall & Zardet, 1987).

Faced with these stakes and influences, socio-economic researchers are increasingly being called in to advise companies on how they can redefine their sales-related systems, with the ultimate goal of enhancing performance. The underlying premise is that a *socio-economic management* system can change an organization's sales practices and improve sales performance, from both an economic and social perspective (Barth, 1994).

THE STRUCTURING ROLE OF SOCIO-ECONOMIC INTERVENTION

Socio-economic intervention (Savall, Zardet, Bonnet, & Moore, 2001) is the process by which the *intervener-researcher* sets up socio-economic management in the target enterprise. It is important to highlight some of the main principles behind this implementation as they form the basis for the changes which the enterprise will experience. Overall, the approach is an integrating, instrumented process: integrating because it applies to the enterprise as a whole; instrumented because it sets up *management tools* that affect the management of time and skills, the monitoring of activities, and the definition of objectives in all aspects, from the most global and strategic level to the individual at his or her workstation. It covers the enterprise in its entirety and, from the start, imbues all those within the enterprise with this view.

Socio-economic intervention (Savall, 2003a, 2003b; Savall & Zardet, 1995; Savall, Zardet & Bonnet, 2000) starts with an analysis based on inward-looking marketing that will introduce consultants in a position of *active listening* that will last throughout the intervention period. This active listening process starts, and this is essential, right from the *negotiation* phase, even at the initial contact. Using open questioning techniques and by reformulating the questions, the socio-economic scientist leads the client, in small steps, to formalize expectations and requirements, which are often different from the initial idea and premise held at the outset. The procedure of "questioning/extracting information/processing the information/standing back/submitting proposals/building-up solutions/ implementation" starts at the initial contact between the socio-economic scientist and the client and is repeated throughout the intervention period (Barth, 2001). The objective of *synchronization* the socio-economic intervention and *socio-economic management* also plays a very important role in the sales and marketing function, often considered to be "outside" the enterprise and somewhat out of phase with it.

The ability to implement these management tools has an enormous impact on people (in this instance, the sales staff) who often consider themselves, with considerable frustration, to be executives without strategic responsibility, even in the short term. The tool enables individuals who see themselves as "thinkers" to make decisions and to act, and those who see themselves as "doers" to conceive and plan. Finally, the principle of coconstruction which is the basis for the whole process of introducing socio-economic management imbues all those in the enterprise with a continuous desire to re-engineer and share information.

ENHANCING MARKETING AND SALES PRACTICES

The chapter summarizes interventions in nine enterprises, identifying and examining changes in the sales and marketing function and their effect on the personnel responsible for undertaking those functions. The analysis examines the direct, indirect, consequential and diffusion effects of introducing socio-economic management on the sales and marketing practices of these enterprises. The study design is based on "before" and "after" snapshots of the sales and marketing functions of the companies, taking into account the introduction of work methods and **management tools** intended to bring about changes in their marketing and sales practices.

Intervention Analysis

The following criteria were used as a basis for company selection for the current study: (1) implementation of socio-economic management; (2) adequate information about their internal operation *before* the introduction of **socio-economic management**; (3) a wide range of enterprise sizes and sectors of activity; (4) enterprises having various degrees of commitment to the appropriation of the socio-economic method; (5) different sales and marketing organizations, with a range of market sensitivities; and finally (6) available personnel for providing information. The analysis is based on nine companies that met these criteria. Table 18.1 provides an overview of these enterprises.

Methodology

As a way of establishing a baseline of information, one of the objectives of the research was to determine the state of the sales practices of these nine enterprises prior to intervention by socio-economic researchers. The **socio-economic diagnostic** method was applied in two phases: the **mirror effect**, recreating the **dysfunctions** mentioned by personnel during the qualitative interviews, and **expert opinion**, presenting the analysis drawn up by the socio-economic intervener-researcher. This approach highlighted a large number of the actual dysfunctions mentioned by the staff of the companies and served as the basis for constructing the interpretation of the statements made by those involved.

The next step was to identify the dysfunctions related to sales and marketing that emerged during the socio-economic analysis (e.g. measuring the dysfunction effect on turnover, the risk in terms of image and opportunities for the enterprise). The basic working material consisted of thousands of *fieldnote quotes* (more than 3,000) relating to **dysfunctions** in the sales and marketing area, which were collected in the general analyses that were undertaken individually by the socio-economic researchers on the team. A *transverse analysis* was produced, bringing together these *fieldnote quotes* as **key ideas**, subthemes, and finally as themes. This method does not aim to ensure that the basic data and results are statistically representative; rather the focus of the method is to capture the types and range of problems in the sales and marketing function.

Table 18.1. Summary Characteristics of the Nine Companies in the Study

	A	B	C	D	E	F	G	H	I
Sector of activity	Catering	Engineering	Telecommunication service	Banking	Insurance	Food industry	Sheet metalworks	Supermarkets	Food industry
Activity	Fabrication and distribution of selected food products	Forging Mechanical engineeering	Sales, installation and maintenance of telecommunications systems	Personal loans	General insurance (fire, accident, sundry risks)	Production and distribution of fancy breads	Sheet metal work	Supermarket (hyper, super, DIY, home, cafeterias)	Production and distribution of biscuits, snacks and cakes
Status	Private limited company (SA) subsidiary of a large group	Subsidiary of a large international group	Limited liability company (SARL), subsidiary of a telecomm cooperative	Subsidiary of an American financial holding company	Mutual	Holding company	Subsidiary	Holding company with franchises	Private limited company (SA)
Annual turnover (€M)	85	40	21.5	30.5	610	137	23	275	259
Total personnel	1,000	400	350	150	4,000	1,000	100	1,300	2,700
Sales department	shops reception	a sales department	a sales department	2 regional branches with sales department	8 sales regions 184 agencies	8 sales regions	a sales department	hypermarkets supermarkets specialist shops	a sales department 7 regional branches
Marketing department	yes (2 persons)	no	1 person	no	yes (at Head Office)	yes (at Head Office)	no	no	yes

Source: © ISEOR 2004.

Results of Transverse Analysis

The key ideas were grouped into four main themes: sales function, the information system, the **internal environment**, and the **external environment**. The quotes that follow are *verbatim* extracts from the interviews.

Sales function. The data indicated that sales personnel had a very short-term, operational view of the sales function, which was expressed through rejection of administrative tasks (e.g., writing of reports, debriefing of visits, purchase orders management). Emphasis was placed on the need to "occupy the ground" and to "sell." The sales force saw themselves only as salespeople, despite the fact that more and more people in the sales force were being asked to *manage* sales. This responsibility implies more team work, an extended vision of the profitability of sales, organizational ability and an awareness of **productivity**. Yet, the sales force, in general, did not appear to embrace this broader, managerial orientation. For example:

> Selling is done in the field … [and] the more information you have to pass back, the more cumbersome the system (Sales/supermarkets [H])

> Instead of following-up leads, we have to write reports. It's maddening. (Sales force analysis/food industry [I])

The analysis also pointed to problems managing customer relations, which were seen mainly as a hindrance that created a heavier work load. It was also evident that the salespeople often formed strong friendships with *"their"* customers and were not always able to maintain the distance necessary for **effectiveness**.

The sales staff also found it very difficult to pass on their sales skills to others, which evoked the "sales gene" philosophy. Conventional wisdom in the firms appeared to be that "hard skills" such as information technology and mechanical engineering could be taught, but that sales people were "born" with the requisite talent. Thus, job-related training was largely limited to learning "on the job," often through working in pairs. Yet, as the following quotes suggest, the sales staff had a number of unmet training-related needs.

> I don't know where I go wrong in negotiating with buyers. (Sales force analysis/food industry [F])

> We are highly autonomous salesmen. So far as customers are concerned, we are on our own for judging whether things went well, whether we made a good impression. (Reception analysis/catering [A])

Finally, the classic methods for motivating the sales staff reflected internal competition and independence. The salesmen were asked to be managers, but they were paid a percentage of their sales. They were asked to work as a team, but the work itself was organized as a competition across individuals. As long as the sales staff is treated as being in a "world apart"—one foot in the enterprise and one foot outside—it is unlikely to be effectively integrated into the enterprise. Consequently there are many dysfunctions caused by such *synchronization* failures, with associated costs and frustration.

> The incentive scheme doesn't provide enough motivation; the ideal would be to have 15 to 25% bonus. (Sales force analysis/Food Industry [I])

> The bonuses don't provide enough motivation for the sales staff and some managers. We'd get a lot more sales if the sales staff had a stake in the sales. (Warehouse analysis/Catering [A])

> The assessment system is too rigid. There ought to be other [*evaluation*] criteria, such as the attitude to suppliers and customers. (Horizontal analysis/ telecom service [C])

Sales information system. The analysis also pointed to a lack of information, particularly feedback from the sales force. This void was explained by the lack of awareness of the importance of the information gathered by each person and required for the enterprise's databases. The sales force was particularly at fault with respect to customer information, which is largely explained by a combination of the lack of training and reliance on the "sales gene" (discussed above), a lack of acknowledgement for providing information, and the effort required to motivate the sales force.

> I ask the salesmen for reports, but I don't get them. (Sales force analysis/ food industry [F])

> There is no feedback from the sales force. (Information analysis/sheet metal works [G])

Along similar lines, the information that was circulated to the sales staff was often criticized as being too much, useless, and badly timed. Although an information system was in place, it was not dynamic and did not facilitate decision making.

> Sales promotions are ready, but we don't find out until the last minute. (Sales force analysis/food industry [F])

> There are serious communication problems between marketing, fabrication and sales. Promotions are announced for products that have not been made or don't exist. (Sales force analysis/food industry [I])

Another reason for the *dysfunction* of the enterprise's information systems was the juxtaposition of local information systems without common language or data interchange, which led to incomprehension, frustration and a waste of time, energy and money.

> An information database would be useful. We don't know how the regional branches compare or where France stands with respect to its competitors. (Sales force analysis/food industry [I])

The information systems, which were regarded both as sirens and scapegoats, also played an aggravating role in many of the dysfunctions.

> Statistics are indigestible, they should be summarized. Very little coming out of the information systems can be used directly. (Horizontal/sales force analysis/Food Industry [I])

Finally, the firms also had difficulty in managing the information coming from the outside. This inability led to a reliance on subcontractors for surveys and an "understanding" of their environment. The enterprises, and in particular the marketing department (where there was one), relied more on such subcontractor reports than on day-to-day realities (i.e., the facts as experienced by the sales force in the field). Moreover, this subcontracting was often ad hoc, so that the enterprises used fixed, often obsolete, representations of their environment. This destabilized the salespeople who were unable to reconcile what they were told by the organization with the reality on the ground with which they were familiar.

Internal environment. The initial analysis revealed that there was often compartmentalization between departments, which resulted in breakdowns in communication and isolation of production, research and development and sales departments. This communication breakdown was aggravated by the dispersion of the sales force.

> There ought to be coordination with the salesman and between the departments, which there isn't. (Agency analysis/telecom service [C])

> Everyone acts for themselves. There ought to be communications between sales, marketing and production. (Horizontal/factory analysis/food industry [I])

> Internal communication is fine at the head office, but it is more difficult in the sales regions. (Human resource analysis/food industry [F])

> There are not enough regional meetings, the management is there for conferences, but that's all. The divisional manager is rarely seen and that's not right. (Sales force analysis/food industry [F])

It was also noted that the decentralization of sales was badly managed, which resulted in poor distribution of job functions resulting in complaints and frustration on the part of the sales staff. This failure could be explained by an insufficient sales management strategy and *resources* constraints. The different sales forces were frustrated about not being involved in drawing up and implementing the strategy. They complained that they felt "left on the sideline."

> It is not easy to involve the sales department in the factory requirements. (Sales analysis/engineering [B])

> I'd like to visit the factory. It's important to know your tools and be able to explain how the product is made. (Sales force analysis/food industry [I])

External environment. The *interaction* between the internal and external environment reflected difficulties in responding to external pressures and managing the internal and external images of the companies. The people who made up the different sales forces were also acutely aware about the deficiencies in enterprise-customer relations, including issues surrounding the staff, the product, associated services, and the agency premises.

> The quality of customer service is deteriorating, which causes a certain discontent among customers. (Agency analysis/telecom service [C])

> When the goods are delivered, half is sometimes missing. That's happened a lot during Epiphany, you lose the sale but it's also a question of image and deteriorating customer relations. (Sales force analysis/food industry [F])

> It's easy to take the easy way. Instead of orienting requirements toward the products we know to increase profitability, we create special products. (Reception analysis/catering [A])

> We have difficulty formulating the requirements. We create a lot of five-legged sheep. (Information analysis/sheet metal works [G])

Finally, the enterprises, through their sales departments, had ideas for implementing actions to co-ordinate the *external* and *internal environment*. Two types of proposal emerged: (1) updating *"hidden" products and/or services*, which could be used as a basis for development strategies; and (2) an outline for partnerships with organizational processes going beyond typical customer/supplier relations.

> It's all very well livening up sales, but it's the claims department that keeps the clients. (Agency analysis/insurance [E])

We think we have a very good product, but we don't make enough of it: delivery, the trimmings, the smile with which it's served, etc. (Reception analysis/catering [A])

Creating an Innovative Internal-External Marketing Program

Socio-economic management proposes a renewed vision of the corporate marketing function, that is, the enterprise's relation to its market. It proposes anchoring that function in the organization, creating a hybrid of marketing and sales functions, defining the market as a group of drivers and varied clients, and questioning the enterprise's boundaries. Each employee becomes aware that he or she can play a role, regardless of position, in the relationship of the enterprise to its clients.

The Sales Function Becomes Responsible for its Own Strategy

Socio-economic management reduced the compartmentalization of the enterprise into functions and the division of work into operational and functional departments (into marketing and sales force in the field which was of interest to us). This redistribution of tasks relied on setting up *synchronized*, managed *decentralization* of the strategy with two driving forces: empowerment of individuals and the use of *socio-economic management tools*.

The intervention repositioned the marketing department as an *internal-intervener* and introduced innovative sales strategies based, for example, on "hidden" products/services, other products/ services that emerged from the enterprise's activity, or efforts that contributed to a more sustainable exploitation of current products. For example, in one of Company H's furniture stores, employees working in the warehouse pointed out during a *focus group* meeting that they had several occasions a day to assist clients with small parts that had been broken during furniture assemblage, worn out or lost. They gave clients spare parts taken from broken or unsaleable furniture. The idea arose to set up a "small parts fair" every Saturday morning, where all small parts would be available for purchase at reasonable prices.

One of the consequences of this decentralization was the repositioning of the use of external marketing consultants. By acquiring a deeper understanding of requirements, the companies were able to reduce the need for outside consultancy and took over marketing activities that were previously subcontracted outside the enterprise (e.g., market surveys), leading to considerable savings.

An *integrated training* program was identified as one of the driving forces behind this reorientation. It was based on structuring, communicating and building on actual sales knowledge from the field that went well beyond the "tricks of the trade" and "cookbook" approaches that too often characterize sales training. For example, setting up *competency grids* permitted Company A's shops to update, organize and homogenize the competency and know-how of sales personnel in the shops. Not only did the personnel have clear job descriptions, but they understood that their missions were not limited to selling. Thus, tasks concerning product control, store decoration, strict application of hygiene *regulations* and so forth were included in every job description. Sales-training action in a catering branch of Company A was designed around the fact that, too often, sales persons stuck close to customer demands, provoking serious production disturbances and high additional costs, for rarely-optimum customer satisfaction. The objective of the training was to lead sales persons to better "pilot" customer demands by being better acquainted with product lines and by mastering reformulation techniques in the negotiation process.

> Marketing missions were returned to the sites—merchandising, consumer quality rating, advertising gadget budgets, selling by management techniques, etc. (Sales analysis/food industry [F])

> Basically, socio-economic management returns the business to those who do it. (Management analysis/catering [A])

Finally, the basic foundation for an effective sales and marketing management function were identified, laying out the entire sales process from initial contact to completion of the order. Qualitative, quantitative and financial indicators were kept up to date by the salesmen to steer this new mission.

Developing an Integrated, Integrating Information System

A second transformation was noted in the companies' information systems, with *developments* occurring both internally (e.g., between departments) as well as between the enterprise and its external environment. The feedback and extraction of information from the *external environment* was a definite challenge for the organizations, particularly with respect to the competition and customers. The sales force became a real spearhead and the ultimate means of acquiring information about company performance and wider environment dynamics.

Computerization and information technology were noted for enhancing the companies' ability to more effectively manage the dispersion, distance and reticence of the sales force, not only as a resource but also for training. The management and adaptation of an internal/external information system rectified many of the dysfunctions identified during the analysis. It enabled a real market survey to be set up in real time at reduced cost by mobilizing the sales force to extract and feedback information on the particular environment. One of the marketing units in the nine-company sample was reoriented as internal intervener, assuming the role of co-ordination and information processing. Sales and marketing operations started to be monitored using the negotiated objectives within *periodically negotiable activity contracts* (*PNAC*).

With the decentralization introduced through socio-economic management, there was greater individual responsibilization. This permits a more active *piloting* of externalized missions: the enterprise is capable of formulating its demands and can better deal with the external advisor. Company F, for example, thus reintegrated the marketing functions that traditionally had been externalized, by having them piloted by an internal "cell" of several employees. The *PNAC* was utilized as a tool for mobilizing *human potential* for the development of a decentralized marketing function, by including several objectives in every *PNAC* involving a marketing action adapted to the employee's post (e.g., information transmission, quality surveillance, delivery service *improvements*).

The *PNAC* is a tool for stimulating individuals; it is therefore a very powerful lever for developing actions concerning the market and clients. In Company H, for example, one of the objectives for cashiers was "managing a more pleasant customer welcome."

> There is an improvement in the caterer/sales problem. A contract that is reviewed regularly helps the reorganization; the arbitrary creation of new products had to be stopped. (Sales analysis/catering [A])

> There is a twice yearly meeting with the sales force and the various departments it deals with to discuss problems. (Information analysis/sheet metal works [G])

> The product flow manager is responsible for the final delivery date. He is the person the salesman has to deal with. (Management analysis/sheet metal works [G])

> The network feeds back information on problems relating to products, obsolescence, etc. (Sales/network analysis/insurance [E])

The Perception and Instrumentation of Continuity Between the External and the Internal Environment

A central environmental theme that emerged centered on the changes in the way the companies in the study perceived their environment. It was noted that, after introducing the decentralization strategy and structured, reorganized information systems, the firms with a sales function (chiefly Companies A, C, F, H and I) had a different view of their environment. This new awareness reflected a continuum, a seamless environment from inside to outside, and realization of the role that those in sales could play in this *integration* (Savall & Zardet, 2005).

Improvements were noted in communication within the enterprises and between the enterprises and the outside world. The quality of this communication was enhanced and was characterized by interactive responses and the creation of a common language. *Socio-economic management* was the basis for formalization and new work methods. This new perception of the external environment enabled the sales staff in particular to better cope with what they had identified as difficulties, helping to reduce stress and improve *effectiveness*.

There was also an increase in the awareness of the external environment, particularly with respect to the competition and customers, which facilitated improvements in service, delivery and innovation. Employees, whose functions were sometimes far removed from the sales department, discovered or rediscovered customers and, by so doing, their sales vocation, even if it was not formally part of their daily activity.

The organization-wide reorientation to "this selling thing" was a key to improvements in service quality. The relationship with the market and customers was now seen as vital for the entire enterprise, and was no longer restricted to a select few specialists. Customers were seen from all points of view, no longer just as a sale. The most striking example was workers from Company B becoming aware that they were manufacturing parts for people traveling, and not just anonymous elements of some airplane cockpit; it was enough to display photographs of airplanes with their passengers on workshop walls.

In Company A, many actions converged toward apprehending the client as a whole, in order to better adapt the catering-service offer (the stakes of a marriage or a reception go far beyond mere food services). A very simple action illustrates this strategy: the "home tasting service," devised after observing that customers have little appetite for tasting products at inappropriate hours. The sale was certainly an important part of the relationship, but not the only part, as was too often believed to be the case.

Once every 6 weeks, there is a meeting with all the regional directors, the network director and the managing director. We discuss what has been well done; it's very effective. (Sales/network analysis/banking [D])

The analysis has made it easier for some people to make themselves heard. (Management analysis/supermarkets [H])

The analysis is one of the advantages of the socio-economic intervention as an *internal marketing* action. (Management analysis/sheet metal works [G])

Socio-economic intervention is a means of releasing internal tensions. (Sales analysis/insurance [E])

The results are found in the way of working together. There is a new way of planning, reaching agreement, forecasting. This struck me with a year's hindsight. (Sales analysis/catering [A])

The basic effect of the socio-economic approach is making those involved responsible. As they open their eyes, they will see the consumer. (Management analysis/catering [A])

The awareness with respect to sales and the environment has changed a lot. (Information analysis/sheet metal works [G])

The analysis has highlighted the concept of the customer; that has made people think. (Management analysis/sheet metal works [G])

Assessing the Implementation of Socio-Economic Management

Socio-economic assessment is a dynamic process. It is the phase of the intervention that evaluates the actions already taken and builds a platform for new proposals and next steps. There is no "right" moment for a *socio-economic evaluation*; it is carried out when the assessor decides to do so in agreement with his or her contacts in the enterprise. The basic principle is to draw up a comparison of the change in practices from the start of the socio-economic intervention (using the "snapshot" of the *mirror effect* of the analysis) up to the time of the assessment.

The socio-economic evaluation is based on identifying the *direct effects* (attributed to the action evaluated within the strict field of intervention), *indirect effects* (effects that have a visible relationship with the action evaluated but which do not fall within the scope of the declared objectives), *diffuse effects* (attributed to the action evaluated but outside the strict scope of the intervention), and *induced effects* (diffuse effects appearing later) of implementing the actions evaluated.

These effects are assessed **qualitatively** and **quantitatively**, including **financial** analysis. The results of the assessment are governed by the information collected (within the time allowed) and also by the extent to which this information is able to stimulate the people concerned. The assessment is based on interviews with persons who have direct or indirect experience of the action assessed. These interviews are semidirected (comprehensive notes are taken) and based on the search for "positive" achievements, even if residual or negative effects are never criticized. In an intervention strategy, the assessment is a step intended to remobilize all active forces by allowing them to see the *progress* that has been made. As much as possible (generally taking account of the constraints of the time for recreating and accessing the information) qualitative, quantitative and financial *"economic balances"* are drawn up, depending on the amount of detail in the information collected. These balances are used to measure what has been gained by the intervention, what has been lost, and the cost of implementing the intervention.

In the present study, the assessment covers the long-term effects of introducing *socio-economic management* on the sales function of the nine companies. The following procedure was applied. Personnel from the nine enterprises were re-interviewed to assess any changes that they felt socio-economic management brought to the enterprise's functions. These interviews lasted between 2½ to 6 hours. The interviewees were chosen for their involvement in this sales function, even if they were not directly part of a sales department at the time of the intervention or at the time of the assessment. A specific guide was created for carrying out these interviews. Particular attention to three general problems: (1) static difficulties classified as "hobby horses" (fixed ideas), (2) "taboos" (subjects to be avoided), and (3) "contentious" topics that can distort understanding during discussion (Savall & Zardet, 1987). There were also historical difficulties "arising from bias introduced in the interpretation of information by current events, memory or anticipation; by this means the information collected at a given time is confused with information from different periods" (Savall, 1986). Our aim was to know the possible biases as they are a fruitful, fundamental source of information when deciphered as "bias becomes a friendly, benign dragon" (Patton, 1978).

Enhancing Financial Performance

All the assessments carried out at the end of the actions showed an increase in performance, even if it was initially only a reduction in *dysfunctions* within the enterprise, which did lead to a consequent reduction in management costs. These savings are a financial springboard for reorienting the activity (see Table 18.2).

Proposals for Enhancing the Sales and Marketing Function

The basic objective of intervention by a socio-economic scientist is to change the way the firm is managed to gain better understanding of how it operates. In the work described in this chapter, the following examples

**Table 18.2. Example of Actions and Related
qQfi Results Evaluation and Financial Effects**

Actions Implemented	Qualitative Assessment of the Effects	Quantitative Assessment of the Effects	Financial Assessment of the Effects
• *Integration* of marketing within the structure of the enterprise (management team and regional managements and marketing department with sales force)	• Remotivation of the sales force	• Reduction in *personnel turnover* by 7% (estimated overhead)	• Reduction in costs of recruitment and training on recruitment [€35,000]
• New sales force organization to reduce *dysfunctions* related to distribution	• Fewer delivery failures and fewer part deliveries	• 30% down	• Reduction of €6,100 x 30% x 7 regional agencies [€12,800]
• Acquisition of reflexes for exploiting existing products	• Drop in cannibalization on shelves	• Reduction of drops in sales due to cannibalization	• Opportunity cost
• Involvement of sales regions in drawing up the marketing strategy	• Remotivation of salespeople, strategy closer to "reality on the ground"	• Not assessed	• Not assessed

Source: © ISEOR 2004.

were identified as key elements that were central to redeploying the strategic capacity of the sales function.

Knowledge About the External Environment

Lack of insight and knowledge about the environment typically results in a lack of awareness about the customer, the competition and other relevant external forces. We defined it as the perception and activation by each person of the role he may play in building up the image and developing sales in the short, medium and long term in his enterprise. This knowledge can also develop in the internal environment with the aim of building strategies or simply discovering or updating *hidden products or services*.

Indeed, one too often thinks that commercial performance is the sole responsibility of sales personnel or other employees who are in direct contact with customers. The accountant plays an equally important commercial role when, for example, he or she sends reminder letters at the right time, with the right tone, stating well-adapted demands (the case of Company E). Proper task-scheduling in workshops (Company B) is also a way of respecting delivery guarantees, which also guarantees positive commercial impact.

A simple example of this was studied in Company A. A cake was broken in the "laboratory" while it was being made by a technician lacking training. There was, of course, loss of raw materials, delivery delay for the client, *risk* of losing earnings since the client refused to change the selected product or to wait, loss of image to the client and an additional 11 other clients (through information that is transmitted if dissatisfied), and finally, deterioration of the atmosphere between the laboratory and the shop, an unpleasant atmosphere that can offend the clientele and turn them away.

Porosity of the Enterprise

This is characterized by microcommunication between each member of the enterprise and the external environment, and the existence of "grey zones" that appear at the interfaces (e.g., sales surface). This highlights the role of those operating in these grey zones. Setting up socio-economic management results in the enterprise no longer being perceived as a closed entity, with an "inside" and an "outside." This was achieved in Company E by setting up factory visits for schools and local associations, to *draw clients* into the enterprise. In another example, Company I created a cookie tasting panel for workers, noting that workers often were unacquainted with the products they manufactured and did not offer them as gifts.

Company B simply noted the client's name on all production orders "to avoid disconnecting the workshop from the exterior."

Sales Potential

Each item of information, decision or act by each member of the enterprise contains "sales potential" (Hanson, 1959). The sales potential of each act, decision and item of information can be conceptualized as the capacity to generate a differential (positive or negative) in the environment relating to the image, the opportunities or revenues of the enterprise, either directly or indirectly. The sales potential of an individual is not directly proportional to his formal power in the enterprise or to how close he or she is to the marketing or sales departments. This is why Company H worked with its cashiers, shaping them into "shop images," observing that this relationship determined the last impression clients took away from the store. Company D and Company E trained telephone receptionists to be truly welcoming to clients, thus improving the store's image by redirecting them to the person who could best deal with their requests. The commercial role of the secretary typing estimates was stressed in Company A, underscoring the consequences of typing errors concerning the vintage year of wines for large, costly receptions.

Vital Sales Function

This expresses what we define as the "entrepreneurship" of each member of the enterprise. In any organization it is important to realize that all organizational members, as appropriate, have a part to play in all the functions of the enterprise. To a varying degree, each member is involved in marketing, production, research and development and maintenance. Each has, of course, a primary function, which he or she carries out most of the time and to a greater degree, but cross-functional activities are a reality of organizational life. In the present study, for example, Company F included heads of production on *negotiation* teams that visited clients. Two effects resulted—(1) production heads became aware of their commercial role and (2) also provided better expertise in responding to clients during negotiation. In Company H, store managers were named "Number 1 sales person" of their city's outlet, with instructions to join various local associations or training *projects*.

CONCLUSION

The sales function, which is a vital part of an enterprise, is highly complex and difficult to understand. The performance of the whole enterprise depends on the performance of the sales function. It is, therefore, essential to have a full understanding of the factors that control sales-related performance and how it might be improved. The chapter explored the potential contribution that socio-economic management could make in dealing with this problem. The study allowed us to go to the heart of the *functioning* of an enterprise and revealed three main points: (1) the importance of mastering the decentralization of the sales function within the enterprises so that it becomes everyone's business; (2) the importance of using effective *management tools* that could be easily understood and used by all members of the enterprise; and (3) the absolute need for each person to take account of the external environment of the enterprise so that all organizational members can become a strategic player.

Socio-economic intervention is not easy to carry out. It can be a time-consuming process, requiring total commitment on the part of the *intervener-researchers* and unflinching single-mindedness to steer toward the initial objectives negotiated with the enterprise. This is a price the consultant as scientist has to pay to penetrate to the heart of the complex structure of an enterprise (Savall & Zardet, 2004) and to help it to change. The mastery of this structuring process, initiated at the first contact between the parties (the decision-makers in the enterprises and the intervener-researchers) determines the initiation of the change and creates the foundation for its quality.

By inducing *active listening* by the personnel in the enterprise (from first contact through analysis and implementation) and its shared constructive approach between the consultant/scientist and the enterprise, socio-economic intervention creates a "process-based" permanent interaction between the enterprise and its particular environment. It leads to an overall view of the internal and external context of the enterprise.

Thus, *socio-economic management* permits renewing the conception of marketing and sales functions in organizations, by endowing all individuals that populate them with an awareness of their "commercial power"— with the means to implement it. The process entails knowing, selecting, informing, convincing, "seducing" clients and prospects, the only true judges of the enterprise's performance, and the only true guarantees of its *survival*.

NOTE

1.   As an example, a focus on sales force performance is reflected in nearly all the articles published in the *Journal of Personal Selling and Sales Management*, a key academic magazine on sales force management.

REFERENCES

Barth, I. (1994). *Proposals for an innovative internal/external marketing strategy, theory, practice and experimental results.* Unpublished doctoral dissertation, University of Lyon, France.

Barth, I. (2001). *The sale of consulting: Proposal for a structured and structuring negotiation process.* Paper presented at the ISEOR, University Lyon 2 and HEC Conference, "Knowledge and Value Development in Management Consulting." Lyon, France.

Hanson, N. R. (1959). *Patterns of discovery.* Cambridge, MA: University Press.

Patton, M. Q. (1978). *Utilization-focused evaluation.* Beverly Hills, CA: Sage.

Rich, G. A., Bommer, W. H., McKenzie, S. B., Podsakoff, P. M., & Johnson, J. L. (1999). Apples and apples or apples and oranges? A meta analysis of objective and subjective measures of salesperson performance. *Journal of Personal Selling and Sales Management, 19*(4), 41-52.

Savall, H. (1974, 1975). *Enrichir le travail humain dans les entreprises et les organisations* [Work and people: An economic evaluation of job enrichment]. Paris: Dunod.

Savall, H. (1986). *Le contrôle de qualité des informations émises par les acteurs des organisations* [The quality control of information emitted by organizational actors].

Paper presented at the ISEOR Conference "Qualité des informations scientifiques en gestion," Lyon, France.

Savall, H. (1987). Les coûts cachés et l'analyse socio-économique des organisations [Hidden costs and the socio-economic analysis of organizations]. *Encyclopédie du management* (pp. 599-628). Paris: Economica.

Savall, H. (2003a). An updated presentation of the socio-economic management model. *Journal of Organizational Change Management, 16*(1), 33-48.

Savall, H. (2003b). International dissemination of the socio-economic model. *Journal of Organizational Change Management, 16*(1), 107-115.

Savall, H., & Zardet, V. (1987). *Maîtriser les coûts et les performances cachés: Le contrat d'activité périodiquement négociable* [Mastering hidden costs and performances: The periodically negotiable activity contract]. Paris: Economica

Savall, H., & Zardet, V. (1995). *Ingénierie stratégique du roseau, souple et enracinée* [Strategic engineering of the reed, flexible and rooted]. Paris: Economica.

Savall, H., & Zardet, V. (2004). *Recherche en sciences de gestion: Approche qualimétrique. Observer l'objet complexe*. Unpublished English translation: *Research in management sciences: The qualimetric approach. Observing the complex object*. Paris: Economica.

Savall, H., & Zardet, V. (2005). *Tétranormalisation: Défis et dynamiques* [Competitive challenges and dynamics of tetra-normalization]. Paris: Economica.

Savall, H., Zardet, V., & Bonnet, M. (2000). *Releasing the untapped potential of enterprises through socio-economic management*. Geneva, Switzerland: International Labor Office-ISEOR.

Savall, H., Zardet, V., Bonnet, M., & Moore, R. (2001). A system-wide, integrated methodology for intervening in organizations: The ISEOR approach. In A. F. Buono (Ed.), *Current trends in management consulting* (pp. 105-125). Greenwich, CT: Information Age.

GLOSSARY

Seam-Related Concepts and Terminology

The Glossary contains references to terms and concepts used in the various chapters, and specific keywords with a brief definition.

Absenteeism: One of the five indicators of hidden costs, both the consequence of certain dysfunctions and the cause of other dysfunctions (see chapters 1, 2, 5, 10, 13, 14, 17).

Academicism: See chapter 1.

Active listening: Active listening is founded in interaction, a heuristic approach composed of three phases: (1) pure (nonjudgmental) listening, (2) reformulation and validation, (3) followed by listening anew (see chapter 18).

Actor: *External* actor designates a person pertaining to the organization's relevant external environment. *Internal* actor designates a person belonging to the organization's internal environment (e.g., owner, director, employee, volunteer) (see chapter 1).

Alignment: See chapters 1, 6, 16, 17.

Applied research: Scientific research that is focused on the effective implementation of discoveries or models to be tested (see chapter 11).

Assessment of achievements: See chapter 7.

Behavior grid: See chapter 8.

Boomerang-effect: Reaction of the external environment to the externalization of costs (externalized costs) forcing an organization to transform them into internal costs (internalized costs) (see chapter 15).

Socio-Economic Intervention in Organizations: The Intervener-Researcher and the SEAM Approach to Organizational Analysis, pp. 421–436
421

Budgeted action plan: Priority action plans implemented during the initial set-up phase of SEAM that serve as the basis to developing budgeted PAPs, which include the budget necessary to carry out priority actions (see chapter 3).

Charges: See chapters 1, 3, 6, 7.

Chronobiology: Study of the life rhythms of a person or organization. The chronology of socio-economic intervention respects both the natural rhythms of the organization where the intervention takes place and certain intervention-process rhythms necessary for effectiveness (see chapters 1, 3).

Clean-up: Maintenance or renovation of the material or intangible objects that make up an organization (e.g., structures, procedures and behaviors) which suffer from inexorable deterioration over time. Clean-up is one of the three efficiency factors (see HISOFIS and synchronization) and one of the main sources of reduced dysfunction and hidden costs (see chapters 1, 3, 14).

Cluster: A cluster is a team composed of internal actors with their hierarchical superiors. Clusters are core elements in the collaborative-training structure, which is part of the Horivert process (see chapters 1, 3, 4, 5, 12).

Cognitive interactivity: Interactive process (between intervener-researcher and company actors) of knowledge production through successive feedback loops, with the steadfast goal of increasing the value of significant information processed by scientific work (see chapters 1, 11).

Cohesion: Level of cooperation among actors within a team or an enterprise, which is a source of sustainable performance according to SEAM (see chapters 1, 2, 6, 7, 12, 14, 16).

Collaborative delegation: Entrusting an operation to a delegate, while ensuring that the delegate has the necessary means to carry it out, which include pedagogical support from the instructing party on behalf of the delegate.

Collaborative learning: See chapter 1.

Collaborative training: Theoretical and practical training of managers in a zone of responsibility in the organization, led by the hierarchical superior, adapting training content to the real work situation (see chapters 1, 2, 4, 5, 6, 7, 17).

Communication: All types of information exchange among actors, formal and informal, top-down and horizontal, frequent and rare, regular and irregular, professional and nonprofessional content, of major importance or not (see chapters 4, 10).

Communication-Coordination-Cooperation (3C): Referred to as **3C**, this concept covers all communication, coordination and cooperation as defined in the present glossary (see chapters 1, 2, 3, 5, 6, 7, 8, 11, 12, 13, 14, 15, 16, 17).

Competence: Theoretical and practical knowledge held by an actor, which is implemented in the exercise of his or her professional activity (see chapters 4, 12, 13, 16).

Competency grid: Synoptic tool displaying the competencies currently available in a team and their concrete deployment (see chapters 1, 2, 5, 6, 7, 8, 11, 12, 13, 14, 15, 16, 18).

Competitiveness: See chapters 1, 7, 8, 11, 12, 13.

Confidence: Positive relationship between individuals or teams, which facilitates cohesion (see chapters 1, 3, 8, 11).

Contradictory intersubjectivity: Technique for creating consensus based on the subjective perceptions of different actors, in order to create more "objective" grounds for working together (see chapters 1, 11).

Conversion of hidden costs into value-added: See chapter 2.

Cooperation: Characterizes the exchanges between actors that permit (1) defining a common operational or functional objective to be carried out within a determined period and (2) defining the game rules of the cooperation (see chapters 1, 2, 3, 4, 5, 6, 7, 8, 11, 12, 13, 15, 16, 17).

Cooperative delegation: Assigning missions for operations and tasks to a collaborator or a team, combined with integrated training action by the hierarchical superior to guarantee adequate competency and leeway for individual initiative; see also *collaborative delegation* (see chapters 1, 12, 13).

Coordination: Information-exchange frameworks between actors with the goal of attaining an operational objective or completing a functional activity (see chapters 1, 3, 4, 5, 6, 7, 10, 11, 13, 17).

Core group: See chapters 1, 6, 11.

Cost estimate: See chapters 1, 10.

Creation of potential: Action that will have positive effects on future financial years, mainly composed of intangible investments (see chapters 1, 2, 3, 4, 7, 11, 12, 13, 14, 16, 17).

Creation of value-added: Goal of internal human activity, responding to the needs of external actors (stakeholders) (see chapters 1, 3, 7, 12, 13).

Decider-payer: Individual in the intervention process with the legitimate power to commit the necessary resources and make change decisions (see chapter 1).

Deontology: See *ethics* (also see chapter 1).

Destruction of value-added: Waste of economic resources due to useless expenditures or noncreation of value-added by internal actors (see chapter 1).

Development: Qualitative and quantitative transformation of the enterprise, its activities, its structures, its behaviors and its results, including both economic and social performance (see chapters 1, 2, 3, 4, 5, 6, 7, 8, 10, 11, 12, 13, 14, 15, 16, 17, 18).

Development-management: See chapter 15.

Diagnostic: See *socio-economic diagnostic* (see also chapters 6, 7, 9).

Direct productivity gaps: One of the five indicators of hidden costs indicating that the volume of production of goods or services is inferior to the standard volume that could be expected (see chapter 15).

Directly productive time: Human time devoted to the completion of tasks, operations or activities directly impacting the immediate operational activity of the organization. According to socio-economic theory, directly productive time is only efficient if coupled with indirectly productive time (see chapter 2).

Discriminating value-added: See chapter 7.

Dysfunction: Consequence of interaction between an organization's structures and the behaviors of internal and external actors. They are described by actors in terms of discrepancies with reference to orthofunctions (see chapters 1, 2, 3, 4, 5, 6, 7, 8, 9, 10, 11, 12, 13, 14, 15, 16, 17, 18).

Dysfunction basket: Method grouping together dysfunctions under three to six relevant themes to facilitate the focus-group's search for solutions during the project phase (see chapters 1, 3, 4, 11).

Economic balance: See chapters 1, 5, 8, 12, 13, 15, 18.

Economic performance: Rational short and long-term utilization of the organization's resources, in keeping with socio-economic rationale, which is multi-dimensional and takes into account psychosociological and anthropological criteria in defining the ultimate goal of economic action (see chapters 2, 6, 7, 12, 15, 16, 18).

Effectiveness: Degree to which a predefined objective is attained (see chapters 1, 3, 4, 5, 6, 7, 8, 10, 12, 14, 15, 18).

Efficiency: Judicious utilization of resources engaged to attain a result (see chapters 1, 3, 4, 5, 6, 8, 11, 12, 14, 15, 16).

Effort/benefit balance for actors: Tool that synthesizes the principal change efforts as well as the principal advantages to be drawn by the various categories of actors involved in a socio-economic innovation project (see chapter 11).

Engineering: See chapters 1, 3, 8, 9, 10, 11, 12, 17, 18.

Ethics: Fundamental values embodied in a code of good conduct shared among actors when cohesion is adequate (see chapters 1, 14, 17).

Evaluation: Measure of the result obtained through a real activity with regards to an objective that was defined *a priori* (see chapters 1, 2, 3, 4, 5, 7, 11, 12, 13, 14, 16, 17, 18).

Evaluation phase: Fourth stage of the socio-economic intervention on the improvement-process axis (following the diagnostic, the project and the implementation phases). The evaluation is qualitative, quantitative and financial in nature, and expresses progress in social and economic performance (see chapters 1, 13).

Excess salary: Component of the cost of certain dysfunctions corresponding to wage differential due to a higher-paid employee performing a task that should have been performed by a lesser-paid employee (downward shift of functions) (see chapters 1, 9, 11, 14, 17).

Exhaustive fieldnote quotes: See chapter 1.

Experimental research: Scientific research that submits ideas for factual experimentation[1] (see chapter 11).

Expert opinion: See chapters 1, 3, 4, 7, 10, 15, 18.

Extension phase: See *territorial extension phase* (see also chapters 1, 3, 4, 5).

External environment: All economic actors outside the organization (stakeholders) whose action directly or indirectly affects the organization (see chapters 2, 5, 7, 18).

External intervener: Professional consultant from the external environment implementing a set of actions involving the modification of structures and behaviors of the organization (see chapters 11, 12).

External validation: Capacity of a theoretical model to be applied to other contexts.

Externalized cost: Cost generated by the organization and imposed upon its external environment; see also *boomerang-effect*.

Field for scientific observation: See chapter 1.

Fieldnote quotes: Handwritten notes accurately reflecting the ideas expressed by actors; an essential technique for the intervener reflecting the quality of his or her active listening (see chapters 1, 6, 15, 16).

Focus group: See chapters 1, 2, 3, 4, 6, 7, 10, 11, 12, 13, 14.

Functioning: The observable operation of an organization is a mixture of ortho-functions and dysfunctions (see chapters 1, 4, 5, 14, 18).

Game rules: Professional activity, that is a game of actors (i.e. theater, sports), where the rules of the game, which are more or less respected, are directly related to the quality of the performance (see chapters 1, 4, 6, 7, 11, 12).

Gears: Made up of human behaviors and frameworks that make procedures and processes of activity and communication-coordination-cooperation function correctly (see chapter 5).

Generic contingency: Epistemological principle introduced by the socio-economic theory that, while recognizing the operational specificities of organizations, postulates the existence of invariants that constitute generic rules embodying core knowledge that possesses a certain degree of stability and "universality" (see chapter 1).

Generic key idea: See chapter 1.

Generic knowledge: Core, decontextualized knowledge (see chapter 1).

Heuristic: A process that produces knowledge by processing factual information through "intelligent groping" in which the search for solutions incorporates, step by step, the rules for discovering relevant information (see chapter 8).

Hidden cost: Destruction of value-added, which is partly or completely left out of a company's accounting information system, consisting of both surplus expenditures and opportunity costs that affect the relevance of decision-making processes (see chapters 1, 2, 3, 4, 5, 7, 8, 9, 10, 11, 12, 13, 14, 15, 16, 17, 18).

Hidden cost-performance (**theory**): Full title of the theory of hidden costs; an interaction exists between such costs and performance (i.e., reducing hidden costs is a performance and reduced performance is a cost) (see chapters 4, 12).

Hidden infrastructure: An organization's infrastructure composed of modes of operation and management that are not visible to external actors (see chapter 2).

Hidden performance: See chapter 1.

Hidden products/services: The spontaneous functioning of an organization often generates microproducts that have a value but are not sold by the enterprise (see chapter 18).

HISOFIS: Humanly Integrated and Stimulating Operational and Functional Information System: HISOFIS-type information stimulates the receiver, individual or group, and incites the receiver to engage in decisive action, which implies the expenditure of human energy. HISOFIS is one of the three efficiency factors (see synchronization and clean-up) and one of the main sources of reduced dysfunctions and hidden costs. OFIS (Operational and Functional Information System) represents the total volume of information emitted and circulated in an organization. HIOFIS (Humanly Integrated OFIS) is the subset of information assimilated by the receivers. HISOFIS (Stimulating HIOFIS) is the part of HIOFIS that stimulates human action, source of performance (see chapters 1, 4, 13).

HISOFISGENESIS: Generating HISOFIS effects (see chapter 6).

Horivert (*architecture* or *intervention* or *model* or *method* or *process*): Socio-economic innovation intervention in an organization composed of two simultaneous actions: (1) a **HORI**zontal action, a diagnostic of dysfunctions with the board of directors and the management team, as well as setting up socio-economic management tools, followed by a horizontal project; and (2) a **VERT**ical action: in at least two departments including vertical diagnostics. Vertical action involves the line personnel as well (see chapters 1, 2, 3, 4, 5, 6, 7, 8, 12, 13, 14, 16, 17).

Horizontal action (or *intervention*): See chapters 1, 2, 3, 4, 7, 11, 12, 17.

Horizontal diagnostic: First phase of the intervention, in parallel with the first training sessions in socio-economic management tools, consisting in drawing up an inventory of major dysfunctions through one-on-one interviews with every senior and middle manager (see chapters 1, 2, 3, 4, 7).

Horizontal project: Action projects responding to the dysfunctions that were revealed during the horizontal diagnostic at the beginning of the Horivert process, setting the stage for improving policies and general procedures (see chapters 1, 4, 7).

Hourly contribution to margin (or *value-added*) *on variable costs* (*HCMVC* or *HCVAVC*): See chapters 1, 2, 3, 5, 8, 11.

Human factor: The socio-economic theory replaces the materialistic and utilitarian concept of the human "factor" with that of human "potential" (see chapter 1).

Human potential: The all-important factor for creating economic value lies in the individual, who can give or withhold his or her energy based on the quality of the informal and formal bond with the organization. This is measured by the HCVAVC indicator (hourly contribution to value-added on variable costs) (see chapters 1, 3, 12, 15, 16, 17, 18).

IESAP: Internal/external strategic action plan (see chapters 1, 2, 5, 7, 8, 11, 12, 13).

Immaterial investment: See *intangible investment* (see also chapter 15).

Immediate result: Results under the headings of visible costs and visible products that are identified and assessed by the actors; often referred to as short-term economic results, such as they appear in the company profit and loss statement (see chapters 1, 2, 3, 11, 12, 13, 14, 16, 17).

Immediate-result indicators: See chapters 1, 13.

Implementation (of SEAM)): See chapters 1, 3, 4, 6, 8, 10, 11, 14

Improvement-process axis: See *(SEAM) three key forces of change* (see also chapter 11).

In-depth intervention (or *integration*) *phase*: See chapters 1, 2.

Indirectly productive time: Human time indirectly impacting the immediate operation activities of the organization. According to socio-economic theory, indirectly productive time is insufficient for certain categories of actors (e.g., line personnel) and excessive for others (e.g., integrated training, communication-coordination-cooperation, evaluation of actions) (see chapter 2).

Industrial injuries: one of the five indicators (see work accidents) (see also chapter 1).

Infrastructure: Framework that organizes and codetermines the quality, coherency, effectiveness and efficiency of the organization's operations, such as synchronization-stimulation frameworks (HISOFIS) and "clean-up." These frameworks are not readily visible and are often omitted when carrying out decisive actions. Dysfunctions are usually lodged in the six "infrastructure" domains: working conditions, work organization, time management, communication-coordination-cooperation, integrated training and strategic implementation (see chapters 1, 2, 7, 8).

Initial set-up (of socio-economic management or SEAM): See chapters 1, 5, 7.

Intangible investment (or *immaterial* or *intellectual*): Cost of creation of potential of an intangible nature (e.g., integrated training sessions, focus-group cost), should be considered as a profitable investment rather than a recurring expenditure in accounting (see chapters 2, 17).

Integral performance: See chapters 1, 11.

Integral quality: Quality of management, operations and products (goods and services), needed to satisfy the needs of internal and external actors (stakeholders) (see chapters 1, 4, 7, 14).

Integrated epistemology: See chapter 1.

Integrated training: Training carried out by an actor in the immediate environment (e.g., the hierarchical superior) involving an inductive pedagogical contribution to local work situations and capitalizing on expertise acquired through experience by explicitly capturing it in an integrated training manual (see chapters 1, 2, 3, 6, 8, 11, 13, 14, 15, 18).

Integrated training manual: See chapters 1, 13, 15.

Integrated training plan: See chapter 7.

Integration: See chapters 1, 2, 3, 5, 12, 14, 18.

Integration phase: See chapter 1.

Interaction: Dynamic process that is constructed in iterative fashion through mutual influences exerted by actors on one another. These influences cause modifications of operations and socio-economic performance. These combined modifications enable actors to pass from operational State 1 to a new operational State 2 (see chapters 2, 3, 5, 6, 15, 18).

Intercompany intervention (or *action*): In the multismall business Horivert, collaborative training in the implementation of SEAM tools is carried out in groups of four to six small enterprises (senior manager and two supervisors) (see chapter 14).

Interface: Critical zone between one or several groups, spaces or domains, which should be managed to ensure effective communication between them.

Internal environment: All active resources in the organization, e.g., human potential, actors, producers of activity and value-added (employees, owners or volunteers) (see chapters 12, 18).

Internal marketing: Term often employed to designate the techniques of information dissemination among the personnel concerning the life of the organization; also applied to in-depth, active listening (socio-economic diagnostic) and subsequent setting-up of innovative actions that respond to employees' actual expectations (socio-economic innovation project) (see chapters 2, 18).

Internal validation: Demonstrates the internal coherency of the links established between the different variables or hypotheses leading to the construction of a theory.

Internal/External Strategic Action Plan (IESAP): Specifies the enterprise's 3 to 5-year strategy, with regard to both key external actors (e.g., clients, suppliers) and key internal actors (e.g., from the CEO to workers). It is updated once a year to take into account change in the enterprise's relevant external and internal environments (see chapters 1, 2, 5, 7, 8, 11, 12, 13, 16).

Internal-client: SEAM considers personnel at all levels as clients whose needs for improved work life conditions should be satisfied (see chapter 15).

Internal-intervener: Individual in the organization collaborating in the implementation of intervention actions (see chapters 1, 3, 4, 7).

Internalized cost: Cost generated by the external environment and imposed on the organization.

Intervener-researcher: Person utilizing the results of his or her interventions to nourish research and scientific knowledge, to help to carry out further interventions (see chapters 3, 6, 7, 8, 16, 18).

Intervention agreement: Contract negotiated by the external intervener with the enterprise that requests a socio-economic intervention. The contract specifies the precise intervention architecture, the number of persons involved, and the schedule of services to be provided (see chapters 1, 4, 5).

Intervention architecture: Horivert schema that shows the distribution of actors and intervention services (see chapters 1, 6, 7, 11).

Intervention-research: Concept of management research, which is close to "action-research," that implies the frequent presence of the researcher within the enterprise in order to ensure systematic observation of management situations under study. This epistemological option acknowledges that the researcher is clearly engaged in his or her research strategy and coconstructs knowledge with the actors "observed" (see chapters 1, 6, 7).

Intracompany intervention (or *action*): In the multi-small business Horivert, the socio-economic diagnostic and project, as well as the implementation of SEAM tools, are carried out inside each one of the four to six enterprises that make up the group (see chapter 14).

Introducer: The person who introduces the external intervener into the client enterprise, but who does not have the power to negotiate the intervention (cf. decider-payer) (see chapter 1).

Itemized operations chart: Tool that breaks down an operation into stages. This tool serves as a medium for training, as a reminder, and for communication among actors who take part in the operation.

Maintenance phase: Follows the socio-economic management set-up and in-depth intervention phases; consists in regularly restoring good practices (see chapters 1, 2, 5, 17).

Management tool: Dynamic tools that mobilize human potential and stimulate internal actor behavior and, at times, external actor behavior (see chapters 1, 2, 4, 5, 6, 7, 12, 13, 14, 15, 18).

Management-tool axis: See *(SEAM) three key forces of change* (see also chapters 1, 11).

Mediator: One of the three fundamental functions of the intervener in the socio-economic intervention method. It entails developing cooperation among actors to build teamwork (see chapter 14).

Metamorphosis: Progressive, radical and sustainable transformation of an organization's operations, performances, structures and actor behaviors (see chapter 1).

Metascript: See chapters 1, 9.

Methodologist: One of the three fundamental functions of SEAM interveners: methodological and pedagogical support to actors so that they might discover by themselves or invent their own solutions (see chapter 14).

Method-product: Part of the selling method for intangible products (example: consultancy), corresponding to an effective technique (e.g., diagnostic, project, assessment and collaborative training are all method-products of socio-economic intervention) (see chapters 1, 4, 11).

Minidiagnostic: See chapter 14.

Mirror-effect: See chapters 1, 3, 4, 6, 9, 10, 11, 12, 13, 16, 17, 18.

Misunderstanding: Category of dysfunctions related to incomprehension on the part of actors concerning strategic choices or procedures. The solution is found in clarifying and communicating, rather than transforming (see chapters 1, 5, 6, 15).

Multismall Business Horivert: Scaled-down Horivert process applied to a group of four to six small enterprises to reduce the intervention cost and to stimulate emulation amongst them (see chapters 1, 14).

Negotiation: Contradictory dialogue engaged between actors to arrive at an agreement that combines the compatible interests of the parties, through a dialectical game that is ultimately positive (win-win) (see chapters 1, 2, 3, 4, 5, 6, 8, 9, 10, 11, 12, 14, 17, 18).

Negotiation phase: Preliminary, complimentary phase of socio-economic intervention that aims to make potential client organizations understand socio-economic management and the rigorous nature of its intervention methodology (see chapter 1).

Noncreation of potential: One of the six components of hidden costs, namely a possible or planned investment of an intangible nature that was never carried through (see chapters 1, 11, 16, 17).

Non-dit (unvoiced comments): The second constituent of the expert opinion that finalizes the diagnostic. The intervener reports major dysfunctions he or she perceived, but that were not mentioned by actors during the diagnostic interviews (see chapters 1, 3, 4, 7, 11, 16).

nonproduct: See chapter 1.

nonproduction: See chapters 1, 11, 12, 14, 16, 17.

nonquality: One of the five indicators of hidden costs corresponding to quality defects or lack of quality (see chapters 1, 2, 11, 13, 15, 16).

Objective-product: Major issues pointed out by the decider-payer during the negotiation phase preceding the intervention. They constitute the explicit goals that serve as references throughout the entire process (see chapters 1, 11).

OMSP (Objective-Method-Service-Product) tool: Synthetic tool summing up objective-products, method-products and service-products which constitute the specification manual at the center of the intervention agreement (see chapters 1, 11).

Orthofunctioning: Functioning desired by the organization's internal and external actors. Orthofunctioning is a relative notion, a flexible reference that admits a certain degree of variability over time. It is useful for defining *grosso modo* the direction of progress actions for the enterprise (see chapters 1, 4).

Overconsumption: Component of the cost of certain dysfunctions corresponding to the quantity of products or services over-employed resulting from the regulation of dysfunctions (see chapters 1, 8, 11, 12).

Overtime: Time in excess of a set limit. Component of the cost of certain dysfunctions corresponding to the time spent in regulating dysfunctions (see chapters 1, 3, 4, 7, 10, 11, 12, 14, 15).

Periodically negotiable activity contract (PNAC): Management tool that formally states the priority objectives and the means made available for attaining them, involving every employee in the enterprise (including workers and office employees), based on a biannual personal dialogue with the employee's direct hierarchical superior (see chapters 1, 2, 3, 7, 8, 11, 12, 13, 14, 16, 18).

Personal assistance: Every collective collaborative-training session in the "clusters" is followed by a one-on-one interview with every manager and senior manager to help him or her adjust SEAM tools to his or her own personal function (see chapters 1, 3, 4, 5, 7, 8, 11, 12, 13, 14).

Personnel (staff) turnover: See chapters 1, 3, 8, 11, 12.

Pertinent environment: See chapter 1.

Piloting structure: See chapter 1.

Piloting: Refers to piloting acts that expend human energy, piloting team cooperation and immaterial management tools (see chapters 1, 2, 6, 8, 11, 12, 13, 16, 17, 18).

Pivotal ideas (idées forces): Family or group of key-ideas. It is a key-idea of a higher generic level (see chapters 1, 3, 11).

Plenary executive committee or plenary steering committee: See ***steering group*** (see also chapter 6).

Plenary group: See chapters 1, 6, 11.

Policy-decision axis: See chapter 11.

Poorly executed tasks: One of the deep causes of numerous dysfunctions corresponding to neglected tasks, tasks not carried out, executed behind schedule or lacking quality (see chapters 13, 14).

Priority action plan (PAP): A coordinated inventory of actions to be accomplished within a period of six months to attain priority strategic objectives, once priorities have been defined and tested for feasibility. These actions are partly generators of dysfunctions (current project on the external environment) and partly reducers of dysfunctions (action on the internal environment) (see chapters 1, 2, 3, 4, 5, 6, 7, 8, 10, 11, 12, 13, 14, 16, 17, 18).

Productivity: Capacity to produce material and immaterial goods appreciated by actors, clients or users for a given period (see chapters 1, 2, 3, 5, 6, 7, 8, 11, 12, 13, 15, 17, 18).

Progressive spiral: A continuous process of cumulative improvement.

Project: See *socio-economic project* (see also chapters 1, 2, 3, 4, 5, 7, 8, 10, 11, 12, 13, 14, 15, 17, 18).

Project leader: Hierarchical leader in charge of conducting the improvement process of socio-economic performances in his or her team, the CEO, department head or workshop head, depending on the situation (see chapters 1, 6, 12).

Project phase: See chapters 1, 4, 11, 13.

Provisional intervention schedule: See chapter 11.

Proximity management (or *supervising management*): The proximity management concept considers that the immediate (or direct) hierarchical superior plays an important role in good working conditions, daily communications and development of competency among his or her co-team members, due to direct and frequent interaction with them (see chapter 3).

qQfi Evaluation: Evaluation combining qualitative and quantitative and financial assessment and measures.

Qualimetrics: See chapters 1, 11, 14.

Qualitative diagnostic: All dysfunctions expressed during a horizontal diagnostic or during the qualitative part of a vertical diagnostic (see chapters 1, 15).

Regressive spiral: A continuously-amplified negative process.

Regulation: See chapters 1, 2, 4, 6, 11, 13, 14, 18.

Regulation of dysfunctions: See chapters 1, 13, 15.

Resolution chart: A tool filled out during a meeting and distributed immediately afterwards to all participants, providing responsibility for and encouraging action steps (see chapters 1, 8, 13).

Resources: The organization's physical and human means, material and immaterial, utilized in company activities and applied to its products (goods or services) (see chapters 2, 3, 4, 5, 6, 7, 8, 11, 12, 14, 15, 16, 17, 18).

Results assessment: See chapters 1, 10.

Risk: See chapters 1, 2, 3, 5, 11, 12, 13, 15, 16, 18.

Root causes (of dysfunctions): Deep, primary causes of dysfunctions of which actors are rarely conscious; they are revealed in the *non-dit* (unvoiced comments) constituent of the expert opinion (see chapters 1, 4, 9, 14, 15).

Rooting phase (or *process* or *approach*) of socio-economic management (or of SEAM intervention): See chapters 1,7.

Scheduling: Designing a precise activity program for all actors in cooperation with their partners; also an activity piloting indicator (see chapters 1, 3, 4, 5, 6, 7, 10, 11, 14, 18).

Scientific consultancy: Advice to organizations based on scientific knowledge drawn from experimentation; it is also a rigorous method of management research (intervention-research) (see chapters 1, 2).

SEAM set–up (phase): See chapter 1.

SEAM star: See chapters 1, 3.

(SEAM) three key forces of change: The socio-economic intervention method consecutively follows three axes: improvement-process axis, management-tool axis and policy-decision axis (see chapter 1).

(SEAM) trihedral: See *(SEAM) three key forces of change* (see also chapters 1, 17).

SEAMES® software (or *SEAM Expert System software*): Expert system developed by ISEOR that processes over 3,500 types of dysfunctions identified in 1,200 organizations in 34 countries through SEAM interventions (see chapters 1, 4, 8, 12).

Security-management: See chapter 15.

Self-analysis of time: See chapters 1, 13.

Self-analysis of time grid: A simple tool enabling an actor to become conscious of the ineffective and inefficient structure of his or her use of time, which causes professional-life quality and under-performance problems (see chapters 1, 6, 11).

Service-product: Actions requiring competency on the part of interveners or actors in application of a "method-product" of the OMSP tool (see chapters 1, 4, 11).

Set-up phase: See *initial set-up* and *(SEAM) set-up (phase)* (see also chapters 1, 7, 17).

Short, medium and long term survival-development: See *survival-development* (see also chapter 1).

Social performance: Multiple sources, notably of a psycho-physio-sociological nature, of satisfaction for actors, clients or producers of the organization in their professional life (see chapters 1, 2, 3, 5, 13, 15, 18).

Socio-economic competitiveness: Capacity to withstand economic competition from the market without weighing too heavily on the social component.

Socio-economic diagnostic (phase): Diagnostic of dysfunctions in an organization and their destruction of the value-added (hidden costs) they cause (see chapters 1, 2, 6, 8, 10, 11, 12, 13, 15, 16, 18).

Socio-economic evaluation: Stage of the socio-economic innovation process that involves evaluating impacts of the implemented improvement actions on the social performance and economic performance of the organization and its actors (see chapters 1, 5).

Socio-economic innovation process (or *action*): Enables change to be progressively assimilated during the experimentation phase, notably through the diagnostic of the enterprise's dysfunctions; the innovation project includes improvement actions for both social and economic performance, their implementation and the evaluation of obtained results. See also *socio-economic project*.

Socio-economic intervention: Simultaneously engaging actions within an enterprise on all its structures and human behaviors, by intervening in the six domains of dysfunctions. The enterprise's operations are thus dealt with as a whole, in order to facilitate the emergence of sustainable, effective and innovative solutions (see chapters 1, 2, 3, 4, 5, 6, 7, 8, 10, 11, 12, 13, 14, 18).

Socio-economic intervention axes: See *(SEAM) trihedral* (see also chapter 17).

Socio-economic management (SEAM): Increased participation and dynamism on the part of the enterprise's or organization's entire personnel and by the development of all human know-how and competency, in the strategic pursuit of both improved social (qualitative) and economic (quantitative) performance, the benefits of which are more or less distributed among internal and external actors (stakeholders) (see chapters 1, 2, 3, 4, 5, 6, 7, 8, 11, 12, 13, 14, 17, 18).

Socio-economic management control: Piloting of the organization's short-, medium-, and long-term economic and social performance through qualitative, quantitative and financial evaluation of visible and hidden costs and performance. It contains a self-control component carried out by operational personnel and an external control component carried out by management control specialists (see chapters 1, 3, 5, 7, 13).

Socio-economic management tools: See chapters 1, 2, 3, 4, 6, 7, 8, 11, 12, 13.

Socio-economic marketing: See chapters 1, 3, 5, 7.

Socio-economic organization: See chapters 2, 3.

Socio-economic performance: Combines social performance and economic performance, the development of one based on the development of the other, neither being exclusively attained to the detriment of the other (see chapters 1, 3, 7, 9, 13, 14, 15).

Socio-economic project (phase): Second stage of the socio-economic innovation process. The goal is to seek solution baskets (improvement actions) that respond to dysfunction baskets. A socio-economic project is an ensemble of socio-economic innovation actions (see chapters 1, 3, 4, 6, 8, 11, 12).

Socio-economic strategy: See chapters 1, 7.

Socio-economic theory (of organizations): An original theory experimented in 1,200 enterprises and organizations, developed by combining knowledge from the social and economic sciences in order to improve sustainable performance (see chapters 1, 2, 3, 4, 11).

Steering committee (or group): See chapters 6, 7, 14.

Strategic alert indicators: Indicators that make it possible to detect the probable critical state of an organization within a timespan of 3 to 5 years.

Strategic ambition: Distance separating an organization's current strategic situation and its intended strategic situation; it may be dependent on or independent of the organization's current strategic force (see chapters 2, 7).

Strategic development: Sustainable effort that focuses on medium- and long-term objectives without maximizing nor sacrificing short-term results (see chapters 1, 2, 13).

Strategic energy: Intangible resource enabling actors to develop their strategic force and capacity to transform their strategic positioning (see chapter 1).

Strategic force: Capacity of an enterprise to transform material and intangible resources into activities that enable the enterprise to improve its strategic situation, namely its capacity for survival-development and its power to negotiate with its internal and external environment (stakeholders) (see chapters 1, 2).

Strategic implementation: Concrete actions to attain the organization's strategic objectives, broken down and discussed at every hierarchical level, in a synchronized, transversal effort uniting various internal and external partners. It involves explicitly defining coherent strategic objectives and scheduling the necessary means to accomplish them (e.g., material, human, intangible) (see chapters 1, 2, 3, 4, 11, 12, 15).

Strategic interactivity: See chapter 1.

Strategic objectives: Explicit, recognized objectives that the organization wants to attain within 3 to 5 years, in order to guarantee its survival-development (see chapters 7, 17).

Strategic patience: See chapter 2.

Strategic piloting: Includes implementation and application of the strategy decided upon by the senior management, as well as evaluation of differentials between obtained results and projected objectives, through a heuristic process of "intelligent groping" (see chapter 13).

Strategic piloting indicators: See *strategic piloting logbook* (see also chapters 1, 5, 7, 12, 13, 14, 16, 17).

Strategic piloting logbook: A tool composed of qualitative, quantitative and financial indicators enabling efficient communication-coordination-cooperation among the members of the management team, enabling them to accomplish decisive actions in operational and strategic activities (see chapters 1, 6, 7, 12, 13).

Strategic threat: A particular state of uncertainty among internal company actors caused by change in the external environment and to a lesser degree in the internal environment, with the subsequent possibility of the organization's partial or total disappearance, or the irreversible obsolescence of an actor's current job (see chapters 3, 4, 7, 14).

Strategic variable: An aspect of the internal or external environment, economic or social in nature, to be taken into account in the strategic analysis when preparing a strategic decision and its implementation.

Strategic vigilance (or *intelligence*): See *vigilance* (see also chapters 2, 5, 7, 13).

Structure: Relatively stable elements of an organization permitting actors to carry out their activities, interacting with their behaviors and their competencies (see chapters 1, 3, 4, 6, 8, 11, 12, 14, 17).

Superstructure (*organizational*): A visible and relatively stable component of the organization (see chapters 1, 2).

Survival: See chapters 1, 3, 5, 6, 8, 12, 14, 16, 18.

Survival-development (*capacity*): Sustainable state of short-, medium- and long-term survival; the development component facilitates the medium and long-term survival of an organization (see chapter 1).

Sustainable (*global* or *integral*) *performance*: Socio-economic performance that can be evaluated or appreciated by the principal stakeholders (see chapters 1, 3, 8, 13).

Sustainable development: See chapters 1, 5.

Synchronization: Coordination in real time or within a very short interval. Synchronization aims to ensure compatibility between activities and decisions of the organization's various entities, divisions departments, units, teams and individuals in view of achieving the objectives and goals of that organization. It is one of the three efficiency factors for strategic and operational piloting of organizations and their performances (cf. HISOFIS and clean-up). It is one of the main sources of reduced dysfunctions and hidden costs (see chapters 1, 2, 3, 4, 6, 14, 18).

Synchronized decentralization: Transferring the initiative of the decisive act to the responsibility level where its implementation will be launched, while setting up game rules (communication-coordination-cooperation) that ensure its compatibility with actions from other zones of responsibility and with the strategic piloting of the entire organization (see chapters 1, 2, 3, 5, 7, 12).

Task group: See chapters 1, 3, 4, 6, 7, 11, 13.

(Territorial) extension (phase): In large enterprises, this second phase enhances the first socio-economic management set-up phase by extending and applying it to additional departments or plants (see chapter 1).

Theater of SEAM: SEAM considers a management situation as a "theater piece," metaphorically speaking (see chapters 1, 9).

Therapist: One of the three fundamental functions of SEAM interveners. It consists in producing *mirror-effects* through active-listening of company actors (see chapter 14).

Three key forces of change: See *(SEAM) three key forces of change* or *(SEAM) trihedral* (see also chapter 1).

Time management: Temporal organization of work that cultivates automatic reflexes of activity planning and scheduling: prevention time, regulation time, preparation time, execution time, control time, improvement time, development time and strategic piloting time (see chapters 1, 2, 3, 5, 6, 7, 8, 9, 11, 12, 14, 15, 16, 17).

Value-added: The very goal and *raison d'être* of an organization (private or public), useful to its societal environment. It is measured by company accounts: the sales figure or revenue minus purchases and other external expenditures (see chapters 1, 2, 3, 5, 6, 7, 8, 12, 13, 14, 15, 16).

Vertical action: In the Horivert intervention process, this action involves every department, its entire hierarchical line of command from the CEO to shop-floor (line) personnel (see chapters 1, 2, 3, 4, 7, 12, 16, 17).

Vertical diagnostic: Inventory of dysfunctions expressed by all department actors, calculation of hidden costs with the cooperation of management, and analysis of competency grids for all actors (see chapters 1, 3, 4, 7, 13).

Vertical project: Concrete project of actions on the dysfunctions discovered during the vertical diagnostic; developed by the focus group and piloted by management with the assistance of line personnel in task groups (see chapters 1, 4, 7, 13).

Vigilance: Active surveillance of the organization's internal and external environment to extract information useful for the efficient strategic and operational piloting of the organization (see chapters 2, 8, 12).

Visibility: The property of an object, action, decision, information item, material or intangible good, as well as a phenomenon of being perceivable by the actors. Visibility influences representation, comprehension, behavior and relevance of decisions (see chapters 6, 7, 11, 12).

Vital sales function: Socio-economic theory places the exchange of resources among actors at the center of human and social activity; the socio-economic intervention develops the sales function among all actors of the organization, public or private, commercial or nonprofit (see chapters 1, 3, 7, 18).

Work accidents: See *industrial injuries* (see also chapter 1).

Work organization: Includes the distribution of missions, functions, operations and tasks, their more or less specialized assignment to internal actors, entities and individuals, and the rules governing the relationships among them (see chapters 1, 2, 3, 6, 7, 10, 11, 15).

Working conditions: Both physical working conditions (e.g., workspace, annoyances from the physical environment, physical or mental strain of the work, security) and the technological conditions and constraints of the work (e.g., equipment, tools available) (see chapters 1, 2, 3, 6, 8, 9, 10, 11, 13, 14, 15, 17).

NOTE

1.   Borrowed from Claude Bernard, *Introduction to the Study of Experimental Medicine* (New York: Dover, 1957).

ABOUT THE AUTHORS

Isabelle Barth is a professor at the University of Metz and at the Institut d'Administration des Entreprises, University Jean Moulin Lyon 3. She is a researcher at the ISEOR and holds a PhD in management sciences from the University of Lyon. She heads two master's programs: sales administration and commercial administration. Her current research interests focus on sales, commercial management and ethics.

Emmanuel Beck is an associate professor at the Institut d'Administration des Entreprises, University Jean Moulin Lyon 3, and a researcher at the ISEOR. He heads the "Social Audit" master's program. He holds a PhD in management sciences from the University of Lyon. His research centers on socio-economic management and interculturality.

Philippe Benollet is an associate professor at the Institut d'Administration des Entreprises, Université Jean Moulin Lyon 3, and a researcher at the ISEOR. He focuses his teaching on logistics, following a career as production manager in industrial enterprises. He holds a PhD in management sciences from the University of Lyon. His current research field focuses on socio-economic performance in cultural organizations.

David M. Boje holds the Bank of America Endowed Professorship of Management at New Mexico State University. He is described by his peers as an international scholar in the areas of narrative, storytelling, postmodern theory and critical ethics. He has published nearly 100 articles in journals, including such top-tier journals as *Management Science, Administrative Science Quarterly, Academy of Management Journal, Academy of Management Review* and *Organization Studies*.

Marc Bonnet is a professor at the Institut d'Administration des Entreprises, University Jean Moulin Lyon 3, and the deputy director of the ISEOR. He heads the "Industrial Security-Environment Certification" master's program. He holds a PhD in management sciences from the University of Lyon and carries on his research, in particular, in the fields of hidden costs and occupational disorders.

Anthony F. Buono, series editor, has a joint appointment as professor of management and sociology at Bentley College, and is coordinator of the Bentley Alliance for Ethics and Social Responsibility. He holds a PhD with a concentration in industrial and organizational sociology from Boston College. His current research and consulting interests focus on the management-consulting industry, organizational change, and interorganizational alliances, with an emphasis on mergers, acquisitions, strategic partnerships, and firm-stakeholder relationships.

Yue Cai is an assistant professor of management at the University of Central Missouri. Her teaching and research interests include global strategic competence, storytelling strategy, business consulting, and the reconciliation of the predominant western strategy paradigm with the ancient traditions of the east. She received her PhD from New Mexico State University

Laurent Cappelletti is an associate professor at the Institut d'Administration des Entreprises, University Jean Moulin Lyon 3 and a researcher at the ISEOR. He heads the "Management Engineering Consultancy" master's program and holds a PhD in management sciences from the University of Lyon. His research interests are organizational and financial audit and corporate performance and value.

Vincent Cristallini is an associate professor at the Institut d'Administration des Entreprises, University Jean Moulin Lyon 3 and a researcher at the ISEOR. He heads the "Team and Quality Management" master's program. He holds a PhD in management sciences from the University of Lyon. His research is centered on personnel management and professionalism.

Miguel Delattre is an associate professor at the University of Lyon. His PhD in management sciences was devoted to organizational performance development. As a researcher at ISEOR, he has conducted intervention-research in organizations. His current research work focuses on the management of collaborative organizational systems and performance development in loosely-structured organizations.

Margarita Fernandez Ruvalcaba is a professor at the Universidad Autonoma Metropolitana de Mexico (UAM) and an associate researcher with ISEOR. She received her PhD in management sciences from the University of Lyon. Her current research interests focus on organizational theories and entrepreneurship.

Nouria Harbi is an associate professor at the University of Lyon. She holds a PhD in computer sciences from the University of Lyon. A researcher at the ISEOR, she conducts on-going research in artificial intelligence and data warehouses.

Randall Hayes received his MBA and PhD from the University of Michigan. He is currently professor of accounting and codirector of the Institute for Management Consulting at Central Michigan University. His research is primarily in the areas of business valuation and consulting techniques.

Mark E. Hillon is an assistant professor of management at the University of Central Oklahoma. His primary research explores the translation of theory into practice in strategy and business consulting, as well as the related qualimetric assessment of the impact of intervention research on company performance. His teaching attempts to make business education relevant and useful to his students' future careers. He received his PhD from New Mexico State University and University Jean Moulin Lyon 3.

Nathalie Krief is an associate professor at the Institut d'Administration des Entreprises, Université Jean Moulin Lyon 3, and a researcher at the ISEOR. She heads the master's program "Management of Public Organizations and Decentralized Administrations." She holds a PhD in management sciences, and her research focuses on hospital management, with an emphasis on reform in that sector.

Lawrence Lepisto is professor of marketing and codirector of the Institute for Management Consulting at Central Michigan University. He received his PhD from Pennsylvania State University. His research interests lie in consumer research and management consulting.

Debra McGilsky is a professor of accounting at Central Michigan University. Prior to her PhD from Michigan State University, she worked on tax and audit engagements at Arthur Anderson & Co. Her research interests include taxation, management consulting, curriculum development and assessment of learning.

Rickie A. Moore is a professor of entrepreneurship and management at EM LYON and an associate researcher with ISEOR. He holds a PhD in management sciences from the University Jean Moulin Lyon 3. His current research and consulting focus on organizational performance and performance measurement, with special interest in performance dilemmas and paradoxes.

Michel Péron is emeritus professor at the University of Paris 3 Sorbonne Nouvelle. He received his PhD from the University of Lyon. He is a researcher at the ISEOR and the CERVEPAS. His research interests lie in cross-cultural management, corporate ethics and the history of economic ideas.

Henri Savall is a professor at the Institut d'Administration des Entreprises, University Jean Moulin Lyon 3, where he is the director of the Centre EUGINOV (Ecole Universitaire de Gestion Innovante) and of the socio-economic management master's program. He is the founder and director of the ISEOR Research Center. He holds a PhD in economic sciences and management sciences from the University of Paris. His current research interests are socio-economic theory, strategic management, and tetra-normalization new program.

Henri M. Talaszka started as accountant auditor and CPA. He received an AMP from Harvard Business School in the 1980s and was hired as financial manager and then CEO of a European subsidiary of a large American agri-food company. In the 1990s, he chaired several service and agri-food companies. He is currently operating a management consulting firm franchised by the ISEOR.

Olivier Voyant is an associate professor at the Institut d'Administration des Entreprises, University Jean Moulin Lyon 3 and a researcher at the ISEOR. He heads the "Audit and Operational Management" master's program. He holds a PhD in management sciences from the University of Lyon. His research focuses on corporate governance and democracy in the workplace.

Véronique Zardet is a professor at the Institut d'Administration des Entreprises, University Jean Moulin Lyon 3 and Codirector of the ISEOR Research Center. She heads the "Research in Socio-Economic Management" master's program. She holds a PhD in management sciences from the University of Lyon. In 2001 she received the Rossi Award from the Academy of Moral and Political Sciences (Institut de France) for her work on the integration of social variables into business strategies. Her research is centered on the conduct of strategic change and the improvement of socio-economic performance in private enterprises and public services.